AF182184

Invisible Bicycle

# Technology and Change in History

*Series Editors*

Adam Lucas
Steven A. Walton

VOLUME 15

The titles published in this series are listed at *brill.com/tch*

# Invisible Bicycle

*Parallel Histories and Different Timelines*

*Edited by*

Tiina Männistö-Funk
Timo Myllyntaus

BRILL

LEIDEN | BOSTON

Cover illustration: The corner of Liisankatu Street and Unioninkatu Street in Helsinki, Finland, 1970.
Photo: Eeva Rista. Source: archive of Helsinki City Museum.

The Library of Congress Cataloging-in-Publication Data is available online at http://catalog.loc.gov

Typeface for the Latin, Greek, and Cyrillic scripts: "Brill". See and download: brill.com/brill-typeface.

ISSN 1385-920X
ISBN 978-90-04-28996-3 (hardback)
ISBN 978-90-04-28997-0 (e-book)

Copyright 2019 by Koninklijke Brill NV, Leiden, The Netherlands.
Koninklijke Brill NV incorporates the imprints Brill, Brill Hes & De Graaf, Brill Nijhoff, Brill Rodopi, Brill Sense, Hotei Publishing, mentis Verlag, Verlag Ferdinand Schöningh and Wilhelm Fink Verlag.
All rights reserved. No part of this publication may be reproduced, translated, stored in a retrieval system, or transmitted in any form or by any means, electronic, mechanical, photocopying, recording or otherwise, without prior written permission from the publisher.
Authorization to photocopy items for internal or personal use is granted by Koninklijke Brill NV provided that the appropriate fees are paid directly to The Copyright Clearance Center, 222 Rosewood Drive, Suite 910, Danvers, MA 01923, USA. Fees are subject to change.

This book is printed on acid-free paper and produced in a sustainable manner.

# Contents

## PART 3
## *Bicycle in the Practices*

# Preface

*Timo Myllyntaus*

The bicycle is an extraordinary phenomenon of the industrial era. Its popularity began when the German Baron Karl von Drais launched his *Laufmaschine,* a kick bicycle, the harbinger of the modern bicycle, at Mannheim in the summer of 1817. Although only a handful of its technological modifications have been revolutionary, the bicycle has shown significant resilience by charming one generation after another. While it has retained its simple structure, it has spread widely over the globe to various cultures. Despite having originated more than two hundred years ago, millions of people still use the bicycle daily: it remains one of the primary means of everyday transport. Nevertheless, the bicycle is not just a vehicle but has various other connotations as well: It is a model of technology adjusting to changes in society and its demands, a signal of societal values and lifestyles, a political statement, a toy, a hobby, sports equipment, status symbol and research topic for historians and social scientists. During its lifetime, the bicycle has had its ebbs and flows and yet it has managed to change its societal image and start new fashions with novel cultural associations.

This edited volume is the result of long-term research and cooperation concerning bicycle history for the authors concerned. For more than a decade, these authors have presented their papers on this theme and met each other at various conferences. Symposia organized by the *International Committee for the History of Technology* (ICOHTEC) have been one important forum for us. At ICOHTEC's symposium held in conjunction with the ICHST in Manchester, UK in 2013, we held our largest joint session. The authors have met at many other conferences and gradually the idea of this book was developed. Some of the essays in this volume are elaborated versions of previous conference papers, while some are the result of brand new research.

Despite its technological simplicity, the bicycle has managed to prove its socially complex character. Previous publications in bicycle history have brought up a variety of viewpoints from which we can examine the bicycle and its history. With this book, we present a multitude of aspects of the bicycle and furthermore aim to open new angles upon its history. This volume does not provide just another contribution in the history of the bicycle's technological development: the focus of the book is to examine the integration and disintegration of the bicycle in society. The broad geographical spread of the essays extends the examination to the international level. In the process, we ask a number of questions: How noticeable has the bicycle been in various societies in different periods? Which roles have the bicycle played in society?

Bicycle history provides a peephole to the values and conventions of societies at the time of study. History has witnessed various public debates that have arisen around bicycles and cyclists. Actually, the bicycle has often seemed bigger in spirit than its physical dimensions might give a reason to suppose. Bicycles have on one hand helped to solve many problems of everyday life but on the other hand they have also ignited numerous public debates. For different reasons, bicycles have also been regarded as a nuisance or even a threat to the social order. Attempts to discriminate against bicycles and cyclists have initiated reactions and a struggle to avoid marginalization. Since its heyday during the post-World War II period, over the past fifty years cycling has developed as a counterculture to automobility.

While this edited volume aims to bring up new approaches and viewpoints to bicycle history, it also puts the field in a broader societal context. The authors claim that car drivers have not only considered cyclists undetectable but historians have also often ignored their significance. Bicycles are regarded as too commonplace to be recognized by mainstream historians. In contrast, bicycle historians consider cyclists as an informal social movement, which substantially contributed to changing the world. For example, in the early twentieth century, bicycles broadened the micro-cosmos of everyday life for various population groups, enabling them to be more mobile and to work, shop and socialize within much larger areas than before. We hope that this edited volume will contribute to a new, broader and more diversified bicycle history.

Formal and informal networking has supported the compilation of this edited volume. We are grateful for several commentators and examiners of our earlier manuscripts; we have learned a lot from their feedback. The book has also benefitted from the financial support by the *Academy of Finland* to our research projects, such as "Gendered Technology of Everyday Life in the Past" (project no. 110034). Some of the authors in this volume have gained encouragement from the prizes awarded to them by scholarly societies, such as the Society for the History of Technology (SHOT). Bicycle history has had a close relationship to the history of technology, but as this edited volume indicates, it is today also expanding towards many other fields.

# Illustrations, Graphs and Tables

## Illustrations



## Graphs



## Tables

## Note on Contributors

*Peter Cox*

is Professor of Sociology at the University of Chester, UK. He chairs the international research network "Scientists For Cycling" for the European Cyclists' Federation and was a founder member of the Cycling and Society Research Group in 2004. Recent research has focused on social movements and cycling activism. His publications include *Cycling and Society* (with D. Horton and P. Rosen, 2007)*; Moving People: Sustainable Transport Development* (2010); *Cycling Cultures* (2015) and *Cycling: Toward a sociology of Vélomobility* (2018).

*Martin Emanuel*

is a historian of technology who received his PhD from Royal Institute of Technology, Stockholm in 2013. He is affiliated with the Dept. of Technology Management and Economics, Chalmers University of Technology in Gothenburg. Emanuel's expertise is at the intersection of mobility, urban, and environmental history, including tourism history. His research has focused on wide ranging historical aspects of cycling culture, urban planning and traffic management. He is author of *Trafikslag på undantag: Cykeltrafiken i Stockholm 1930–1980* [*Excluded through Planning: Bicycle Traffic in Stockholm 1930–1980*] (2012) and co-editor and author of *Cycling Cities: The European Experience* (SHT 2016; 2nd edition MIT forthcoming 2018).

*Tiina Männistö-Funk*

received her PhD from the University of Turku, Finland in 2014. She is a historian of technology currently working as a researcher at the Chalmers University of Technology, Sweden. Her research deals with everyday practices and situations around such technologies as the bicycle, gramophone, photography, street pavements and copying machines. She has published on bicycle history in *Technology and Culture, Transfers*, ICON, *Cycle History*, *Tekniikan Waiheita* and edited volumes.

*Timo Myllyntaus*

is Professor of Economic and Social History at Turku School of Economics, Finland. His research interests stretch from economic and social history to environmental history and the history of technology. He has published several monographs and more than 100 scientific articles. His edited volumes include *Thinking through the Environment: Green Approaches to Global History,* (Cambridge 2011). He is the past president of the International Committee for the History of Technology (ICOHTEC).

*Nicholas Oddy*
is Head of Design History and Theory at Glasgow School of Art. He is Chair of the Cycling History and Educational Trust and has acted as consultant for Cycles and Cycling Memorabilia for Phillips and Bonhams Auction Rooms. He was one of the speakers at the First International Cycling History Conference in 1990 and has been active in its organization since. Recent publications include "This Hill is Dangerous" in *Technology and Culture* (2015) and "Bicycle Histories, they have a past, but do they have a future?" in *West86* (2017).

*Harry Oosterhuis*
is a historian and sociologist who teaches at the Faculty of Arts and Social Sciences of Maastricht University. He is also affiliated to the Huizinga Research School for Cultural History in Amsterdam. His research focuses on the cultural and social history of psychiatry and mental health care, of sexuality and gender, of health and citizenship, and of bicycling and cycling policies. His books include *Stepchildren of Nature: Krafft-Ebing, Psychiatry, and the Making of Sexual Identity* (2000), *Verward van geest en ander ongerief: Psychiatrie en geestelijke gezondheidszorg in Nederland (1870–2005)* (2008) and *Health and Citizenship: Political Cultures of Health in Modern Europe* (2014/2016).

*M. William Steele*
is Professor Emeritus at the International Christian University in Tokyo, Japan. He specializes in the social and cultural history of Japan in the late nineteenth century. His major publication is *Alternative Narratives in Modern Japanese History* (2003). His recent publications focus on mobility and environmental issues, including the history of rickshaws, bicycles, and automobiles in modernizing Japan.

*Manuel Stoffers*
is assistant professor of history at the Faculty of Arts and Social Sciences of Maastricht University. His research focuses on the history of collective ideas, meanings and mentalities, and most recently includes the history of cycling cultures. He has published on cycling history in the *Journal of Transport History, Mobility in History, Transfers, Cycle History,* BMGN*/The Low Countries Historical Review* and several edited volumes. In 2009, he launched an online international cycling history bibliography (www.fasos-research.nl/sts/cyclinghistory).

*Sue-Yen Tjong Tjin Tai*
is a historian of technology and business historian. She received a PhD from Eindhoven University of Technology for her dissertation *Connecting Small Firms for Innovation. Roles of Trade Associations and the Dutch Rijksnijverheidsdienst,*

*1900–1940*. She is currently studying the role of the Raleigh Bicycle Company in the glocalization of cycling.

*Frank Veraart*

is an assistant professor in the history of technology at Eindhoven University of Technology. His research is focused on the historical development of sustainability issues. His research spans from the history of mobility, to spatial planning and computing. Among his publications are *Cycling Cities: The European Experience* (Eindhoven, 2017) and *Well-Being, Sustainability and Social Development: The Netherlands 1850–2050* (Berlin, 2018).

CHAPTER 1

# Introduction: The Historical Production of the Invisible and Visible Bicycles

*Tiina Männistö-Funk*

## 1 Introduction

"Sorry mate, I didn't see you." This line, delivered by car drivers to cyclists after collisions or close shaves, has been deemed so typical in England that it has merited an acronym of its own: SMIDSY. In some situations, however, cyclists seem too visible and too easy to notice. In his 2013 BBC 'Future' column, Tom Stafford argues that car drivers do not hate cyclists only because they are annoying, but because they offend the moral order of the road by merely being there.[1] Questions of the societal and cultural visibility and invisibility of the bicycle seem extremely topical: our book endeavours to introduce historical insight into these debates. We approach the origins and developments of the bicycle as a mode of mobility, as a socio-technical product, and as a cultural category. We argue that taking the history of the bicycle into account allows us to understand why the bicycle and cyclists struggle to gain visibility in practice, although the bicycle is arguably the single most prominent symbol of the most recent discussion on, and plans for, a new urban culture. In this introductory essay, I will first discuss invisibility and visibility as theoretical concepts, before broadening the scope to representations of cycling and to the content of this volume.

In certain respects, the bicycle has a high visibility in the contemporary Western world. One after the other, major European cities have declared the promotion of the bicycle as one of their tools in fighting traffic congestion, air pollution and health issues. Not all have agreed with this development. In some public discourses, the bicycle has become synonymous with specific political or cultural identities in the views of both the supporters and adversaries of bicycle promotion. In January 2016, in an interview with the German newspaper, *Bild*, Poland's foreign minister, Witold Waszczykowski, accused the previous centre-right government of following a left-wing agenda of mixing

1 Tom Stafford, "The psychology of why cyclists enrage car drivers", BBC *Future*, 12 February 2013. http://www.bbc.com/future/story/20130212-why-you-really-hate-cyclists.

© KONINKLIJKE BRILL NV, LEIDEN, 2019 | DOI:10.1163/9789004289970_002

cultures and races, while striving for "a world of bicyclists and vegetarians" that would support renewable energies and reject religion. From his point of view, such visions were in conflict with traditional Polish values and a disease that must be cured.[2]

While the bicycle is made visible in the most recent urban agendas and political discourses, actual cyclists, as a group and as individuals, experience invisibility when negotiating roads and traffic. Fighting for the "Right to the Road"[3] can be conceived as a protest for more visibility. This fight has a long history, as James Longhurst has shown in his recent book, *Bike Battles*.[4] In the introduction to her recent popular history of the bicycle in America, Margaret Guroff frames her book as an answer to the car drivers who ignore or bully her on the road and deny her right to use the roads.[5] White ghost bikes set up at the roadside serve to make visible traffic accidents that have resulted in the death of a cyclist. Many challenges in bicycle research and promotion concern different invisibilities. These reach from the level of whole cycling periods, nations and cultures being invisible, to the disappearance of bicycles from many urban planning and construction projects. They also cover the often fatal invisibility of an individual cyclist in the traffic amongst cars. We argue that these blind areas make it considerably more difficult to understand and to promote cycling successfully.

In media and film studies, cultural invisibility refers to the under-representation of minorities and marginalized groups in media products.[6] In the social sciences, social invisibility refers to the sense of invisibility and social disconnectedness experienced by individuals of minority groups systematically

---

2 The original quote from Waszczykowski: "'Wir wollen lediglich unseren Staat von einigen Krankheiten heilen, damit er wieder genesen kann', rechtfertigte er das Vorgehen seiner Regierung gegen staatliche Medien: 'Dort wurde unter der Vorgängerregierung ein bestimmtes linkes Politik-Konzept verfolgt. Als müsse sich die Welt nach marxistischem Vorbild automatisch in nur eine Richtung bewegen—zu einem neuen Mix von Kulturen und Rassen, eine Welt aus Radfahrern und Vegetariern, die nur noch auf erneuerbare Energien setzen und gegen jede Form der Religion kämpfen. Das hat mit traditionellen, polnischen Werten nichts mehr zu tun.' Polen-Minister verteidigt Mediengesetz." *Bild*, 3 January 2016.

3 See Glen Norcliffe, *Critical Geographies of Cycling: History, Political Economy and Culture* (Farnham: Ashgate, 2015).

4 James Longhurst, *Bike Battles: A History of Sharing the American Road* (Seattle: University of Washington Press, 2015).

5 Margaret Guroff, *The Mechanical Horse: How the Bicycle Reshaped American Life* (Austin: University of Texas Press, 2016), 7–8.

6 Amanda Hoynes, "Mass Media Re-Presentations of the Social World: Ethnicity and 'Race'", in *Media Studies: Key Issues & Debates*, ed. Eoin Devereux (Los Angeles: Sage, 2007), 168.

overlooked by the wider public.[7] This experience of social invisibility bears similarities with the experiences of cyclists in urban areas. In these areas, the often sporadic and discontinuous infrastructural solutions, coupled with the behaviour of more dominant modes of mobility, combine to make the riding experience of the cyclist one of being overlooked and neglected. At the same time, the bicycle as a technological product is so commonplace and domesticated that it may easily go unnoticed. Bicycle technology has been in use for a long time and is not surrounded by the same kind of industrial innovation hype as that which currently characterizes automobile technology.

Daniel Miller has talked about the humility of everyday material objects—their surprising ability to disappear from our sight. This is also the core problem of writing the history of the most commonplace of cycling practices—they have been nearly invisible to their contemporaries and thus remain difficult to trace.[8] David Edgerton has proposed a history of technology-in-use that would make "a whole invisible world of technologies" visible and lead to "a rethinking of our notion of technological time, mapped as it is on innovation-based timelines."[9] As Edgerton discusses technology-in-use, Mikael Hård and Andrew Jamison have discussed the cultural appropriation of technologies as a key to alternative technological narratives. This appropriation can be studied with respect to the levels of discourses, organizations and practices.[10]

In this volume, we will be examining the different and parallel timelines of the bicycle, showing the ways it has become invisible and visible over time. We shall also be considering how it has mixed and matched with other technologies, economic and political tendencies, and cultural valuations. The different chapters are divided into three parts, loosely following Hård's and Jamison's model of three levels. The first part provides insights into the interparts of the various historical timelines of the bicycle and the concrete realities supporting or questioning these. In the second part, political, economic and bureaucratic institutions are considered as appropriators of the bicycle. In the third part, the focus is directed at those practices that have influenced the visibility of the bicycle at national, local or individual levels. In this introduction, I will first give an example of the invisibility and visibility of the bicycle in

7 Mikael Carleheden, Carl-Göran Heidegren, and Rasmus Willig, "Recognition, Social Invisibility, and Disrespect", *Distinktion: Scandinavian Journal of Social Theory* 13, no. 1 (2012).

8 Daniel Miller, *Stuff.* (Cambridge: Polity Press, 2010), 155.

9 David Edgerton, *The Shock of the Old: Technology and Global History since 1900* (London: Profile Books, 2006), xi.

10 Mikael Hård and Andrew Jamison, *Hubris and Hybrids: A Cultural History of Technology and Science* (New York: Routledge, 2005), 13–14.

popular works of science fiction, to illustrate how the bicycle is both seen and not seen, and then examine how this connects with the wider questions of the bicycle's role in contemporary societies. I will then discuss the invisibilities and visibilities of bicycle historiography and the reasons for them. Finally, the possibilities opened up by parallel bicycle histories are contemplated. The introductory article concludes with an overview of this volume's chapters and their perspectives.

## 2 Of Zombies and Bicycles

In July 2011, the American humour website Cracked.com published a list of six "technologies conspicuously absent from sci-fi movies". The first item on the list is the bicycle "in every post-apocalyptic film". Such films usually feature people who are constantly on the move, either looking for rescue, searching for resources or escaping murderous creatures, such as zombies. Strangely, as the article states, the only options ever considered are walking or using a motorized vehicle:

> [T]hat means either walking over insane stretches of possibly radioactive desolation or fighting other people for gas. That's just the way it is, though, because if the whole world has gone to shit, how else are you gonna get around? So, what's missing? How about grabbing a bike? In most of these films, there always seems to be a gap between having a vehicle and gas and being shit out of luck, as if no other possibility existed. Why don't they ride bikes? Did all the zombies eat them? Did the nukes somehow specifically target bicycles but miss all the cars? Bikes are cheap, fast and easy to maintain, plus they require no fuel and they're freaking everywhere—literally the only reason we can think of for why they are never used in these films is that they would look kinda ridiculous.[11]

Although written in jest, the article makes one wonder about the invisibility of the bicycle in science fiction and fiction at large. Searching for bicycles in this genre very soon makes it clear that they are few and far between, scattered over decades: perhaps the best-known example from literature is the surreal

11 David Christopher Bell, "6 Technologies Conspicuously Absent from Sci-Fi Movies", *Cracked.com*, 27 July 2011, http://www.cracked.com/article_19325_6-technologies-conspicuously-absent-from-sci-fi-movies.html.

postmodern novel, *The Third Policeman*, by the Irish writer, Flann O'Brien. The novel was written between 1939 and 1940 and published posthumously in 1967. It includes characters obsessed with bicycles,[12] which reflects the central role the bicycle had in the Irish countryside of the late 1930s and early 1940s.[13] It presents cycling as an affective, mystical and erotic activity that transforms the cyclist partially into a bicycle and, in turn, gives the bicycle human qualities.[14]

Among classical science fiction, we find Avram Davidson's *Or All the Seas with Oysters* that won the Hugo Award for Best Short Story in 1958. It tells of ominous, self-repairing bicycles that organically develop from safety pins and coat-hangers in a 1950s American bicycle shop.[15] Interestingly, Jeremy Withers also highlights some other examples from the 1950s, such as Robert A. Heinlein's novel *The Rolling Stones* (1952) and Poul Anderson's novella *A Bicycle Built for Brew* (1958), that similarly portray bicycles as potent and agentic, countering the era's dominant ideology of automobility.[16] Such literal examples can give us glimpses of the resistance that has existed alongside growing motorization, but which has been rendered nearly invisible by the story of modernity and the car's victorious ascendancy.[17]

Kim Westwood's 2011 novel, *The Courier's New Bike*, follows the adventures of a young bicycle courier, who cycles the streets of a dark, future version of Melbourne in a post-pandemic future, where energy, fuel and freedom are scarce.[18] Since 2013, a small publisher, Elly Blue, has produced anthologies of "feminist bicycle science fiction" named *Bikes in Space*, which was originally intended as a humorous one-off publication, but has since then grown into a

12 Flann O'Brien, *The Third Policeman* (London: MacGibbon & Kee, 1967). Flann O'Brien is one of a number of literary pseudonyms of Brian O'Nolan.

13 Flore Coulouma, "Cycling in Circles: Flann O'Brien's Free-wheeling Stories in *The Third Policeman*", *The International Journal of the History of Sport* 29, no. 12 (2012).

14 Mark Byron, "Modernist Wheelmen", in *Flann O'Brien & Modernism*, ed. Julian Murphet, Rónán McDonald, and Sascha Morrell (New York: Bloomsbury, 2014).

15 Avram Davidson, "Or All the Seas with Oysters", *Galaxy Science Fiction*, May 1958.

16 Jeremy Withers, "Bicycles across the Galaxy: Attacking Automobility in 1950s Science Fiction", *Science Fiction Studies* 44, no. 3 (2017). Withers also points out that bicycles appear already in H.G. Wells's *The War of the Worlds* (1898) as one of the means to try and escape the Martians, that Wells incorporated bicycles also in some of his futuristic war stories, and that from the latter part of the twentieth century there is environmental science fiction that includes bicycles.

17 See Peter Norton, *Fighting Traffic: The Dawn of the Motor Age in the American City* (Cambridge, MA: MIT Press, 2008).

18 Kim Westwood, *The Courier's New Bicycle* (Sydney: Harper Voyager, 2011).

series.[19] The examples from the field of film are even fewer than from literature. In a famous scene from the 1982 family science fiction film, *E.T.*, the suburban American youths who are the film's main protagonists outrun the police cars pursuing them as the small alien makes their bicycles fly. More recently, this combination of suburban childhood imagery and the freedom given by the bicycle has been used to create nostalgic 1980s feelings in the science-fiction horror series *Stranger Things*.[20] And we should not forget the British Superman parody "Bicycle Repair Man" in Series 1 of *Monty Python's Flying Circus*, first aired in 1969.

Works of science fiction tell us a lot about the attitudes and understandings of their era, both social and technological.[21] It seems that the bicycle has not inhabited a large space in our collective imagination, at least when we imagine technologies that shape our present day and future. There are multiple probable reasons for this: the bicycle is usually thought of as a rather low-tech device. It is categorized as "old" technology and marginalized as a means of transportation in most places around the world. The history of its marginalization is at the same time the history of technological, economic, social, cultural and environmental developments that have been central in shaping the world as we know it today. Therefore, even the idea of a major societal, economic and environmental meltdown is probably not enough to make the invisible bicycle visible in popular visions.

In the first episode of the first season of the horror series, *The Walking Dead*, released in 2010, we see the main protagonist, Rick Grimes, grab a bicycle and ride away from a badly mangled zombie that is trying to reach out to him. But we only see him riding for one short take on the sidewalk, before abandoning the bicycle and continuing on foot.[22] In the cyberpunk television series, *Dark Angel*, originally released in 2000–2002 by Fox, the lead character is a genetically modified soldier who works as a bicycle messenger in post-apocalyptic Seattle. Her cycling constitutes a central theme in the series, but is also used

19 Elly Blue, ed., *Bikes in Space: A Feminist Science Fiction Anthology* (Portland, OR: Microcosm, 2013); Elly Blue, ed., *Bikes in Space: Volume 2: More Feminist Science Fiction* (Portland, OR: Microcosm, 2014); Elly Blue, ed., *Pedal Zombies: Thirteen Feminist Bicycle Science fiction Stories* (Portland, OR: Microcosm, 2015).

20 *Stranger Things*, written and directed by the Duffer Brothers (Netflix, 2016), video-on-demand.

21 Thomas Haigh, "Technology's Other Storytellers: Science Fiction as History of Technology", in *Science Fiction and Computing: Essays on Interlinked Domains*, ed. David L. Ferro and Eric G. Swedin (Jefferson: McFarland, 2011).

22 *The Walking Dead*, season 1, episode 1, "Days Gone By", written and directed by Frank Darabont (AMC, 2010), television.

to highlight her outsider status as a participant of the marginalized messenger culture.

A Hugo Award-winning short story, Bruce Sterling's "Bicycle Repairman" from 1996, centres on an owner of a bicycle repair shop in an anarchist squatter community of a dystopian corporate future.[23] This story is an exception among the few works that place the bicycle in a future setting, as it includes discussions of bicycle technology. The protagonist is a bicycle enthusiast who wants to develop an intelligent inertia brake, chip-driven and designed to feel like a boost in the muscle power of the cyclist, unlike the battery-power of the electric bikes. He contrasts bicycles with computers that are black boxes with no big visible working parts, but ends up implying that cyclists prefer their bicycles invisible in another sense:

> People were strangely reticent and traditional about bikes. That's why the bike market had never really gone for recumbents, even though the recumbent design had a big mechanical advantage. People didn't like their bikes too complicated. They didn't want bicycles to bitch and complain and whine for attention and constant upgrading the way that computers did. Bikes were too personal. People wanted their bikes to wear.[24]

The reference to recumbent bicycles is very interesting, as they have been the most futuristic bicycles around during the whole history of the bicycle. Manuel Stoffers describes in his chapter in this volume the visions of the 1970s' and 1980s' Human Powered Vehicle movement that promoted the recumbent bicycle as the modern, scientific means of future personal transport. The bicycle repairman's explanation as to why this vision never became reality proposes that there is something about the fundamental characteristics of the bicycle that makes it a humble, invisible technology of everyday life. However, there are also highly visible bicycle cultures, such as Japanese bicycle racing, which is also portrayed and reproduced in manga and anime.[25] In this volume, we propose that this humbleness and invisibility, as well as the areas of high visibility, are historically produced and linked to many complex, overlapping and parallel developments that call for scrutiny. These developments help us to understand the social and cultural challenges the bicycle and cyclists face today.

---

23 Reprinted in Bruce Sterling, *A Good Old-fashioned Future* (New York: Bantam Books, 1999), 188–228.

24 Sterling, *A Good Old-fashioned Future,* 203.

25 One example of a popular bicycle racing manga is the series *Yowamushi Pedal,* written by Wataru Watanabe. Between 2008 and 2016 it was published in 47 volumes and produced into an anime series as well as a television drama film.

## 3 Typical Timelines and Beyond

For more than two decades now, bicycle history has been an active field of study, academically situated mainly inside the history of technology, but with another strong line originating from half- or non-academic organizations for bicycle history. The historiography of the bicycle and cycling contains a multitude of diverse studies, ranging from detailed accounts of technological development or bicycle racing to social histories of cycling and theoretical approaches on bicycle use and innovation. Despite the rather large volume of research and the high quality of many books and articles on bicycle history, the actual scope covered has often been surprisingly limited.

Three limiting tendencies can be detected when we study the timelines of bicycle history. Firstly, bicycle historiography is heavily biased towards the very early days of cycling, giving less attention to other eras. Moreover, the typical timeline in the studies has tended to follow a predictable trajectory: the bicycle boom in the late nineteenth century that is followed by a loss of status. Then comes the mass-cycling of the first half of the twentieth century, which is followed by the post-war all-time-low of cycling. Finally, there appears a "cycling renaissance," starting in the 1970s. Last, but not least, the history of motorized traffic has often forced its own timeline on bicycle history, so that the bicycle has been staged as the predecessor of the car in the early period, and a poor man's substitute for a motorized vehicle later on. The current decade has witnessed an increasing number of studies scrutinizing the story of the bicycle's rise and fall from new perspectives, as well as broadening the variety of subjects in bicycle history. This volume is intended as a contribution to these perspectives.

Even as we want to question the typical timelines of bicycle history, we acknowledge that there are valid and apparent reasons for their existence. The bias towards nineteenth-century cycling is perhaps the most striking, but also the easiest to explain.[26] A good example of this bias is David Herlihy's 2004 work, *Bicycle: The History,* one of the rare attempts to chart the whole history of the bicycle in one book. Three-quarters of this book are dedicated to nineteenth-century cycling, and the remainder is left to cover the entire twentieth century.[27] Herlihy writes extensively on the early and

---

26 For more on this bias, see Manuel Stoffers, Harry Oosterhuis, and Peter Cox, "Bicycle History as Transport History: The Cultural Turn", in *Mobility in History: Themes in Transport: T2M Yearbook 2011*, ed. Gijs Mom, Peter Norton, Georgina Clarsen, and Gordon Pirie (Neuchâtel: Presses Universitaires Suisses, 2011), 266–67, 273.

27 David V. Herlihy, *Bicycle: The History* (New Haven and London: Yale University Press, 2004).

mid-nineteenth-century history of *draisines* and velocipedes. He then moves on to consider the subject that has enjoyed the strongest interest of bicycle historians for decades: the sporty, fashionable, Western upper and middle class riding of the high wheelers in the 1870s and the 1880s, and the safety bicycles of the 1880s and the 1890s.

Glen Norcliffe has analyzed late nineteenth-century bicycle use in Canada, and especially what he calls "the mania" of the 1890s, as a phase of industrial modernity that helps us to better understand the whole logic of modernity. After the end of the Canadian bicycle boom in 1898, the bicycle becomes, in his view, a commonplace and boring matter.[28] This might seem a rather harsh judgment, but if one is mainly interested in cycling as a fashion and a novelty of the wealthy few in the urban, Western, industrialized settings, it is a plausible conclusion. Such a point of view has led to the overrepresentation of the "bicycle booms" or "bicycle crazes" in research, visible also in Herlihy's book. Like Norcliffe, many others have analysed the culture and social relations of these upper and middle-class circles through the study of Victorian cycling and the Victorian representation of the bicycle, sometimes also comparing it to the beginning of a different cycling era in the early twentieth century.[29] Countries such as Great Britain, Germany, the United States and Canada have dominated bicycle history, and this has also had an impact on the timelines perceived as typical in the history of the bicycle and cycling.

When James McGurn published one of the earliest general histories of the bicycle in 1987, he wanted to draw attention to the discarded role of the bicycle in social history.[30] Even now, almost thirty years later, the cycling that has been most commonplace and popular is still the cycling we know least about, and

28 Glen Norcliffe, *The Ride to Modernity: The Bicycle in Canada, 1869–1900* (Toronto, Buffalo, and London: University of Toronto Press, 2001), 247–56.

29 E.g., Daniel Di Falco, "Prometheus fährt Rad: Die Mythen der industriellen Moderne in der frühen Fahrradwerbung", in *Bilder vom besseren Leben: Wie Werbung Geschichte erzählt*, ed. Daniel Di Falco, Peter Bär, and Christina Pfister (Bern: Haupt, 2002); Anne-Katrin Ebert, "Liberating Technologies? Of Bicycles, Balance and the 'New Woman' in the 1890s," ICON 16:1 (2010); Sarah Hallenbeck, *Claiming the Bicycle: Women, Rhetoric, and Technology in Nineteenth-Century America* (Carbondale: Southern Illinois University Press, 2016); Phillip Gordon Mackintosh, "A Bourgeois Geography of Domestic Cycling: The Responsible Use of Public Space in Toronto and Niagara-on-the-Lake, 1890–1900", *Journal of Historical Sociology* 20, no. 1–2 (2007); Rüdiger Rabenstein, *Radsport und Gesellschaft: Ihre sozialgeschichtlichen Zusammenhänge in der Zeit von 1867 bis 1914* (Hildesheim, München, and Zürich: Weidmann, 1991); Christopher S. Thompson, "Bicycling, Class, and the Politics of Leisure in Belle Epoque France", in *Histories of Leisure*, ed. Rudy Koshar (Oxford and New York: Berg, 2002).

30 James McGurn, *On Your Bicycle: An Illustrated History of Cycling* (London: John Murray, 1987), 9.

the eras, as well as the places, that witnessed the most widespread social significance of the bicycle, are the ones least studied. In many Western countries, the first decades of the twentieth century introduced the first dissemination of mass-produced bicycles outside the wealthiest classes. These decades also saw the beginnings of the broad, everyday use of the bicycle that would in many places make the bicycle the most common vehicle on the road for several decades. Although adult bicycle use declined in America at the dawn of the new century, in Europe, the interwar years meant soaring bicycle sales and ubiquitous bicycle use. In European urban spaces, the bicycle became and stayed the dominant form of transportation from the First World War to the beginning of the 1960s.[31] There are quite practical reasons behind the lack of historical research on this widespread cycling. Already during the first decades of the twentieth century, the uses and the users of the bicycle changed in ways that have made them more difficult to study historically. Although more numerous than in the nineteenth century, early twentieth-century cyclists left behind far less explicit traces of and documents about their cycling. While the fashionable cyclists of the late nineteenth century made a public spectacle of their cycling and left an easily researchable paper trail of bicycle club archives, bicycle journals, newspaper reports and picture documents, the early twentieth-century cyclists drew little attention and produced fewer documents in comparison.[32]

Once cycling became a common and important part of everyday life in many industrialized countries, there were cyclists everywhere, but from a historical perspective they are mostly anonymous, almost invisible. In a parallel way, Harry Oosterhuis and Manuel Stoffers have pointed out that there has been a relative lack of interest in bicycle history in Denmark and the Netherlands, the two leading bicycle countries of the Western world. In their opinion, this may be because of the self-evident and ingrained nature of cycling and bicycles in the everyday life of these countries. Therefore, cycling calls for less reflection and debate than in countries where the status of cycling and cyclists is more problematic and a subject of controversy.[33] Thus, we can detect a certain invisibility of the most common everyday form of cycling, both historically and

31 Ruth Oldenziel and Mikael Hård, *Consumers, Tinkerers, Rebels: The People Who Shaped Europe* (Hampshire: Palgrave Macmillan, 2013), 131.

32 Nicholas Oddy, "Cycling's Dark Age? The Period 1900–1920 in Cycling History", in *Cycle History 15: Proceedings of the 15th International Cycling History Conference*, ed. Rob van der Plas (San Francisco: Van der Plas and Cycle, 2005), 79–86.

33 Harry Oosterhuis and Manuel Stoffers, "A Strong Presence, but a Weak History: The Bicycle in Dutch Historiography", in *Cycle History 21: Proceedings of the 21st International Cycling History Conference*, ed. Andrew Ritchie (Birmingham, UK: Cycling History, 2012), 166–71.

geographically, whereas the relatively small number of enthusiastic cyclists that view cycling as a lifestyle or a hobby have often attracted more attention.

## 4 The Rise and Fall Revisited

The history of rise and fall, often seen from the perspective of an anticipated renaissance in our time,[34] is the currently dominant narrative of bicycle history. It makes us see similar kinds of bicycle timelines—defined by economic shifts and motorization—everywhere we look. This generalization may prevent us from considering the actual diversity of twentieth- and twenty-first-century bicycle history. After a closely similar cycling fashion in all of the Western industrial countries in the late nineteenth century, the twentieth century saw large differences forming between countries and areas in the popularity of cycling, bicycle infrastructure and cultural meanings of the bicycle.

The most profound difference appeared between the United States and Europe. In the United States, adult cycling effectively disappeared at the beginning of the twentieth century, as bicycles came to be perceived as a vehicle for children[35]. Although the 1930s saw a rise in cycling as a leisure activity, the role of cycling in American life remained small. According to Herlihy, in the mid-1930s, the United States had a total of 2–3 million bicycles, whereas Germany had 15 million, Britain and France both 7 million and the Netherlands 3 million, the last number roughly translating into one bicycle for every second Dutch citizen.[36] A closer look also reveals significant differences between European countries. For example, in the Netherlands, the bicycle achieved a strong position as a vehicle of the whole nation, whereas in Germany, its cultural, social and political role was contested.[37]

A history of rise and fall can be a practical and compact way of giving some basic information to a reader not familiar with the history of the bicycle. It is also relatively useful in sketching the major developments of cycling in the leading bicycle countries. However, it also presents pitfalls. From this perspective, subtle differences, alternative stories and local developments seem trivial or become difficult to discern. It can also easily make all of the diverse forms of and reasons for bicycle use and non-use appear as monolithic reactions

34 See, e.g., Tim Birkholz, "Die stille Revolution—Das Fahrrad kommt zurück", in *Das Fahrrad: Kultur, Technik, Mobilität,* ed. Mario Bäumer (Hamburg: Junius, 2014).

35 See Longhurst, *Bike Battles*.

36 Herlihy, *Bicycle*, 322–28.

37 Anne-Katrin Ebert, *Radelnde Nationen: Die Geschichte des Fahrrads in Deutschland und der Niederlanden bis 1940* (Frankfurt am Main: Campus, 2010), 419–25.

to economic up- and downturns or the rise of the car. In more recent times, this generalizing trend has been counteracted by historical and contemporary studies that concentrate on the cycling culture and policy of specific cities.[38]

We can also apply postcolonial criticism to such Western-centric historical periodization and universalist historical narratives that bring with them the tendency of comparing the rest of the world to the timelines based on developments in a few Anglo-Saxon countries.[39] The Eurocentric problematic is apparent, for example, in comments that make comparisons between the idea of the current Western bicycle "renaissance" and the contrasting preference for cars in "classic bicycle nations such as China and India".[40] This frames the Chinese and Indians as followers of the same old Western timeline, just dragging behind in progress, and renders the domestic bicycle histories and timelines of these countries effectively invisible.

In the case of India, the timeline of bicycle history was strongly influenced by the struggle for independence under British colonial rule. The bicycle was one important instance of *swadeshi*, promoting indigenous industries, production and habits. Prior to the 1950s, India had to rely mainly on imported bicycles, but the bicycle became nonetheless associated with *swadeshi* ambitions, which increased its popularity and availability. In the 1960s, it was among the first modern consumer goods used also by low-caste rural Indians.[41] In China, there were solo bicycles in the late nineteenth century, but they were left in the shadow of rickshaws and pedicabs before becoming the dominant form of urban transportation in the 1950s. In the twenty-first century, local governments have favoured motorized traffic over bicycles. In this context, it is perhaps unsurprising that sales of e-bikes have soared.[42]

The example of China demonstrates the need to analyse the use of the bicycle in relation to the whole spectrum of transportation and mobility, a point

38 See, e.g., Martin Emanuel, *Trafikslag på undantag. Cykeltrafiken i Stockholm 1930–1980* (Stockholm: Stockholmia, 2012); Ruth Oldenziel, Martin Emanuel, Adri Albert de la Bruhèze, and Frank Veraart, *Cycling Cities: The European Experience* (Eindhoven: Foundation for the History of Technology, 2016); Bernard Tschofen and Yonca Krahn, *Velo: Erkundungen zu Zürcher Fahrradkultur* (Zürich: University of Zürich, 2016).

39 For more on this line of criticism, see, e.g., Dipesh Chakrabarty, *Provincializing Europe: Postcolonial Thought and Historical Difference*, new ed. (Princeton: Princeton University Press, 2007), 6–16.

40 Mario Bäumer, "Vom auf und ab des Fahrrads—Fahrradmobilität im Wandel", in *Das Fahrrad: Kultur, Technik, Mobilität*, ed. Mario Bäumer (Hamburg: Junius, 2014), 157.

41 David Arnold, *Everyday Technology: Machines and the Making of India's Modernity* (Chicago: University of Chicago Press, 2013), 96–107.

42 Edward J.M. Rhoads, "Cycles of Cathay: A History of the Bicycle in China", *Transfers: Interdisciplinary Journal of Mobility Studies* 2, no. 2 (2012).

illustrated in this volume by Peter Cox's chapter that also further discusses the problematic nature of periodization. The advance of motorization has shaped the timeline of bicycle history in many ways. After the bicycle fashion of the 1890s was over, wealthy Westerners turned with similar enthusiasm towards the car. The discourse of mankind's victory over time and space was also directly relocated from the bicycle to the car at the beginning of the twentieth century.[43] In many European countries, touring organizations that had been promoting bourgeois cycling transformed into automobile clubs.[44] However, this is not the whole picture.

In Britain, the Cyclists' Touring Club and motorists stood in open conflict over the use of the roads during the interwar years.[45] At the same time, many early motoring enthusiasts continued cycling and celebrated the links between motor and pedal traffic, not the differences.[46] Accordingly, comparison with the car are often accurate when writing the history of the bicycle, but not in the narrow framing of the bicycle as a precursor of the car, as the development is often characterized in general transport histories.[47] In a closer analysis, the relationship between the bicycle and the car forms a much more complex picture than a simple timeline of growing motorization and diminishing cycling would suggest. The dominance of the car and the marginalization—invisibility—of the bicycle was in Europe not a result of widespread motorization, but a phenomenon preceding it, as recent studies have made clear.

In the 1920s and 1930s, local authorities and urban planning experts throughout Europe carried out traffic solutions that prioritized cars and excluded or marginalized cyclists. This happened at a time when the bicycle was overwhelmingly the most popular mode of transport and in many places carried more people than all other modes of transport together. Although car sales increased in many countries dramatically, the overall number of cars remained low. In the whole of Europe, there were seven bicycles to one car in the 1930s. The United States, where there were seventeen cars to one bicycle, was an anomaly in the industrialized world. Despite this, European governments and experts saw cycling as an old-fashioned, outmoded means of transportation that needed to be tamed or removed, as cyclists were considered an

43 Di Falco, "Prometheus fährt Rad", 92–96.

44 Oldenziel and Hård, *Consumers, Tinkerers, Rebels*, 147.

45 Peter Cox, "'A Denial of Our Boasted Civilisation': Cyclists' Views on Conflicts over Road Use in Britain, 1926–1935", *Transfers: Interdisciplinary Journal of Mobility Studies* 2, no. 3 (2012): 4–30.

46 Carlton Reid, *Roads Were Not Built for Cars: How Cyclists Were the First to Push for Good Roads & Became the Pioneers of Motoring* (Washington: Island Press, 2015).

47 Stoffers, Oosterhuis, and Cox, "Bicycle History as Transport History", 269, 273–74.

undisciplined nuisance to motorized traffic.[48] In many places, traffic planning practically excluded the bicycle from its inception, as cycling was not perceived as actual transport.[49] This historical exclusion is still prone to make cycling in many cities of the Western world an act of poaching, as Michel de Certeau has named the tactics of individuals finding their way in environments defined by strategies of institutionalized power.[50]

The bicycle falls between transport planning's main divide of quick and slow traffic, as it might be both or neither. This is a major cause of conflict between cyclists and other road users. It is also remarkable how the solutions to this in-between problem differ from one country and region to another, and how they have been affected by different groups of cyclists and non-cyclists. For example, in America, the decision not to build separate bikeways during the last decades of the twentieth century, but to place cyclists on highways, was not only supported by highway engineers, but also by club cyclists who saw the bikeways as an obstacle to high-performance road cycling.[51]

Motorization and cars have influenced the history of the bicycle in many ways, but not simply by eroding the popularity of the bicycle. A growing number of cars is not the reason for the invisibility of the bicycle. Problems caused by motorized traffic have actually increased bicycle visibility, at least on the level of policy and publicity. This was also true of earlier bicycle advocacy. The rise of bicycle activism in the 1970s predated the oil crisis and was caused by a myriad of environmental concerns and other problems linked to increased car use and dependency.[52] Bicycle advocacy in the Western world can also be conceived as the outcome of these problems, and it has directly or indirectly inspired a large volume of bicycle research, a subject discussed in Peter Cox's chapter in this volume. Even at the level of practices, the dominance of cars can have both obstructive and conducive effects on cycling. The relative

48 Martin Emanuel, "Constructing the Cyclist: Ideology and Representations in Urban Traffic Planning in Stockholm, 1930–70", *Journal of Transport History* 33, no. 1 (2012); Ruth Oldenziel and Adri Albert de la Bruhèze, "Contested Spaces: Bicycle Lanes in Urban Europe, 1900–1995", *Transfers: Interdisciplinary Journal of Mobility Studies* 1, no. 2 (2011); Oldenziel and Hård, *Consumers, Tinkerers, Rebels*, 145–61.

49 Johanna Kallioinen, *Pyöräilyn institutionaalinen asema liikennesuunnittelussa*, VATT-keskustelualoitteita 267 (Helsinki: VATT Institute for Economic Research, 2002).

50 Michel de Certeau, *The Practice of Everyday Life*, trans. Steven Rendall (Oakland: University of California Press, 1984), xii–xix.

51 Bruce D. Epperson, *Bicycles in American Highway Planning: The Critical Years of Policy-Making 1969–1991* (Jefferson: McFarland, 2014), 158–59.

52 Peter Cox, "Cycling, Environmentalism and Change in 1970s Britain" (presentation, Conference on Mobility and Environment, Kerschensteiner Kolleg, Deutsches Museum, Munich, 13–14 February 2015).

advantage in speed that the bicycle has over the car in many large cities—a state of affairs that has, for example, enhanced the use of bicycle couriers—is largely caused by the number of cars clogging the streets.

## 5 Parallel Bicycle Histories

The evaluation and use of history has played an important role in cycling practices throughout the existence of the bicycle, as Nicholas Oddy demonstrates in his chapter in this volume. Bicycle histories have been used as early as the 1870s to mark technological progress, and in the twentieth century also to stage the bicycle as a vehicle of the past. Such present-day phenomena as the Tweed Run present a nostalgic use of bicycle history as a form of fashionable urban lifestyle. As Oddy argues, the mountain bike's key to popularity was its ability to reinvent the bicycle outside the typical bicycle history timeline of utility riding on the one hand and bicycle racing on the other.

A different kind of reinvention was the small-wheeled unisex bicycle design popular in Scandinavia in the 1960s. It soared above the otherwise plummeting bicycle sales numbers. Inspired by the British Moulton bike (launched in 1962), the Finnish Helkama started to produce Jopo bicycles in 1964 and the Norwegian Øglænd Kombi bicycles in 1967. Unlike the Moulton bicycles, both of the latter turned into long-term commercial successes that continued into the 1970s. An interesting twist in the story took place after the initial production of these bicycles had been stopped. Both Jopo and Kombi developed into national cult phenomena in the 1990s. In 2000, both were re-launched by their original manufacturers. The new Kombi, re-named Komby, with a frame structure similar to the original Kombi, but otherwise with modern design features, was a flop that marked the end of Norwegian domestic bicycle production. The new Jopo, by contrast, made greater use of its retro potential, and its sales figures grew until it became the best-selling bicycle model in the Nordic countries by the 2010s.[53]

Awareness of history is not only useful for manufacturers wanting to create best-selling bicycle models. Studying the history of the bicycle can help us to understand the cycling practices of today and to create an informed basis for policies and strategies. As Harry Oosterhuis argues in his chapter in this volume, the historical trajectories of cycling have a strong steering power

53 Kjetil Fallan, "Kombi-Nation: Mini Bicycles as Moving Memories", *Journal of Design History* 26, no. 1 (2013); Tapani Mauranen, *Hopeasiipi—Sata vuotta Helkamaa* (Helsinki: Otava, 2005).

on people's decisions as to whether to cycle or not to cycle. According to Oosterhuis's analysis, studies carried out for the purposes of bicycle promotion rarely take the historical dimension into account. In this volume, we want to demonstrate the need to look at that dimension on several levels: local, national and transnational. We also want to study different user and non-user groups and analyse the connections the bicycle has had to diverse social, economic, political, technological and cultural currents. In this way, multiple bicycle histories become written and the restrictions of typical bicycle history timelines can be overcome. The bicycle is a global technology, and consequently one produced, used and defined in diverse local, national and transnational cultures and economies. The bicycle has not only acquired different meanings in different places at different times—it has been a different machine in different places and at different times.

Even with the lively interest in studying the history of the bicycle in recent decades, we have only begun to form a picture that takes into account the multiple histories of bicycle use. Whole continents remain virtually invisible in histories of cycling. For instance, there is relatively little internationally published research on Asian cycling and even less on South American and African cycling.[54] Hans Peter Hahn, who has studied present-day appropriations of the bicycle in West Africa, has observed the invisibility of African cycling in research. Despite its significance to local societies, African cycling practices have attracted very limited attention. Depending on the researchers' points of view, they might appear either as not traditional enough or not modern enough to be worthy of consideration.[55]

Concentration on technical development remains a major trend in bicycle histories and tends to leave large parts of the world unexamined.[56] In the Western world, rural areas form a blind area of bicycle research. Bicycles belong to the group of modern, mass-produced consumer goods that have been studied mostly in urban surroundings. Here, the research follows the general tendency of granting urban areas a decisive role in the history of modernity and of ignoring the role of rural areas.[57] However, the potential of the bicycle

54 Paul Smethurst's recent book scrutinizes Japan and China in an illuminating manner, but gives much less attention to India, Southeast Asia and Africa, and none at all to Latin America; Smethurst, *The Bicycle: Towards a Global History* (Houndsmills, Basingstokes, and New York: Palgrave MacMillan, 2015).

55 Hans Peter Hahn, "The Appropriation of Bicycles in West Africa: Pragmatic Approaches to Sustainability", *Transfers: Interdisciplinary Journal of Mobility Studies* 2, no. 2 (2012): 33.

56 See, e.g., Tony Hadland and Hans-Erhard Lessing, *Bicycle Design: An Illustrated History* (Cambridge, MA: MIT Press, 2014), xi–xiii.

57 See, e.g., Kate Murphy, *Fears and Fantasies: Modernity, Gender, and the Rural–Urban Divide* (New York: Peter Lang, 2010), 185.

to change the lives of its users has often been greatest in rural areas, where long distances need to be travelled and public transportation is scarce.[58] Studying neglected areas potentially reveals phenomena that cannot be reduced to the typical timelines and uses of the bicycle, but which open new perspectives and introduce different meanings and practices to the target of study.

For example, in West Africa, an old bicycle is seen as more useful than a brand new one, as it will probably have been modified to fit its surroundings.[59] In a peripheral Western context, rural Finns built their own bicycles out of wood and iron at the turn of the nineteenth and twentieth centuries, not to copy the behaviour of the urban cyclists and not even to ride them to any significant extent, but to demonstrate their superior handicraft skills to the village community.[60] In Sweden, the national Forest Service planned over 3,500 kilometres of bicycle trails to criss-cross the vast expanses of forest in Northern Sweden. Approximately a third of these were built in the 1930s and the 1940s to improve the working conditions of the forest workers.[61] Beside the above-discussed bias towards certain eras and areas, another kind of restriction can be detected. Anne-Katrin Ebert has argued that the fascination with the technical has overshadowed many other characteristics that have been an integral part of cycling, such as playfulness.[62]

## 6 Intersecting Levels of Action and Location

The chapters in this volume have been developed as a follow-up to the bicycle symposium organized by the editors of this volume in July 2013 as part of the 24th International Congress on the History of Science, Technology and Medicine. The symposium brought together researchers from various countries and disciplines, but, most importantly, with diverse perspectives on bicycle history. This book is a result of the diversity that we feel is of essential importance for our understanding of the bicycle, its uses and its history. The articles in this book analyse various aspects of the bicycle and cyclist invisibility and

58 Tiina Männistö-Funk, "The Prime, Decline, and Recalling of Rural Cycling: Bicycle Practices in 1920s' and 1930s' Finland Remembered in 1971–1972", *Transfers: Interdisciplinary Journal of Mobility Studies* 2, no. 2 (2012).

59 Hahn, "The Appropriation of Bicycles", 36–41.

60 Tiina Männistö-Funk, "The Crossroads of Technology and Tradition: Vernacular Bicycles in Rural Finland 1880–1910", *Technology and Culture* 52, no. 4 (2011).

61 Anna-Maria Rautio and Lars Östlund, "'Starvation Strings' and the Public Good: Development of a Swedish Bike Trail Network in the Early Twentieth Century", *The Journal of Transport History* 33, no. 1 (2012).

62 Ebert, *Radelnde Nationen*, 34.

visibility in research and historiography, as well as in historical developments and phenomena in diverse locations.

Our book concentrates especially on the period from the Second World War to the present day, although earlier developments are also discussed in some of the chapters. Our common aim is to highlight the importance of local context in the use of global technology, and to study the role of cultural interpretations, patterns of use, and social and economic significance, as well as policymaking in the history of the bicycle. We also want to ask how cycling has affected matters, instead of only concentrating on the influence of other factors on the bicycle. While the effects of the car are everywhere to be seen, heard, touched and smelled, bicycles leave minimal traces.

The chapters of the book are loosely arranged to follow the tripartite divide of discourses, organizations and practices proposed by Hård and Jamison. The chapters by Peter Cox and Harry Oosterhuis are not explicitly discourse studies, but highlight the connections between discourses and the material world in bicycle history and historiography. They show the actual material effects of the ways cycling and cycling history are discussed and understood. The chapters in the second part of the book, by Martin Emanuels and Sue-Yen Tjong Tjin Tai and Frank Veraart, focus on institutional and industrial actors that in several ways provide a backbone to the historical developments of cycling between the general discursive and basic practical levels. The chapters by William Steele, Manuel Stoffers and Nicholas Oddy constitute the last part of the volume and scrutinize the practical level of bicycle use, both commercial and private.

Some of the phenomena studied in the chapters are transnational with national and local specialities, while others are national or local developments embedded in transnational and global history. Tweed runs, mountain bikes, the Human Powered Vehicle movement, and cycling advocacy are such transnational phenomena, dealt with by Nicholas Oddy, Manuel Stoffers, and Harry Oosterhuis in this volume. National and local studies include Japanese *keirin* races, the Dutch bicycle industry, the history of cycling policy in Stockholm, and British cycling during the era of declining numbers of cyclists. These studies are written by William Steele, Sue-Yen Tjong Tjin Tai and Frank Veraart, Martin Emanuel and Peter Cox.

William Steele shows in his chapter how bicycles, in the form of popular *keirin* races, acquired a decisive role in shaping the future of the nation in postwar Japan. Sue-Yen Tjong Tjin Tai and Frank Veraart study in their chapter the formation of the small-scale Dutch bicycle industry and the specific Dutch bicycle style resulting from active economic and technological exchange with large-scale bicycle manufacturing countries. It is interesting to consider these two chapters in the light of the more general bicycle histories, where Japanese bicycle exports and bicycle-related innovations are often mentioned in the context of their effects on the bicycle market and technology of Europe and America,[63] while the social, economic and cultural effects of the bicycle industry in Japan are left untouched. The Netherlands, for its part, often serves as a point of comparison and scrutiny in histories of cycling due to the social and cultural importance the bicycle has achieved in the country. However, its economic and industrial bicycle history remain largely unknown internationally.

The various chapters also provide concrete examples for revisiting specific points made in typical bicycle history timelines by closely analysing their dynamics in different environments. Peter Cox's chapter argues strongly for the broadening and contesting of the linear artefactual history of the bicycle through studying use-centred histories in different locations. By introducing new perspectives into the 1950s' and 1960s' cycling in the UK, he highlights the importance of national and local scrutiny of such transnational phenomena as the decline of bicycle use during the decades following the Second World War. Martin Emanuel's chapter demonstrates how this is also true in the case of the "cycling renaissance" usually conceived in Western cities as starting in the mid-1970s and which is often seen as a relatively smooth process. When studied at the level of one city, Stockholm, this "renaissance" is revealed as a wave-like development affected by numerous political, cultural and social changes in the local government, which made work on bicycle infrastructure highly non-linear.

When reading the chapters of this volume, one can discover both similarities and differences inside and across all the levels: local, national and transnational. One can also detect a certain resilience in the bicycle cultures described. As Harry Oosterhuis argues in his chapter, local and national user practices and patterns of bicycle use have proven remarkably difficult to change. It is difficult to trigger growth in cycling by external policy measures. However, cycling has also resisted traffic planning and structures that have aimed to exclude it. The same flexibility that causes problems in traffic planning and everyday commuting has allowed the bicycle to exist in niches and find its places beside and among newer and more visible technologies. For example, many large factories in different branches of manufacturing use bicycles as a means of quick and easy transportation around their plants. Bicycles have even been used to operate the largest and perhaps most complex machine in the world: the Large Hadron Collider in CERN is situated in a 27-kilometre-long ring-tunnel, where technicians working on it get around on bicycles. CERN actually has a longer

63 See Herlihy, *Bicycle*, 316, 336, 364–65; Hadland and Lessing, *Bicycle Design*, 131–33, 254, 277.

tradition of utilizing the bicycle: in the 1960s its main data network relied on bicycles that carried data tapes to the computer centre.[64]

As a conclusion to this introduction, we will return to the bicycle in science fiction by creating a small speculative fiction of our own. Let us imagine a world otherwise relatively similar to ours, but with bicycles as the main vehicles of personal transport. When immersing in this thought experiment, we soon discover that it necessitates creating an imaginary world that differs from ours in many ways, not only in the realm of mobility. Our speculative bicycle-run world sounds, smells and looks different. It has different patterns of work and leisure. Structures of cities and rural areas are different.[65] And it must have a different history. Even if it might be tricky to create a plausible alternative history that would have led to the supremacy of bicycles, in the days of heightened awareness of the overuse of natural resources it is relatively easy to imagine futures in which motorized personal mobility has become impossible or very rare. Imagining a future of a human society without bicycles is much more difficult. The bicycle might not inhabit a large space in the collective imagination of either the past or future, but it has already proven itself a device that can survive and adapt to major societal and technological changes.

64 Hazel Morris, "Interesting Facts", LHC Machine Outreach, compiled April 2004, http://cern.ch/lhc-machine-outreach/lhc-interesting-facts.htm.

65 Although maybe not dramatically so: see Colin Pooley, "Landscapes without the Car: A Counterfactual Historical Geography of Twentieth-Century Britain", *Journal of Historical Geography* 36, no. 3 (2010).

# PART 1

## *Discourses and Materialities of the Bicycle*

CHAPTER 2

# Rethinking Bicycle Histories

*Peter Cox*

## 1 Introduction

As studies of the bicycle and cycling come to international recognition as legitimate subjects of academic inquiry in transport studies and beyond, cycling historiography has emerged as a significant theme for discussion. Panels and special sessions at conferences on the history of technology, transport history, and within the broader field of mobility studies all reflect growing cross-disciplinary interest in bicycle history, and consequently pose questions as to the form and argument of its historiography and historiology.[1] Yet all of these studies are relative latecomers to a pre-existing interest in bicycling history outside of the academy.

As with the study of other mobility technologies such as railways and aviation, popular histories, enthusiast publications and academic studies sit side-by-side to comprise a varied body of literature in the United Kingdom.[2] Similarly, historical accounts appear as part of a vibrant user culture.[3] For example, the Veteran Cycle Club (founded in 1955 as the Southern Veteran Cycle Club) set out not only to conserve and ride old cycles, but also to exchange information on the history of cycles and cycling, a task it continues to pursue

---

1 See, e.g., the "Invisible Bicycle" panel (from which this volume is derived) at the 40th Symposium of the International Committee for the History of Technology (ICOHTEC), Manchester, July 22–28, 2013; the roundtable on "Cycling History and Cycling Policies" at History and Future of Intermodal Mobilities: The 10th Annual Conference of the International Association for the History of Transport, Traffic and Mobility (T2M), Madrid, November 15–18, 2012; the sessions on "Velomobilities" and "Cycling Futures" at Networked Urban Mobilities: The 10th Anniversary Cosmobilities Conference, Copenhagen, November 5–8, 2014.

2 Given limitations of space, this chapter is largely confined to a detailed examination of English language sources. Broader issues of historiography are, however, international, and the international dissemination of much English language material through study networks means that significant historiographical issues cross geographical and disciplinary boundaries.

3 Some of this is reflected in regular historical accounts and articles appearing in newsstand magazines such as *Rouleur* and *Cycling Plus*.

© KONINKLIJKE BRILL NV, LEIDEN, 2019 | DOI:10.1163/9789004289970_003

even today through its various publications.[4] The publishing of cycle histories in mass distribution book form has understandably coincided with periods of general public interest in cycling in its many forms. These vary according to whether the interest is in the general social culture of the bicycle, its technological aspects or in cycle sport. The task of this chapter is not to present an overview of the literature, but to examine some of the underlying narratives (re)produced by the ways that histories of cycles and cycling are written.[5]

Book length studies of cycling history are relatively few in number but dominated by accounts of the nineteenth century, exploring developments up to the construction of the safety bicycle in its recognizable "modern" form.[6] Consequently, twentieth-century bicycle history is rather less explored, except partially via the new wave of coffee table picture books on bicycle design, and these take little or no consideration of cycling practices.[7] In fact, the very term 'bicycle history' creates its own narrative of the technology separate from use and users. As academic research in cycling has grown in recent years, assumptions concerning cycling and bicycle history emerge, reflecting and reiterating approaches dating "to the 1970s and 1980s, when the hopes and wishes of cycling advocates inspired cycling historiography".[8] Despite the pre-figurative work of enthusiast networks, cycle and cycling historiology is deeply bound to the political spaces inhabited by cycling—especially in light of the growth of counter-culture environmentalism in the 1970s and of subsequent discourses of sustainability from the late 1980s. This "new master narrative" of cycling history identified by Stoffers and Ebert is very much framed by a declensionist history: dominated by a narrative focused on reductions across Europe in

4 See the club's website http://www.v-cc.org.uk/ for its extensive range of activities and publications.

5 For an overview of the literature, see the online "Cycling History Bibliography" (compiled and maintained by Manuel Stoffers), Maastricht University, Faculty of Arts and Social Sciences, accessed 28 December 2017, http://fasos-research.nl/cycling-history-bibliography-database/.

6 Andrew Ritchie, *King of the Road: An Illustrated History of Cycling* (London: Wildwood House, 1974); Jim McGurn, *On Your Bicycle: The Illustrated Story of Cycling* (London: John Murray, 1987); David V. Herlihy, *Bicycle: The History* (New Haven, CT: Yale University Press, 2004). Publishers' demands on nineteenth-century bicycle historians to provide a complete and universal narrative are partly responsible for this tendency.

7 E.g., Gerard Brown and Graeme Fife, *The Elite Bicycle: Portraits of Great Marques, Makers and Designers* (London: Bloomsbury Sport, 2013); Michael Embacher, *Cyclepedia: A Tour of Iconic Bicycle Designs* (London: Thames and Hudson, 2011). An even bigger selection of similar works concentrates exclusively on racing bikes; see, e.g., Richard Moore and Daniel Benson, *Bike! A Tribute to the World's Greatest Cycling Designers* (London: Aurum Press, 2012).

8 Manuel Stoffers and Anne-Katrin Ebert, "New Directions in Cycling Research: A Report on the Cycling History Roundtable at T2M Madrid", *Mobility in History* 5 (2014): 13.

cycling as everyday transport in the 1950s, followed by a partial renaissance from the 1970s onward.[9]

The argument of this chapter is not to refute these accounts of changing patterns of cycle use, but to consider how the writing of cycling histories constructs norms, how these dominant narratives obscure other stories, and to consider the impact of these underlying discourses on our understanding of cycles and cycling. To do this, I will first consider some of the practical and theoretical issues—both constraints and opportunities—confronting academic research on the topic. Subsequently, the problems of two identifiable patterns in cycling historiography are examined. First the constraints produced by periodization are confronted, in which where time periods are characterized by particular patterns of use. Second, taking a parallel example from environmental history, the place of declensionist narratives is questioned for the manner in which these narratives reproduce both a sense of inevitability and tend towards technological determinism. The final parts of the chapter re-examine cycling in the UK 1950–1970, to illustrate how different emphases might present alternative perspectives on the period.

## 2 Issues of Theory in Cycling Research

One of the first problems encountered when we come to study the bicycle or cycling is that the actual subject itself is often obscure. Although the bicycle appears to be a self-evident object it nevertheless has layers of use and meaning that are not always apparent.[10] Even the act of riding a bicycle (or tricycle, or …) can mean very different things to different participatory constituencies: transport, sport or play. This diversity is also constantly value-laden, reflecting social diversity by class, age, gender and many other markers.[11] Hence the potential schools of academic study and theoretical perspectives that may relevantly be brought to bear on the subject are many and diverse. To study cycling history as history we can choose from a range of approaches informed by, for example, social studies in technology, transport history, economic history, environmental history, sports history and the sociology of sport, together with disciplines as diverse as engineering, sociology and film studies. The emergence of mobilities as a field of study in its own right provides a further

9 Stoffers and Ebert, "New Directions", 13.

10 Luis Vivanco, *Reconsidering the Bicycle: An Anthropological Perspective on a New (Old) Thing* (New York and London: Routledge, 2013).

11 Peter Cox, ed., *Cycling Cultures* (Chester: University of Chester Press, 2015).

layer of complexity.[12] While some mobilities scholarship has been criticized for the weakness of its historical perspectives, it is clear that historical dimensions have become both a vital and lively aspect of current mobilities analysis, and a significant amount of current cycling history operates within networks of mobilities studies.[13] The advantage of framing research into cycling and the bicycle within a mobilities perspective is that it brings with it an inherent expectation of an interdisciplinary approach to study.[14]

One set of difficulties in both interdisciplinary and transdisciplinary studies is that each academic tradition has its own legacy and bias, shaping the particularities of its narrative forms, deployment of sources and expectations. Interdisciplinary work frequently transfers method from one discipline to another, potentially raising questions of underlying epistemological and ontological concern.[15] Transdisciplinary work, destabilizing and breaking down disciplinary barriers can be even more problematic. How do we navigate between disciplines and methods and decide which ways of knowing are prioritized at any given time? Moreover, as academics we must be aware that each school has its own disciplinary demands and norms to uphold, and sometimes reputations to defend.[16] Since studies of cycling can potentially be framed within such a myriad of academic perspectives, it can be hard to correlate studies from differing backgrounds. Further, we need to also acknowledge that whatever discipline(s) and tradition(s) we as academics bring to organize our analysis, the majority of research and publications in bicycle history has been contributed by dedicated amateurs outside of the academy.[17] Before examining the case studies themselves, it is worth thinking about some of the

12 See Mimi Sheller, "The New Mobilities Paradigm for a Live Sociology", *Current Sociology* 62, no. 6 (2014).

13 E.g., Stoffers and Ebert, "New Directions"; Ester Anaya and Santiago Gorostiza, "The Historiography of Cycling Mobility in Spain in the Twentieth Century", *Mobility in History* 5 (2014).

14 Sheller, "The New Mobilities Paradigm"; Gijs Mom, "The Crisis of Transport History: A Critique, and a Vista", *Mobility in History* 6 (2015); Peter Merriman, "Mobilities, Crises, and Turns: Some Comments on Dissensus, Comparative Studies, and Spatial Histories", *Mobility in History* 6 (2015).

15 Basarab Nicolescu, "Multidisciplinarity, Interdisciplinarity, Indisciplinarity, and Transdisciplinarity: Similarities and Differences", in *Minding the Gap: Working Across Disciplines in Environmental Studies*, RCC Perspectives 2014/2, ed. Rob Emmett and Frank Zelko (Munich: Rachel Carson Center, 2014).

16 For parallels, see Emmet and Zelko, *Minding the Gap*; Johan Galtung, *Launching Peace Studies: The First PRIO Years: Strategies Findings Implications* (Grenzach-Wyhlen: TRANSCEND University Press, 2008).

17 A similar pattern is discernible in the case of railways, other diverse forms of road transport, and aviation.

recent trends in the study of cycling, starting with those from social studies in technology.

Through the work of Wiebe Bijker and Trevor Pinch, the bicycle has emerged as an archetypal motif in understanding the social construction of technology (SCOT).[18] Although a major aim of SCOT analyses has been to place technologies within their context of use and users, and to challenge or overturn heroic and linear narratives of invention and dissemination so as to rescue these from technological determinism, this very ambition has had paradoxical effects for our thinking about cycling.[19] First, it has put an emphasis on the technical artefact of the bicycle rather than its users and the spaces and contexts of its use, and, second, despite Rosen's work on 'reframing', the socio-technical "stabilization" of the safety bicycle has been taken up in non-academic accounts as implying either the impossibility of any further development of bicycles or that non-safety bicycle designs represent less valuable or less efficient deviations from a perfected technology. To complement and counter this tendency, a clearly cultural turn in bicycle studies has begun to reconsider bicycle history in relation to transport history, drawing on studies of power, social class, gender and other pertinent factors of social inclusion/exclusion, as well as social and political capital in operation.[20]

The methodological turn in studies of technology from a focus on producers (with a distinct bias towards economic history), to a more cultural approach (in which users come to the fore), has been mirrored in cycling studies. A rapid rise in ethnographic accounts of cycling practices has assisted engagement with the policy and politics of bicycling.[21] Similarly, we are beginning to see the emergence of historical studies that also focus on user accounts and

---

18 Trevor J. Pinch and Wiebe E. Bijker, "The Social Construction of Facts and Artefacts: Or How the Sociology of Science and the Sociology of Technology Might Benefit Each Other", *Social Studies of Science* 14, no. 3 (1984); also Weibe Bijker, *Of Bicycles, Bakelites, and Bulbs: Toward a Theory of Sociotechnical Change* (Cambridge, MA: MIT Press, 1995); Paul Rosen, *Framing Production: Technology, Culture and Change in the British Bicycle Industry* (Cambridge, MA: MIT Press, 2002).

19 This has clearly not been the intention of the authors like Bijker and Pinch, but is linked to the popular dissemination of their work.

20 Manuel Stoffers, Harry Oosterhuis, and Peter Cox, "Bicycle History as Transport History: The Cultural Turn", in *Mobility in History: Themes in Transport: T2M Yearbook 2011*, ed. Gijs Mom, Peter Norton, Georgina Clarsen, and Gordon Pirie (Neuchatel: Alphil, 2010).

21 See Justin Spinney, "A Place of Sense: A Kinaesthetic Ethnography of Cyclists on Mont Ventoux", *Environment and Planning D: Society and Space* 24, no. 5 (2006); Rachel Aldred and Katrina Jungnickel, "Constructing Mobile Places between 'Leisure' and 'Transport': A Case Study of Two Group Cycle Rides", *Sociology* 46, no. 3 (2012).

experiences.[22] Schot and Albert de la Bruhèze note the need for a conjunction of production-oriented and consumer-oriented studies for understanding technology and employ the two poles of user- and producer- influence to map the agency at work in the social construction of technology.[23] In relation to the cycling and the (bi)cycle we need also observe a third and very important factor at work. The bicycle is especially dependent upon the space in which to use it and the surfaces on which it is to be used. Just as the bicycle and rider combine to make a machinic combination, this combination cannot exist without terrain to traverse.[24]

Because the bicycle is a technology that operates in and consumes public space, it is constrained within webs of interaction, social and physical. It is also therefore reliant on the infrastructure of public space and the legal governance of that public space in civil society, and of the public interest. Variations in national legislation pertaining to highways and routeways, their classification, access and use-rights shape relations between traveller and travel. These histories reflect existing power relations of land ownership and the mobility expectations accorded to relative class positions. If we consider mobility as a market, then the forces of production and consumption are joined by, and mediated by, the political and legal regulation of that market. The historiography of the bicycle and of cycling must therefore address the distinction of political differences across a range of territories.

In sum, a comparative use-centred study of cycling is a conjunction of consumer-, producer- and politico-oriented studies. Each of these is a complex of multiple levels of differentiation, and requires us to take into account the classic distinctions of class, gender and ethnicity as they bear on the practice, alongside their divergent forms as they relate to national and regional distinctions. To summarize, there is a pressing need to go beyond simple narrative accounts of cycling and turn our gaze outward from the bicycle to engage more deeply with the broader contexts in which cycling takes place and the multiplicity of forms that cycling takes, not only as transport. From this perspective cycling history becomes a very difficult terrain to negotiate. Simplifications inevitably have to be made in producing a coherent narrative. Nevertheless,

---

22 See, e.g., Bernhard Hachleitner et al., eds., *Motor bin Ich Selbst: 200 Jahre Radfahren in Wien* (Vienna: Metroverlag & Wienbibliotek im Rathaus, 2013).

23 Johan W. Schot and Adri A. Albert de la Bruhèze, "The Mediated Design of Products, Consumption and Consumers in the Twentieth Century", in *How Users Matter: The Co-Construction of Users and Technology*, ed. Nellie Oudshoorn and Trevor Pinch (Cambridge, MA: MIT Press, 2003).

24 In Deleuzian terms, the bicycle rider machine connects with the road/path machine; see Claire Colebrook, *Gilles Deleuze* (London: Routledge, 2002), 56f.

we need to distinguish between forms of simplification that produce heuristic clarity, and those that may serve either to traduce the events they describe or to introduce misleading models.

## 3 Mapping Recent Bicycle Research

In his extensive survey of recent trends in the rapidly expanding field of bicycle research, Harry Oosterhuis identifies an emergent central research question: "why people use or don't use the bicycle for utilitarian purposes and, consequently, how cycling can be promoted".[25] Policy- and practice-linked research not only has a high profile but is also connected to wider dissemination networks through international conferences and national and international lobbying networks.[26]

A dominant motif behind this central research question is recognition of the rapid post-Second World War decline in European bicycle transport, vividly mapped in the ground-breaking work of Albert de la Bruhèze and Veraart, further developed in Oldenziel and Albert de la Bruhèze.[27] Indeed, Oosterhuis begins his analysis by commenting that "the bicycle was surpassed by the car as the dominant mode of individual transport".[28] He argues that there is an explicit reliance on historical analysis within the policy orientation of current research, but specifically historical studies that could be used for the compilation of comparative analysis and to produce *longue durée* perspectives remain relatively thin on the ground. For example, Aldred's overview of post-war British cycling activism is in no sense an inaccurate portrayal of events, yet the very clarity of its simplification obscures some of the more nuanced debates and tensions that might lie beneath the surface of the events described, and the complex processes that led to the development of positions and controversies

25 Harry Oosterhuis, "Bicycle Research between Bicycle Policies and Bicycle Culture", *Mobility in History* 5 (2014): 20.

26 E.g., the ECF (European Cyclists' Federation) with its own Velo-city and Velo-city Global conference series and Scientists for Cycling network (for more information on these, visit the Federation's website at http://www.ecf.com/), or the grassroots-led World Bicycle Forum (http://www.fmb4.org/en/home/).

27 Adri Albert de la Bruhèze and Frank C.A. Veraart, *Fietsverkeer in praktijk en beleid in de 20e Eeuw: Overeenkomsten en verschillen in het fietsgebruik te Amsterdam, Eindhoven, Enschede, Zuid-Oost Limburg, Antwerpen, Manchester, Kopenhagen, Hannover en Basel.* (Den Haag: Ministry of Transport, Public Works and Water Management, 1999); Ruth Oldenziel and Adri Albert de la Bruhèze, "Contested Spaces: Bicycle Lanes in Urban Europe, 1900–1995", *Transfers* 1, no. 2 (2011).

28 Oosterhuis, "Bicycle Research", 20.

cited.[29] But this is due to the lack of available published primary histories, rather than any omission on the part of the author.

The International Association for the History of Transport, Traffic and Mobility (T2M) has consistently sought to support studies of this nature and to identify the research gaps.[30] However, this work is focused on utilitarian cycling. Oosterhuis concludes "Research into utilitarian cycling would benefit from a new approach that attends to national historical trajectories and national bicycle habitus".[31] However, taking this Bourdieusian approach thoroughly, I would add that to understand a national bicycle *habitus* one will have to attend to all of the uses and practices of the bicycle of which it is comprised, alongside the elements of *doxa* that inform actions and conceptualizations in relation to cycles and cycling. Concentration on utilitarian purposes alone disconnects transport uses of bicycles from other potentialities and practices. Little connection is made with cycling studies in the history of sport, or in leisure research. The public imagination of cycling may be profoundly influenced by these non-transport images.[32]

To go even further, perhaps, in this analysis we might need to interrogate even the divisions of cycling into rigid categories of activity (leisure, sport, utility) and more closely consider the multiple meanings and identities attached to diverse practices: to think of heterogeneous and complex 'cyclings', rather than 'cycling'.[33] In this way, we can deconstruct the categories, understanding how they are differently defined and understood across a variety of social and geographic locations, and how they overlap and interact.

Specific research in cycling history as a discrete field of interest, as mentioned above, is dominated by an international enthusiast-led network, and connected through a well-established annual *International Cycling History Conference* series.[34] Since its first meeting in 1980 and through annual

29 Rachel Aldred, "The Role of Advocacy and Activism", in *Cycling and Sustainability*, ed. John Parkin, Transport and Sustainability, Vol. 1 (Bingley, UK: Emerald, 2012).

30 See, e.g., Francis Papon, "Historiographical Needs in the Study of Bicycling Mobility in France", *Mobility in History* 5 (2014); Evan Friss, "Writing Bicycles: The Historiography of Cycling in the United States", *Mobility in History* 6 (2015).

31 Oosterhuis, "Bicycle Research", 35.

32 "Exploring the Relationship between Leisure and Commuter Cycling: Policy Analysis Research Summary, October 2011", Transport for London (TfL), accessed 28 December 2017, http://content.tfl.gov.uk/exploring-the-relationship-between-cycling-leisure-and-utility-trips.pdf.

33 I am grateful to Tiina Männistö-Funk for making this point. For illustration, see also the contributions in Cox, *Cycling Cultures*, especially Angela van der Kloof, "Lessons Learned through Training Immigrant Women in the Netherlands to Cycle".

34 See http://www.cycling-history.org/.

publication of conference proceedings, the ICHC draws together scholarship on bicycle history across international boundaries, time periods and from individual, national networks of study on the bicycle. In its conferences the ICHC has sought to problematize historical metanarratives and to encourage plural histories, but the wealth of information produced, however, remains dominated by micro-level studies. Invaluable though detailed histories of marques, specific events, groups, clubs and organizations; they are only incidentally connected to wider issues of social change. Studies of cycling separated from other mobilities or from other historical factors can be difficult to connect with broader themes or to integrate into macro-level analyses. These lacunae are also visible in the production of general circulation histories of the bicycle and cycling, as discussed below. Early contributions to specialist bicycle history were drawn largely from outside academia. However, the ICHC series has insisted on raising the standard of scholarship and challenged populist assumptions, providing a constant source of detailed knowledge and important correctives to widely circulating myths in bicycle history. The conference today serves as a forum for a diverse range of perspectives and opinions, across backgrounds and disciplines.

In overview, therefore although the field of cycling research is lively and growing, historical dimensions are relatively under-researched. Moreover, cycle and cycling histories suffer from fragmentation. The dominance of policy-related research has produced a reliance on summary historical understandings and it is the problems arising from the simplification of the history of cycling and the bicycle to which we now turn our attention.

## 4 Writing Bicycle History: Periodization

For any historian faced with writing an account of cycle and cycling history, the challenge of making a complex narrative comprehensible is a stern one. Two primary techniques stand out as obvious means by which the complexities may be rendered into a coherent narrative—thematic and chronological. Thematic studies identify organizing themes and construct a narrative around these. For example, Andrew Ritchie's seminal *King of the Road* (1974), is organized around the following chapter headings: "Bicycle Archaeology"; "Amateur Mechanics"; "Velocipedomania"; "The Cult of the Ordinary"; "Tricycle and 'Sociable' Cycling"; "The Search for Safety"; "Women's Liberation"; "A Fact of Everyday Life". These titles illustrate two key points. First and most obvious is the dominance of the nineteenth century. Second is the clearly thematic organization of the work. While the first three chapters chart a chronological development of patterns of invention, the next four map a series of diverse activities and themes that occur synchronically. The obvious advantage of this approach is that it allows both diachronic and synchronic accounts to be incorporated. Multiple synchronic narratives allow elucidation of diverse experiences.

A second general approach may be to organize the entire study chronologically: there are obvious reasons to adopt a generally chronological approach, especially when addressing a non-specialist audience. Within this there are also logical reasons to divide time into relevant blocks, if only for the very justifiable purpose of creating readable chapters. Chronologies allow easy navigation through changes, perhaps at the risk of linearizing the histories.

Problems arise, however, when the two methods are conjoined. In brief, combining delineated time blocks with specific themes results in a periodized history, divided into separate eras in which, for each chronological period, one specific narrative identity or set of processes is identified as archetypal. Not only do specific time periods become reduced to single narrative motifs, but also the overall image is of a unified, linear history constructed from the unitary narratives of each successive era. Plurality of experience and practice becomes subsumed in the formation of a monolithic and singular history.

To illustrate we can take a relatively recent publication, *Voyages à vélo, du vélociped au Vélib*, produced to accompany an exhibition of the same name in Paris 13 May–14 August 2011.[35] Lavishly illustrated, with text by the noted historian Catherine Bertho Lavenir, and with a preface by Paul Fournel, author of *Besoin de Vélo*, one of the few works to reflect seriously on the experience of riding. The text is ordered into five chapters: "L'age du vélocipede 1812–1880"; "La bicyclette et le loisir bourgeois 1880–1914"; "La culture poulaire du vélo 1918–1945"; "La bicyclette au temps de l'automobile 1945–80"; and "Renouveau du cyclism 1980–2010". The text itself provides as complex and insightful analysis of bicycle use across the centuries as could be hoped for from a brief (122 page) guidebook for a general audience. And yet, the structure in which the illustrations and discussion sit is problematic on a number of levels.

While a periodized history may be an effective communication device, especially in the context of narratives written for a non-specialist audience, or even as the basis of a single study, this device makes an awkward foundation for writing long perspectives on cycling history. In this latter role, it serves to conflate and to oversimplify: sometimes misleadingly, sometimes dangerously erasing significant elements, particularly those of geography, space and power.

35 Catherine Bertho Lavenir, *Voyages à vélo, du vélociped au Vélib* (Paris: Paris Bibliotheques, 2011).

Additionally, this simplification can re-introduce a sense of historical determinism. Changes in cycling and cycle use become almost inevitable products of the passing of time, irrespective of the forces mobilized around processes of historical change. In the case of bicycle historiography, an overarching discourse emerges through oversimplified summary accounts and which then serves as a master narrative to distort existing events, and to erase the operations of power. The contributions that cycle use has made to historical change become invisible, cycling becomes a passive object shaped by circumstance. Given that cycling research has such close links with bicycle policy today, these erasures are potentially deeply troubling.

At the broadest conceptual level, periodization naturalizes processes of historical change.[36] One era, characterized by a single motif, gives way to another with little indication of the forces and processes involved in change. Ascribing specific dates to a particular theme is a necessarily arbitrary process. Moreover, the date of a changing pattern in one country may not be the same as another. Even relatively small time delays between different territories where similar patterns are discernible can reveal also the roles of price control mechanisms, wage differentials across social classes, the relative power of retail markets and capital investment in different territories, for example. Although *Voyages à Vélo* is a specifically French history, its territorial specificity is lost in the broad sweep of the themes. Similar universalism is to be seen even more explicitly in other accounts, where images and examples are drawn from a range of national origins. Events lose their originating geography and become reported as universal trends. Conversely, in an explicitly periodized history, narratives from a range of particular places and times become simplified into a general trend.

By becoming more geographically specific in analysis, cycling histories can produce stronger comparative studies. It may be that through such studies particular typologies can be discerned by writer and reader, and consequently related to other social, political or economic patterns. However, the idea of a singular 'bicycle' history, assuming the universality of the technology, acts only to elide territorial differences. Even a critical text such as this, by making its argument through a re-examination of bicycling patterns in the UK, runs the risk of perpetuating the dominance of US/Anglo North European normativity in cycling studies. A comparative stance also would enable more

36 This insight is at the heart of much postcolonial critique of historiography. Key examples can be found in the work of Ashis Nandy; see, e.g., Vinay Lal, ed., *Dissenting Knowledges, Open Futures: The Multiple Selves and Strange Destinations of Ashis Nandy* (New Delhi: Oxford University Press, 2000).

serious engagement with non-European histories of cycling, which while tied in through international trade and travel, produce distinct narratives in different territories, reflecting individual national fortunes and international relations. International trade patterns, the role of tariffs and imperial/colonial imperatives are accompanied by more cultural factors of use patterns, practices and imagery, and demand transnational studies as well.[37] Pluralism is a necessary dimension in the production of cycling histories.

Another valuable form of comparative study is the examination of cycles and cycling in relation to other technologies and practices. This does not only apply to relations to other *transport* technologies, but in a broader socio-historical framework. As mentioned above, current interest in policy-relevant research stresses cycling as transport, but this leads to a relative lack of examination of the leisure/sport nexus and the parallels with other forms of mobile leisure. Rethinking bicycle historiography one might ask why particular uses, occurrences, events and user groups are more privileged than others in the accounts we weave. What political agendas might be hidden within this selectivity? To take two examples raised in other chapters in this volume, how does emphasis on the nineteenth century bicycle affect the way we see riding today? Or how does the proliferation of writing on cycle sport impact upon an agenda of inclusivity and mundane riding to replace car use? As Timo Myllyntaus has written of environmental history, "studying history means making choices, defining and framing topics".[38] Writing histories requires transparency in our choices of subject, approach and significance, and it is the evidence of these processes of structuration that remain problematic when narratives imply a singular history of cycles and cycling.

The predominance of interest in the nineteenth century, especially in overview studies, also has the unintended consequence of inscribing invention and novelty as being the most important aspects of bicycle history. Cycling history, that is the history of the *use* of cycles in all their myriad forms, thus becomes separated from the history of the object itself. Confining interest in design innovation principally to the nineteenth century facilitates simplification into a linearized history, which, in turn fits into models of product lifecycle. Rogers's model of the diffusion of innovation, with its bell curve distribution of product innovation and adoption, has become a truism of popular assumptions

37 I am grateful to Ruth Oldenziel for this point.

38 Timo Myllyntaus, "Methods in Environmental History", in *Thinking through the Environment: Green Approaches to Environmental History*, ed. Timo Myllyntaus, (Cambridge, UK: White Horse Press, 2011), 2.

about technologies.[39] When cycle design is projected into this model, the long-term fate of the bicycle is assumed to mimic other modern consumer products. Invention, adoption and diffusion will ultimately be followed by obsolescence and decline as the next innovation product comes on stream.[40] Despite the insistence of historians of technology that this is a flawed way of thinking about technology, bicycle history exhibits a tendency towards teleological narratives, often organized in relation to the narrative of declining European use in the 1950s. It is to this framework that we now turn our attention.

## 5 Declensionist Narratives and Cycling Historiography

The combination of an emphasis in studies on nineteenth century innovation and the decline in cycle use in Europe in the 1950s, coupled with bell curve models of innovation, produces a powerful, if unintended, declensionist dimension into cycling studies. That is, cycling histories must cope with a structural history of progressive decline. Consequently, there is also a powerful impulse in cycling studies to curb this tendency towards pessimism by an overemphasis on positive narratives.

If we turn our attention away from cycling historiography to consider the field of environmental history, similarities are striking. One of the first issues that environmental history struggled with was its relationship to policy. As Opie put it back in 1983, "environmental history is dogged by the spectre of advocacy" an observation for which Oosterhuis's observation on bicycling history (above) could be seen as a rephrasing.[41] Even more revealing than the struggle with advocacy, however, are the discussions that have necessitated its reconsiderations of historiography. William Cronon's 1992 article, "A Place for Stories: Nature, History, Narrative", provided a set of arguments that allowed environmental history to reconsider the shaping of narratives.[42] Put briefly, he argued that every story that historians tell is necessarily a selective process, an

39 Everett Rogers, *Diffusion of Innovations*, 5th ed. (New York: Simon and Schuster, 2003).

40 It is worth noting to which degree recent publications on racing manufacture depend on an evocation of nostalgia and the celebration of "classic" styling (see Oddy in this volume).

41 John Opie, "Environmental History: Pitfalls and Opportunities", *Environmental History Review* 7, no. 1 (1983): 10.

42 William Cronon, "A Place for Stories: Nature, History, and Narrative", *Journal of American History* 78, no. 4 (1992).

exercise of power in choosing which elements to recount for the reader.[43] But more than noting the necessary selection of events, Cronon pointed to common narratives in environmental histories, depicting an historic Edenic period later destroyed by particular interventions—even though the specificities and politics of their narratives, and the blamed interventions, vary.

Although he was critiquing histories of the Great Plains, we might think of how histories of bicycling are framed by similar conceptual frameworks, however covert. A golden age of riding at the dawn of the twentieth century is gradually eroded by the growth of motor traffic, and cycling is finally brought to an entirely marginal state by the inevitable growth of the private car, a situation from which it must be redeemed, by some means. Stories, as Cronon argues, "are intrinsically teleological forms, in which an event is explained by the prior events or causes that lead up to it."[44] In other words, once the storyline is established, changes do not need explanation, they are merely the unfolding of inevitably predestined narrative arcs. The mythic story takes precedence over material histories. It is this inevitability that we challenge here. The echoes of Edenic foundation and subsequent Fall lurking in the background of the stories of cycling are not simply teleological but eschatological in their mimicry of the theological narratives of Christian tradition. Fascinatingly, they then covertly pose the question of how this current fallen state might be redeemed.[45] What form might the intervention take that will restore the lost paradise—is it infrastructure? Is it a new bicycle technology? This may be all far too fanciful, but nevertheless it remains important to understand the power of the (hi) stories that we tell and the manner in which these connect with those nonreflexive understandings within which we culturally operate.

Although declensionist environmental histories can provoke progressive change, as can the opposing impulse of a corrective emphasis on positive narratives, Carolyn Merchant argues that both positions can be too simplistic. Instead, she argues for a dialectic environmental history, emphasizing the excavation of material and power dimensions in the formation of events.[46] Such an approach moves away from the tendency towards teleology noted above,

---

43 Cronon's argument foreshadows Deegan's discussion of the production of cycling maps, which constantly show processes of power and politics in the choice of what should be shown or what not and which elements should be stressed. See Brian Deegan, "Mapping Everyday Cycling in London", in Cox, *Cycling Cultures.*

44 Cronon, "A Place," 1370.

45 Cf. Donald Worster, *The Wealth of Nature: Environmental Histories and the Ecological Imagination* (New York and Oxford: Oxford University Press, 1993).

46 Carolyn Merchant, preface to *Ecological Revolutions: Nature, Gender, and Science in New England*, 2nd ed. (Chapel Hill, NC: University of North Carolina Press, 2010).

and reinstates the politics of change, including their dimensions of class, gender and other social distinctions. Encouragingly, this shift is visible in recent writing on cycling. For cycling history to inform policy, rather than focusing on the decline of bicycle use for transport we might look to include other factors involved in modal shift. Broader perspectives enable clarity on the degree to which the fortunes of any technology are rarely inherent qualities of the technology itself, whether in its rise or decline. Even more, we need to unpick what exactly is going on in times of change to understand the complexities of changes in any given period. Acknowledging the existence of multiple 'cyclings' requires us to rediscover multiple histories. Responding to this consideration to rethinking histories in this manner, the final part of this chapter revisits a selection of events between 1951 and 1971 that perhaps enable another way of thinking about cycling in the UK during this period.

## 6 Decline, Rise or Just Change? Cycling in the United Kingdom, 1951–1971

The absolute decrease in reported numbers of miles travelled by bicycle between the 1950s and the end of the 1960s in the UK is obvious and undeniable (Graph 2.1). Similarly, changes in CTC (Cyclists' Touring Club) membership from a peak of 53,374 in 1950 to a low of 18,564 in 1971 might be taken as a proxy for this decline. However, headline numbers can be misleading. This part will argue that while recorded bicycle mileage declines, we also need to understand other narratives in order to better interpret changes in quotidian transport during this period, and to take note of other stories of cycles and cycling to provide a thicker description of the changing national bicycle *habitus*. Displacement of cycling from everyday transport is one function of a number of changes within British society and of the specific engineering of other changes in local and national transport provision.

In 1948 the anonymous author of *The Cycling Manual* was able to state boldly that, despite current shortages and unavailability of many items and components during these times of austerity, "a new era of cycling history is upon us".[47] Certainly, the reported 23.6 billion-kilometre cycle travel (compared with a total for all motor vehicle of 46.5bn km) indicated that the bicycle accounted for more than the travel distance covered by motorcars, taxis and motorcycles

47 Cycling, *Cycling Manual*, 2nd ed. (1948), xii.

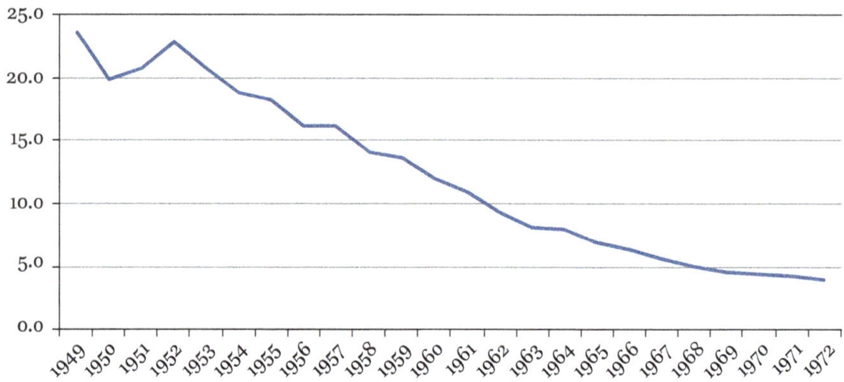


GRAPH 2.1 Distances cycled in the United Kingdom, by year (in billion km)
SOURCE: UNITED KINGDOM DEPARTMENT FOR TRANSPORT

combined.[48] This figure, which corresponds to some 2,500 km per cyclist per annum, suggests that the majority of this travel must be made on a quotidian basis.[49] Similarly, the dramatic decline in distances travelled suggests that it is this regular use that disappears. When coupled with the dramatic rise in motor vehicle numbers (from approximately 4 million to 13 million) and total distances travelled by motor vehicle, both for drivers and passengers during this period, we see not simply substitution, but a considerable increase in new journeys and longer distance journeys. Entirely new mobility patterns are emerging, not just changes in modes of transport for existing journeys.

Yet what we see here is that this sort of everyday distance strongly suggests that the bicycle was being used as a means of getting to everyday employment: and in 1950s' Britain, this is a deeply gendered activity.[50] During the period from 1951 to 1971, the percentage of households with access to a car rose from 14 per cent to 52 per cent. But this is also gendered: even in 1975, only 29 per cent of women in the UK had a driving license.[51] Four fifths of the population had no exclusive access to a car in 1971, and the majority of these were women. The decline in cycling is not necessarily a general phenomenon, but a stripping out of a significant group of male employees (cycling to the workplace) from the

---

48 Department of Transport, "Road Traffic and Speeds" (Table TRS9901), http://www.dft.gov.uk/pgr/statistics/datatablespublications/roads/traffic.

49 Estimate calculated from Department for Transport recorded mileage data and parliamentary estimates of cyclists' numbers; see 190 Parl. Deb. H.L. (5th ser.) (1954) cols. 590–644, http://hansard.millbanksystems.com/lords/1954/dec/21/road-traffic-bill-hl-1.

50 Stephanie Spencer, *Gender, Work and Education in Britain in the 1950s* (Basingstoke: Palgrave Macmillan, 2005).

51 Department for Transport, NTS0201.

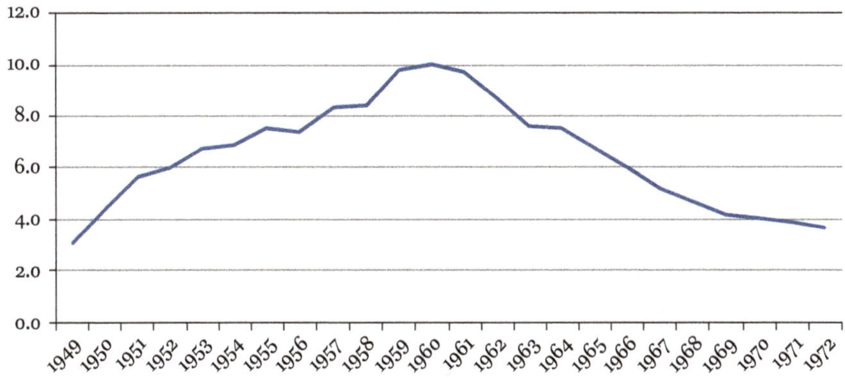


GRAPH 2.2 Distances travelled per year in the United Kingdom by powered two-wheeler (in billion km)
SOURCE: UNITED KINGDOM DEPARTMENT FOR TRANSPORT

sum of those making cycle journeys. A secondary reinforcement of this analysis comes from examination of the growth in motorcycle traffic (including scooters, mopeds and associated vehicles) through the 1950s (Graph 2.2). It almost exactly replicates the fall in cycling journeys.

We may reasonably hypothesize much of the decline in cycling to work during the 1950s involved the substitution by motorcycle journey for the largely male fulltime workforce, with the subsequent growth in car use in the 1960s taking away from both commuting modes.[52] The importance of the scooter was that it ushered in a new model of mobility as a an object of consumption, as manufacturers set out "not just to make a new category of machines but a new category of consumer ... and the conversion of consumption into lifestyle".[53] In other words, as choices became available for affordable and reliable motor cycles and scooters, as in Italy and France, they were taken. Cycling to work had not been a matter of choice but of necessity. The British public images of the bicycle as the poor man's transport, narrated from the 1930s onwards, had some material basis in this period In the 1960s: the car was presented not only as just one other transport possibility, but as the essential ingredient for participation in a rapidly modernizing society. Hence, the double modal shift visible

52 Although women accounted for 45.9 per cent of the overall workforce in 1955, their wages (and only partly because of the much larger proportion of women in part-time work) were approximately 50 per cent of men's during this period. See Stephen Brooke, "Gender and Working Class Identity in Britain during the 1950s", *Journal of Social History* 34, no. 4 (2001).

53 Dick Hebdige, "Object as Image: The Italian Scooter Cycle", in *Material Culture: Critical Concepts in the Social Sciences*, Vol. 2, ed. Victor Buchli (London: Routledge, 2004), 139.

in the Department for Transport data. One must also therefore consider what simultaneous changes in demographic distributions, employment patterns in relation to dwelling location occur during this period, as well as changes going on in a wider analysis of transport.

Road transport is only one part of the mobility equation, however. Railway closures had been in operation prior to the Beeching report of 1964, which recommended complete rationalization of the national rail assets: 2,363 stations to be closed, 266 services withdrawn, seventy-one modified.[54] An earlier programme of closures had been initiated almost immediately after nationalization in 1948, and over 1000 miles of track closed in the first five years.[55] Taking the 1955–71 period overall (as highlighted in the 1974 Independent Commission on Transport), this relatively short period saw a 39 per cent reduction in track mileage open to passenger traffic, a 56 per cent cut in the number of stations, and 54 per cent reduction in passenger capacity.[56] In total, the passenger network contracted by 8,000 miles between 1948 and 1973, the number of stations from over 6,500 to 2,355.[57] The Beeching plan, *The Reshaping of British Railways* (1963) commissioned by Transport Minister Ernest Marples and named after its author, initiated a strategic re-organization of the railways with a series of closures of rural and cross country lines that made the system less of a network.[58] Ostensibly, this was done in order to rationalize, modernize and save money. In retrospect, an ideological agenda hostile to rail transport can be discerned.[59] As Christian Wolmar puts it, "The Beeching report had been commissioned in order to demonstrate that minor railway lines were fundamentally uneconomic and it was hardly surprising that this was its conclusion".[60] Even lines that still ran after the cutbacks might not have stations near communities. The closure of passenger stations in particular, removed the possibility of alternatives to road travel for numerous communities. Without the clear substitution of other public transport provision, plans to provide bus services

---

54 Charles Loft, *Last Trains: Dr Beeching and the Death of Rural England* (London: Biteback Publishing, 2013).

55 Ruud Filarski (in co-operation with Gijs Mom), *Shaping Transport Policy* (Den Haag: SDU Uitgevers, 2011).

56 Independent Commission on Transport, *Changing Directions: The Report of the Independent Commission on Transport* (London: Hodder, 1974), 30.

57 Loft, *Last Trains*.

58 A second report, *The Development of the Major Railway Trunk Routes* (1965), recommended rationalization to only 3,000 miles of key routes for future development; see Loft, *Last Trains*.

59 David Henshaw, *The Great Railway Conspiracy*, rev. ed. (Dorchester, UK: A to B Books, 2013).

60 Christian Wolmar, *Fire and Steam* (London: Atlantic Books, 2007), 284.

for former rural rail link ("bustitution") notoriously failed to materialize in any meaningful form; the only practical option for many was to invest in a private car. As branch lines closed and the network became less dense, communities became increasingly isolated, or saw rail travel become increasingly inconvenient, requiring secondary transport to get to the nearest station. In many cases the car was the only viable option for this as well. That rail travel remained relatively static in this period is remarkable and only comprehensible through the overall increase in travel arising from demographic shifts.

The Beeching plan was actually the second major intervention arising from the initiative of Ernest Marples, the Conservative Transport Minister appointed in 1959.[61] Addressing his Party Conference in 1960, he declared that, "we have to rebuild our cities. We have to come to terms with the car", neatly foreshadowing the conclusions to be reached three years later by Colin Buchanan's report *Traffic in Towns*, which Marples had recently commissioned.[62] Such was the impact of *Traffic in Towns* that a shortened edition was published the following year as a mass-market paperback.[63] In the preface to the shortened volume, Sir Geoffrey Crowther, former editor and chairman of *The Economist*, wrote that:

> to liberate the motor vehicle ... we shall have to make a gigantic effort to replan, reshape and rebuild our cities.... What the Victorians built, surely we can rebuild. Nor is this an unpleasant necessity. Our cities, most of them, are pretty depressing places, and to rebuild them would be a worthwhile thing to do even if we were not forced to it by the motor car.[64]

Edward Heath, leader of the Conservative Party, stated as he opened the 1966 Motor Show, "Of course traffic in towns creates a problem. My approach is not to restrict, to hamper or confine the motorist. Instead, we must learn to cope with the motor car and care for the motorist".[65] The 1970 White Paper (published shortly after Heath's election victory), *Roads for the Future*, laid

61 Marples, owner of 64,000 of the 80,000 shares in Marples Ridgeway, a specialist road-building firm, narrowly avoided scandal for appearing to benefit from government-subsidized road construction contracts. See Mick Hamer, *Wheels within Wheels: A Study of the Road Lobby* (London: John Murray, 1987).

62 *Traffic in Towns* (London: HMSO, 1963).

63 *Traffic in Towns: The Specially Shortened Edition of the Buchanan Report* (Harmondsworth: Penguin, 1964).

64 Geoffrey Crowther, preface to *Traffic in Towns: The Specifically Shortened Edition*, 14–15.

65 Quoted in Patrick Rivers, *Restless Generation: A Crisis in Mobility* (London: Davis-Poynter, 1972), 40.

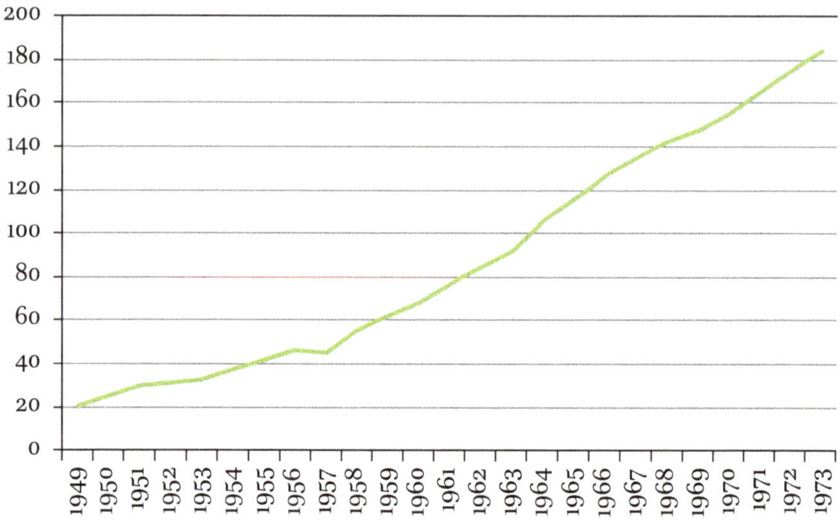


GRAPH 2.3 Car and taxi distance travelled in the United Kingdom, by year (in billion km)
SOURCE: UNITED KINGDOM DEPARTMENT FOR TRANSPORT

out a strategy for a four billion pound investment in trunk roads to double capacity in the next 20 years.[66] Transport modal shift towards the car was a clear political priority, a deliberate reorientation of the transport system and economy (Graph 2.3). If motor scooters and cycles had allowed greater numbers of working class men to replace the bicycle journey to work in the 1950s with one deliberately constructed as more glamorous, as well as opening up the possibility of greater distances between home and work, the restructuring of the 1960s ensured that car ownership became enshrined as a primary means through which participation in newly modernizing Britain was understood.

Unlike provision of rail transport or bus services, government investment in the road network was justified as a public good providing for both private motoring and road goods haulage in the 1960s. What we see is not simply growth in the long-distance road network but also a much broader change in the mobility structures of the UK. As distance travelled multiplies in this period, and longer journeys become normalized through demographic changes and urban restructuring, so the bicycle becomes less of an appropriate tool. Social housing policy in the post-war years, especially the rapid growth of construction of public housing schemes, frequently built on cheaply and quickly available greenfield sites in order "to build the maximum number of houses in

66 Ministry of Transport, *Roads for the Future: The New Inter-Urban Plan for Transport*, Cmnd. 4369 (London: HMSO, 1970), 6–7.

the shortest possible time", had the effect of disrupting and resettling existing urban communities.[67] Although important in improving living standards, this served to break assumptions of immediate proximity between housing and employment.

Only with the 1968 Transport Act did the idea pass into legislative policy that transport was a function of the public good, might legitimately be subject to state subsidy, and "that subsidies might be used as an instrument of wider transport policy".[68] This was despite years of practical subsidy by government to make up for losses in the rail system. Nevertheless, the greater picture of British transport legislation is that it had remained a bastion of laissez-faire public policy.[69] Yet laissez-faire is not simply a non-interventionist policy instrument in a neutral, "natural" environment. It is an ideological tool as much as its opposite, defining particular understandings of the relationship between citizen and state, and concerning the management of inequalities of power and wealth. And any pretence that the growth of motoring in the latter half of the twentieth century was the outcome of laissez-faire public policy is solidly undermined by the evidence.

## 7 Other Cycling Stories: Sport and Industry

If transport cycling in the post-war years was depicted as the poor man's necessity, in both scooter and car promotion, we should also be aware that it was only one competing image of cycling in the 1950s. For British cycle sport the picture looks quite different. During the late 1940s, sport cycling in the UK was in something of a turmoil. The pastime had many thousands of adherents, belonging to cycle clubs throughout the country, but since 1890 the governing body, the National Cyclists Union, had banned mass start racing on public highways, depriving Britain of the spectacle of road-racing and riders of the opportunity to participate in mass events, excepting those run on closed circuits such as airfields and parks. Instead, the main participant racing activity was time trialling: individually against the clock on distances from 10

67 From the Glasgow Corporation mission statement, quoted in David Kynaston, *Family Britain, 1951–1957* (London: Bloomsbury, 2009), 275. See also Tim Butler and Chris Hamnett, "Social Geographic Interpretations of Housing Spaces", in *The Sage Handbook of Housing Studies*, ed. David Clapham, William Clark, and Kenneth Gibb (London: Sage, 2012). The classic contemporary study of community change is Michael Young and Peter Wilmott, *Family and Kinship in East London* (Harmondsworth: Penguin, 1957).

68 Independent Commission on Transport, *Changing Directions*, 144.

69 William Plowden, *The Motor Car and Politics, 1896–1970* (London: Bodley Head, 1972).

miles to 100 miles, plus events of 12 and 24 hours run by a separate governing body (RTTC Road Time Trials Council). The early 1950s, with little traffic on the roads, is remembered by many participants as a golden age for this activity.[70] However, in 1942, with circuit racing courses in short supply and minimal private traffic on the roads, police permission was obtained for a one-off road race from Llangollen to Wolverhampton, and participants, suspended from the NCU, formed a breakaway association (the British League of Racing Cyclists, BLRC), which continued to organize road races, with considerable public spectator support. Although club affiliation to the BLRC meant a ban from the other two associations, sufficient national support was gained to result in the first "Tour of Britain"—the *Daily Express Round Britain Cycle Race*, in 1951. The sponsorship shows the level of interest, and the race rapidly became an annual fixture, the Milk Marketing board taking over sponsorship from 1958.

BLRC teams took part in international cycle sport as well, including sending a team to the Tour de France in 1955. Professional riders, whether on road or track—like Reg Harris, Olympic medal winner and 1954 world sprint champion—became household names, alongside their amateur compatriots, with significant coverage in print and newsreel. Road racing became both spectator and participant sport: to the point where the NCU gave in and amalgamated with the BLRC in 1959 as the British Cycling Federation. Although the number of events and crowds at track meets declined rapidly at the end of the 1950s, through the 1960s the number of events grew, alongside emergent (Pro) stars such as Tom Simpson or later Barry Hoban, while Beryl Burton dominated amateur women's racing, not just in Britain but internationally.[71]

For the sporting cyclist in Britain, the period from 1950–1970 was far from a straightforward picture of decline. Rather the reverse. It saw the growth and establishment of the sport, overcoming divisions so that from 1959 onwards riders could compete in any type of event, while thousands turned out to watch top riders at the annual Tours of Britain or in local circuit races.[72] New images of the bicycle and cycling were being forged in this period. If we

70 See, e.g., Dave Moulton, "Once upon a Time, Britain Had a Bike Culture", *Dave Moulton's Blog*, 29 July 2010, http://davesbikeblog.squarespace.com/blog/2010/7/29/once-upon-a-time-britain-had-a-bike-culture.html. This perspective has been confirmed over a number of years through personal communication from members of various cycle clubs.

71 Burton was a five-time world pursuit champion, two-time world road race champion and domestic British Best All Rounder (a combined distance time-trialling competition) for 25 consecutive years beginning in 1959; see Beryl Burton, *Personal Best* (Horsham: Springfield Press, 1986).

72 Seaside towns such as Morecambe would hold them on the promenade for maximum visibility, even as attendance at track meetings declined.

are considering the ways in which a national bicycle *habitus* is formed, then the growth of sports cycling's image in this era may not be entirely irrelevant to the peculiarity of rising interest in cycling as a sport—spectator and participant—in the UK since 2008. There is certainly a historical precedent, even though the relationship between leisure and utility riding practices is ambiguous.[73]

From the perspective of the industry also, the period was one of tremendous change. Falling sales of utility roadsters in particular—the core machines of the everyday working travel being usurped by other travel possibilities or necessities—required industry to change both production and presentation of the act of cycling. In total, UK bicycle production fell by about one third between 1950 and 1960.[74] However, this stabilized in the mid-1960s, despite the continuing fall in distances travelled.[75] The Moulton bicycle, which came onto the market at the beginning of 1963, reengineered the bicycle.[76] With its small wheels and single frame size design, it was no longer a strictly gendered design product, constructed in Ladies' or Gents' models, but a unisex vehicle, which, with one simple adjustment to a quick release seatpost, could suit a wide range of different riders. The familiar ubiquity of this feature in today's small wheeled and folding cycles is apt to obscure the degree to which it represented a revolutionary re-imagination of the bicycle. It was followed rapidly by small wheeled designs from Raleigh (at the time accounting for over 75 per cent of UK cycle production) and other manufacturers. To counter the success of Moulton, Raleigh embarked on an unprecedented marketing drive in which the bicycle was forged in a new image, that of a lifestyle product.[77] While leisure had always been a major part of imagery of bicycle sales material, this was a significantly different approach—the new sales material repositioned bicycle as "consumer goods, not bits of light engineering".[78]

---

73 Jillian Anable and Birgitta Gatersleben, "All Work and No Play? The Role of Instrumental and Affective Factors in Work and Leisure Journeys by Different Travel Modes", *Transportation Research Part A: Policy and Practice* 39, nos. 2–3 (2005).

74 In 1950, the total production was 3,528,000 units, in 1955 it was 3,526,000 units, and in 1960 2,278,000 units; Paul Rosen, *Framing Production: Technology, Culture and Change in the British Bicycle Industry* (Cambridge, MA: MIT Press, 2002), 73. Rosen's analysis of Raleigh is crucial to understanding the challenges to industry at this time.

75 Tony Hadland, *The Moulton Bicycle*, 2nd, rev. ed. (Coventry: Hadland Books, 2000).

76 Alex Moulton was initially inspired by the manner in which scooters had transformed the way in which motorcycles were understood and presented. See Hadland, *The Moulton Bicycle*.

77 Bruce Epperson, "A New Class of Cyclists: Banham's Bicycle and the Two-Wheeled World It Didn't Create," *Mobilities* 8, no. 2 (2013).

78 Peter Seales, head of Raleigh Marketing in 1973, quoted in Rosen, *Framing Production*, 102.

Even if pressed into adaptation by necessity, the changes made and the reconfigurations embarked upon were successful enough in reframing the bicycle in consumer focus for it to become a new product in successive decades with the rise of BMX and then mountain biking. These and subsequent re-imaginings of the bicycle, its place and use, can arguably be traced back to the interpretive flexibility initiated in this period.[79] That they have had insignificant impact in the use of bicycles as road transport should not overshadow their dramatic importance for histories of cycles and cycling. Traditional road traffic surveys and travel surveys may convey little of the uses for which such cycles may be employed, but that is a problem for metrics. Re-examining the cycles and cycling in the UK between 1950 and 1970 we can see that the picture is a complex one. A profound decline in the numbers of everyday bicycle commuting, certainly. Simultaneously, however, other uses and meanings of cycles and cycling were growing or being established.

When we consider the growth of car ownership and its emergence as the primary transport mode, the emphasis placed on it in UK government policy through the 1960s suggest that car use was not just demand driven, but that demand was created through concerted efforts to reconfigure travel around the private motor vehicle. This despite the fact that it remained available to a minority of households and an even smaller proportion of individuals—largely male—within these households across the whole period. Bicycle sales declined during the same period, but by nowhere near the same proportion as the mileage decreases in cycling.

What changed most were the uses of the bicycle. By the early 1970s, a much stronger narrative of the cycle as a means of fashionable leisure was beginning to be written. Bicycles had become consumer products and production was diversified as was design, opening the way for the boom years for the industry in the later 1970s. The image of the bicycle—an essential part of the national *habitus* of the bicycle—could potentially change from a utilitarian necessity to an object of lifestyle choice, projecting any one of a number of different meanings and messages of identity. Whether one views this as a positive or a negative accomplishment is a separate value judgment. However, the narrative of the bicycle as the poor man's transport did persist (and still does). Sports and leisure uses of the bicycle in Britain are significantly less pejoratively viewed and today, images of these activities are used by advertisers to signify freedom and desirable lifestyle choices.

---

79 Rosen, *Framing Production*, Ch. 1.

## 8 Conclusions

This chapter set out to critically engage with the historiography of cycles and cycling. I have argued that the unintentional impact of certain frequently occurring devices in practices of writing and presenting cycling histories have had unintended consequences, despite the best efforts of their authors. The relationship of current interest in cycling studies to the world of advocacy provides both opportunities for research and danger in relating to historical precedents. I have argued that metanarratives are distinctly problematic and need to be carefully interrogated to establish what elements of diversity, multiplicity and counter-narratives they may conceal. Finally, I have examined one period of UK history, usually depicted as one of general cycling decline, and considered the extent to which that narrative might actually be confined to one very specific group of cycle users. Further, the chapter looked at how those changes were manipulated as part of a broader political strategy. Finally, attention was given to some of the other forms of change, and counter-narratives from other cycling activities. Overall, therefore, the argument is for more pluralized histories of cycling that are both socially and geographically diverse.

CHAPTER 3

# Entrenched Habit or Fringe Mode: Comparing National Bicycle Policies, Cultures and Histories

*Harry Oosterhuis*

## 1 Introduction

From the First World War until the 1960s, the bicycle was a popular means of personal transport all over Europe. After the volume of pedalling traffic peaked in the 1950s, it was rapidly outstripped by motoring. In many countries, the share of the car in the total number of traffic movements (modal share or modal split) would surpass that of the bicycle by around 1960—a development that came about earlier in North America. Cycling seemed out-dated and headed for an all-time low. However, since the 1970s, when countercultural criticism of technocratic car-geared systems and bicycle activism arose,[1] it has regained support among the general public as well as from governments. Worries about energy depletion, environmental and noise pollution, traffic congestion and safety, ill health and obesity, social exclusion and insecure streets, have entailed a re-evaluation of the two-wheeler as a clean, silent, sustainable, healthy, flexible, inexpensive, democratic and humane vehicle. Its modal share increased again, in some countries and cities more sharply than in others, but nowhere did it reach the 1950s level. Over the last two or three decades, national governments and cities throughout the western world, have launched ambitious policy statements and programs aimed at promoting cycling. Apart from students in university towns, the bicycle's popularity increased in particular among young and well-educated residents of cosmopolitan cities. Also, it won a prominent position in the marketing of popular tourist destinations such as Paris, Amsterdam and Barcelona, and even of traditionally bicycle-unfriendly cities such as London and New York.

1 See, e.g., E.C. Claxton, "The Future of the Bicycle in Modern Society", *Journal of Royal Society of Arts* 116, no. 5138 (1968); Ivan Illich, *Tools for Conviviality* (New York: Harper and Row, 1973); Ivan Illich, "Energy & Equity", in Illich, *Toward a History of Needs* (Berkeley: Heyday Books, 1978); André Gorz, "L'idéologie sociale de la bagnole", *Le Sauvage* (September–October 1973); Paul Rosen, "Up the Vélorution: Appropriating the Bicycle and the Politics of Technology", in *Appropriating Technology: Vernacular Science and Social Power*, ed. Ron Eglash et al. (Minneapolis: University of Minnesota Press, 2004).

© KONINKLIJKE BRILL NV, LEIDEN, 2019 | DOI:10.1163/9789004289970_004

All of this has nourished the belief that the western world is witnessing a "bicycle renaissance" and "a veritable bicycle boom".[2] Policymakers, policy-oriented bicycle researchers and cycling activists seem quite optimistic about the possibilities to increase the bicycle's modal share in daily transport for short-distance trips (up to 5 or 7.5 kilometres) by means of infrastructural engineering and programs for bicycle promotion. Bicycle policies have been introduced not only in countries with relatively high cycling levels (Netherlands, Denmark, the Flemish part of Belgium, Germany and Finland, and, to a lesser extent, Sweden, Norway, Ireland, Austria and Switzerland), but also in countries with low volumes of cycling (Great Britain, the United States, Canada, Australia, France, and Italy). By way of contrast, in the eastern part of Europe, where recent trends point to declining levels of pedalling as a consequence of economic growth and fast growing motorized traffic, and also in Spain, Portugal and Greece, bicycle policies, if in place at all, are still in their infancy.

The arguments reinforcing cycling policies are basically similar everywhere, but their implementation as well as actual wheeling levels reveal significant and persistent differences between countries. Around 2000, the bicycle's modal share in passenger transport amounted to 27 per cent in the Netherlands and 20 per cent in Denmark. It varied between 7 and 12 per cent in Germany, Belgium, Austria, Switzerland, Sweden and Finland; between 4 and 5 per cent in Italy, France and Norway; and between 2 and 3 per cent in Great Britain, Canada, Ireland and the Czech Republic. And it stagnated at around 1 per cent in the United States, Australia, New Zealand, Spain, Portugal and Greece. The annual pedalling distance per capita in kilometres fluctuated between 850 and 1,020 in the Netherlands and Denmark; between 250 and 330 in Belgium, Germany, Sweden and Finland; between 140 and 230 in Ireland,

---

2 John Pucher, Charles Komanoff, and Paul Schimeck, "Bicycling Renaissance in North America? Recent Trends and Alternative Policies to Promote Bicycling", *Transportation Research Part A: Policy and Practice* 33, nos. 7–8 (1999); John Pucher, Ralph Buehler, and Mark Seinen, "Bicycle Renaissance in North America? An Update and Re-appraisal of Cycling Trends and Policies", *Transportation Research Part A: Policy and Practice* 45, no. 6 (2011); John Pucher and Ralph Buehler, eds., *City Cycling* (Cambridge, MA: MIT Press, 2012); Trine Agervig Carstensen and Anne-Katrin Ebert, "Cycling Cultures in Northern Europe: From 'Golden Age' to 'Renaissance'", in *Cycling and Sustainability*, Transport and Sustainability, Vol. 1, ed. John Parkin (Bingley: Emerald, 2012); Tim Birkholz, "Die stille Revolution—das Fahrrad kommt zurück", in *Das Fahrrad. Kultur/Technik/Mobilität*, ed. Mario Bäumer and Hamburg Museum der Arbeit (Hamburg: Junius, 2014); Paul Smethurst, *The Bicycle: Towards a Global History* (New York: Palgrave Macmillan, 2015), 3–4, 143. For a tempering of this belief, see, however, Manuel Stoffers and Anne-Katrin Ebert, "New Directions in Cycling Research: A Report on the Cycling History Roundtable at T2M Madrid", *Mobility in History* 5, no. 1 (2014); Smethurst, *The Bicycle*, 144–48.

Italy and Austria, between 70 and 100 in France, Great Britain and Greece; while it did not reach 50 in Portugal and Spain. Whereas all residents of the Netherlands and Denmark, on average, own a bicycle, the same goes for 3 out of 4 Germans; 2 out of 3 Swedes and Finns; about 1 out of 2 Belgians, Italians and Austrians; 1 out of 3 Frenchmen and British; 1 out of 4 Portuguese; and 1 out of 5 Spaniards.[3] Also, the reasons for, and the appreciation of, pedalling show considerable variation. Whereas in countries with high volumes of bicycle traffic, a positive image and utilitarian purposes (commuting to work, school, shops and other activities and destinations) prevail, in countries with low cycling levels, more negative views abound on daily use of the bicycle and pedalling as a leisure time, sportive or childhood activity comes first.[4]

These substantial differences between nations raise several questions. For one thing, what does the so-called "bicycle renaissance" imply, and what is its impact? Is it possible to explain variations in the frequency, purpose and appreciation of bicycle-use on the basis of geographical and climatological conditions, environmental and infrastructural planning, demographic characteristics, habits in mobility, and the image of the bicycle? What is the impact of cycling policies in various countries? Are they effective at all?

This chapter considers these issues on the basis of a meta-analysis of social-scientific and historical bicycle studies as well as policy documents. First, I discuss policy-oriented research into the factors that advance or impede bicycling. Next, I argue that this research and the associated policy plans leave several of the questions unanswered, and that some of their basic assumptions

3 For these and other quantitative data, see J. Dekoster and U. Schollaert, *Cycling: The Way ahead for Towns and Cities* (Luxemburg: Office for Official Publications of the European Commission, 1999), 19; Ulrike Huwer, "Let's Bike: The 10 Point Pedalling Action Programme to Support Cycling", *World Transport Policy & Practice* 6, no. 2 (2000), 41, 43; European Conference of the Ministers of Transport, *Implementing Sustainable Urban Travel Policies: National Policies to Promote Cycling* (Paris: Organisation for Economic Co-operation and Development, 2004), 19–20, 24; Piet Rietveld and Vanessa Daniel, "Determinants of Bicycle Use: Do Municipal Policies Matter?" *Transportation Research Part A: Policy and Practice* 38, no. 7 (2004), 534; Kurt van Hout, *Literature Search Bicycle Use and Influencing Factors in Europe* (Hasselt: Hasselt University/Transportation Research Institute, 2008), 8, 14–18; John Pucher and Ralph Buehler, "Making Cycling Irresistible: Lessons from the Netherlands, Denmark and Germany", *Transport Reviews* 28, no. 4 (2008), 498–99; David R. Bassett et al., "Walking, Cycling, and Obesity Rates in Europe, North America, and Australia", *Journal Physical Activity and Health* 5, no. 6 (2008), 799; Directorate-General for Internal Policies, Policy Department B, Structural and Cohesion Policies, Transport and Tourism, *The Promotion of Cycling* (Brussels: European Union, 2010), 28; "Bicycle Statistics: Usage, Production, Sales, Import, Export", International Bicycle Fund, accessed 28 December 2017, http://www.ibike.org/library/statistics-data.htm.

4 Huwer, "Let's Bike", 43.

should be brought up for discussion. As I will show, policymakers and bicycle researchers largely ignore the historical and national-cultural dimension of cycling. My claim is that this overlooked aspect is highly relevant for explaining international differences in both cycling levels and the effectiveness of policies.[5] This will be demonstrated by comparing diverging cycling patterns among Western countries, and explaining how they are rooted in the past. In this way, my argument seeks to bridge the gap between bicycle policies and the interrelated social scientific research on the one hand and cultural-historical studies of pedalling on the other.

## 2 Determinants of Bicycling

The growing concern for bicycling in the transport policies of many western governments in the last three decades has boosted social-scientific bicycle research in the field of mobility, traffic engineering and urban planning. Quantitative and statistical methods—in particular, measurements of traffic movements and surveys—have been predominant in this research. Central concerns pertain to why people either use or do not use the bicycle (in particular for utilitarian purposes) and how cycling can be facilitated and promoted. Strikingly, most of these studies have appeared in the United States, the United Kingdom and Australia, where bicycle levels are fairly low, while Dutch, German, Belgian, Danish and other Scandinavian scholars also figure prominently. My argument is based on an analysis of more than two hundred published and unpublished research papers[6] and several policy documents, mostly produced in the last two decades.[7]

5 The considerable variations in cycling volumes between regions or cities within countries are beyond the scope of this chapter. See, however, Stoffers and Ebers, "New Directions".

6 Within the confines of this chapter, I can only refer to a selection of these studies.

7 Ministry of Transport, Public Works and Water Management, The Netherlands, *Nota Fietsverkeer 1983. Een volledig beeld* (The Hague: Ministry of Transport, Public Works and Water Management, 1983); European Commission, *Policy and Provision for Cyclists in Europe* (Brussels: Commission of the European Community, Directorate General for Transport, 1989); Ministry of Transport, Public Works and Water Management, The Netherlands, *Beleidsnotitie Masterplan Fiets* (The Hague: Ministry of Transport, Public Works and Water Management, 1991); Federal Highway Administration, the United States, *The FHWA National Bicycling and Walking Study Case Study No. 3: What Needs to Be Done to Promote Bicycling and Walking?* (Washington, DC: Federal Highway Administration, 1992); Federal Highway Administration, the United States, *The FHWA National Bicycling and Walking Study Case Study No. 4: Measures to Overcome Impediments to Bicycling and Walking* (Washington, DC: Federal Highway Administration, 1993); Ministry of Transports and Communications, Finland, *Finland Moving*

*on Two Wheels* (Helsinki: Ministry of Transports and Communications, 1993); Department of Transportation, Federal Highway Administration, the United States, *The National Bicycling and Walking Study: Transportation Choices for a Changing America* (Washington, DC: Federal Highway Administration, 1994); Department of Transport, the United Kingdom, *National Cycling Strategy* (London: Department of Transport, 1996); Federal Ministry of Transport, Building and Urban Development, Germany, *Erster Bericht der Bundesregierung über die Situation des Fahrradverkehrs in der Bundesrepublik Deutschland* (Bonn: Federal Ministry of Transport, Building and Urban Development, 1998); Ministry of Transport, Public Works and Water Management, Directorate-General for Passenger Transport, The Netherlands, *Eindrapport Masterplan Fiets. Samenvatting, evaluatie en overzicht van de projecten in het kader van het Masterplan Fiets, 1990–1997* (The Hague: Ministry of Transport, Public Works and Water Management, 1998); Federal Highway Administration, the United States, *The National Bicycling and Walking Study: Five-Year Status Report by the US Department of Transportation* (Washington, DC: Federal Highway Administration, 1999); Ministry of Transport, Public Works and Water Management, Directorate-General for Passenger Transport, The Netherlands, *The Dutch Bicycle Master Plan: Description and Evaluation in an Historical Context* (The Hague: Ministry of Transport, Public Works and Water Management, 1999); Ministry of Transport, Denmark, *Promoting Safer Cycling: A Strategy* (Copenhagen: Ministry of Transport, 2000); Department of Transport, the United Kingdom, *Ten Year Transport Plan* (London: Department of Transport, 2000); *The Swedish National Strategy for More and Safer Cycle Traffic* (Stockholm: N/A, 2000); European Commission, *Promotion of Measures for Vulnerable Road Users: Measures to Promote Cyclist Safety and Mobility* (Brussels: European Commission, 2001); Federal Ministry of Transport, Building and Urban Development, Germany, *Nationaler Radverkehrsplan 2002–2012: FahrRad! Massnahmen zur Förderung des Radverkehrs in Deutschland* (Berlin: Federal Ministry of Transport, Building and Urban Development, 2002); Ministry of the Flemish Community, Department of Environment and Infrastructure, Belgium, *Ontwerp Vlaams Totaalplan Fiets* (Brussels: Ministry of the Flemish Community, 2002); Ministry of Transport and Communications, Finland, *Towards Healthy, Sustainable Transportation: Implementation of the* WHO *London Charter in Finland* (Helsinki: Ministry of Transport and Communications, 2002); Geneviève Laferrere, *Comparison of National Cycling Policy in European Countries* (Brussels: Association for European Transport, 2002); Norwegian Public Roads Administration, *National Cycling Strategy: Making Cycling Safe and Attractive: National Transport Plan 2006–2015* (Oslo: Norwegian Public Roads Administration, 2003); Department for Transport, the United Kingdom, *Walking and Cycling: An Action Plan* (London: Department for Transport, 2004); European Conference of the Ministers of Transport, *Implementing*; Department of Transportation, the United States, *The National Bicycling and Walking Study: A Ten-Year Status Report* (Washington, DC: Department of Transportation, 2004); Austroads Incorporated, *The Australian National Cycling Strategy 2005–2010* (Sydney: Austroads Incorporated, 2005); Federal Ministry of Agriculture and Forestry, the Environment and Water Resources, Austria, *Masterplan Radfahren—Strategie zur Förderung des Radverkehrs in Österreich* (Vienna: Federal Ministry of Agriculture and Forestry, the Environment and Water Resources, 2006); Interministerial Co-ordinator for Bicycle Policy, *Plan pluriannuel d'actions de l'État en faveur du vélo proposé en 2007* (Paris: Ministry of Ecology, Sustainable Development and Energy, France, 2007); Federal Ministry of Transport, Building and Urban Development, Germany, *Zweiter Bericht der Bundesregierung über die Situation des Fahrradverkehrs in der Bundesrepublik Deutschland 2007* (Berlin: Federal Ministry of Transport, Building and Urban Development, 2007); Department of Transport, the United Kingdom, *A Sustainable Future for Cycling* (London: Department of Transport,

Based on my reading of the research reports, the following six factors that advance or impede cycling can be distinguished: (1) natural conditions; (2) land use patterns and built environment; (3) demography; (4) traffic infrastructure; (5) individual motivation and (6) ingrained and taken for granted collective habits (*habitus*) with respect to mobility. Four of these determinants cannot be changed through direct and purposive human intervention, at least not in the short term: they largely depend on the forces of nature (1) or they have taken on a more or less fixed shape in long-term historical developments (2, 3 and 6). In principle, it is possible for traffic infrastructure and individual motivation to be influenced more or less directly in the shorter term through goal-oriented policy measures. Policy as such—the choices and priorities made and the way it is implemented—can be considered a possible immediate influence on pedalling levels in its own right.

The basic assumption of cycling policy is the more or less optimistic idea that riding bicycles can be stimulated by technical and social design. Policymakers, planning experts, and policy-oriented bicycle researchers feel themselves challenged by two main problems. The first one is that people who do not use a bicycle for personal transport are hampered by material and environmental barriers, such as the dominance of motorized traffic and the lack of appropriate infrastructural facilities or other provisions. The second issue is that such people are not aware of the two-wheeler's benefits because they lack experience with it and have the wrong ideas about it. The engineering and planning approach implies the belief that these problems can be tackled by implementing the appropriate measures based on scientific (in particular quantitative) knowledge and expertise. This way of reasoning also presupposes to a large extent that people's decisions as to whether or not to pedal is mainly based on an individual and rational-instrumental consideration of costs and benefits, and that such a choice can be influenced by adapting the

2008); Swiss Federal Council, *Stratégie pour le développement durable: Lignes directrices et plan d'action (2008–2011)—Guide* (Bern: Federal Office for Spatial Development [ARE], 2008); Department of Transport, Ireland, *Ireland's First National Cycle Policy Framework* (Dublin: Department of Transport, 2009); Australian Bicycle Council and Austroads, *Gearing up for Active and Sustainable Communities: The Australian National Cycling Strategy 2011–2016* (Sydney: Austroads, 2010); Directorate-General for Internal Policies, *The Promotion of Cycling*; Department of Transportation, the United States, *The National Bicycling and Walking Study: 15-Year Status Report* (Washington, DC: Department of Transportation, 2010); Department of Transportation, the United States, *Policy Statement on Bicycle and Pedestrian Accommodation Regulations and Recommendations* (Washington, DC: Department of Transportation, 2010); National Association of City Transportation Officials, *Cities for Cycling* (Washington, DC: National Association of City Transportation Officials, 2010).

physical environment, by promoting a positive image of cycling and by educating them about its advantages.

After a brief discussion of the main results of bicycle research with respect to natural conditions, land use and built environment, and demography, my argument centres on infrastructure, motivation and *habitus* in relation to policies. Finally, I turn to what is generally lacking in bicycle studies and planning: a consideration of the influence of history and (national) culture on pedalling.

## 3 Natural Conditions, Built Environment and Demography

Although considerable differences in altitude and extreme climates (hot as well as cold) seem to be major barriers for cycling, research shows that such natural conditions are not always a decisive factor and perhaps even play a subordinate role. The two-wheeler's modal split is largest in Denmark and the Netherlands, more or less flat countries with a temperate (though also rainy) climate. Despite the cold winters, cycling levels are generally higher in Scandinavia than in several countries with a warmer climate. And even with their icy winters, Canadians on average pedal more often than Americans. Moreover, both the climate and the topography in Ireland, Eastern England as well as the North of France, Germany and Italy are not very different from those in the Netherlands and Denmark, but cycling levels vary substantially among these regions. Swiss and Austrian pedalling volumes are larger than those in several less mountainous areas.[8]

---

8 Mark J. Koetse and Piet Rietveld, "The Impact of Climate Change and Weather on Transport: An Overview of Empirical Findings", *Transportation Research Part D: Transport and Environment* 14, no. 3 (2009); Paul Emmerson, Tim Ryley, and D.G. Davies, "The Impact of Weather on Cycle Flows", *Traffic Engineering & Control* 39, no. 4 (1998); Max Nankervis, "The Effect of Weather and Climate on Bicycle Commuting", *Transportation Research Part A: Policy and Practice* 33, no. 6 (1999); A.J. Richardson, *Seasonal and Weather Impacts on Urban Cycling Trips*, TUTI Report 1–2000 (Victoria, Australia: The Urban Transport Institute, 2000); Christiane Brandenburg, Andreas Matzakaris, and Arne Arnberger, "The Effects of Weather on Frequencies of Use by Commuting and Recreation Bicyclists", in *Advances in Tourism Climatology*, ed. Andreas Matzarakis, C.R. de Freitas, and Daniel Scott (Freiburg: Berichte des Meteorologischen Instituts der Universität Freiburg, 2004); John Pucher and Ralph Buehler, "Why Canadians Cycle More than Americans: A Comparative Analysis of Bicycling Trends and Policies", *Transport Policy* 13, no. 3 (2006); Meghan Winters et al., "Utilitarian Bicycling: A Multilevel Analysis of Climate and Personal Influences", *American Journal of Preventive Medicine* 32, no. 1 (2007); Christiane Brandenburg, Andreas Matzakaris, and Arne Arnberger, "Weather and Cycling—A First Approach to the Effects of Weather Conditions on Cycling", *Meteorological Applications* 14, no. 1 (2007); Muhammad Sabir, Mark J. Koetse, and Piet

There is evidence that spatial and urban characteristics (building and population density, the degree of (sub)urbanization and the dispersion or mixing of functions such as living, working, shopping and leisure, which determine commuting distances, have a greater impact on bicycle use than climate and topography. In general, trips of up to a maximum of 5 to 7.5 kilometres seem to be feasible for utilitarian cycling. The contrast between compact towns and inner cities in Europe and pervasive urban sprawl in North America and Australia partly explains the large differences in bicycle use. Some researchers point out that urban design and land use patterns are not independent variables because self-selection may play a role as well: areas with greater building density and lesser functional dispersion attract people who may choose the bicycle over the car in daily transport. Also, various degrees of urban compactness or sprawl may not only be a cause, but also an effect of higher and lower levels of pedalling and motoring respectively. In general, the influence of existing bicycle volumes and patterns on the built environment and people's willingness to use the bicycle in daily transport tends to be underrated. The increase of automobility has advanced urban sprawl (most obviously in North America and Australia), while high numbers of cyclists (such as in Dutch and Danish and also other European towns) may stimulate higher building densities and the intermixing of economic and social functions. Moreover, there is evidence that the correlation between sizeable cycling volumes and high building density holds truer for smaller towns than for larger cities (over 100,000 residents), which usually have more elaborate public transport networks. There appears to be an inverse correlation, in other words, between the modal share of public transport and that of the two-wheeler in daily mobility.[9]

---

Rietveld, *The Impact of Weather Conditions on Mode Choice Decisions: Empirical Evidence for the Netherlands*, Tinbergen Institute Discussion Paper (Amsterdam: Vrije Universiteit, 2008).

9 Kevan Shafizadeh and Debbie Niemeier, "Bicycle Journey-to-Work: Travel Behavior Characteristics and Spatial Analysis", *Transportation Research Record* 1578 (1997); Tim Schwanen, "Urban Form and Commuting Behavior: A Cross European Comparison", *Tijdschrift voor Economische en Sociale Geografie* 93, no. 3 (2002); Brian E. Saelens, James F. Sallis, and Lawrence D. Frank, "Environmental Correlates of Walking and Cycling: Findings From the Transportation, Urban Design, and Planning Literatures", *Annals of Behavioral Medicine* 25, no. 2 (2003); Robert Cervero, "The Built Environment and Travel: Evidence from the United States", *European Journal of Transport and Infrastructure Research* 3, no. 2 (2003); Robert Cervero and Michael Duncan, "Walking, Bicycling, and Urban Landscapes: Evidence From the San Francisco Bay Area", *American Journal of Public Health* 93, no. 9 (2003); Anne Vernez Moudon et al., "Cycling and the Built Environment, a US Perspective", *Transportation Research Part D: Transport and Environment* 10, no. 3 (2005); John Douglas Hunt and J.E. Abraham, "Influences on Bicycle Use", *Transportation* 34, no. 4 (2007); Jessica Y. Guo, Chandra R. Bhat, and Rachel B. Copperman, "Effect of the Built Environment on Motorized and Non-Motorized Trip Making: Substitutive, Complementary, or Synergistic?",

In some countries, there is a significant correlation between the bicycle's modal split and particular demographic characteristics of the population (age, gender, income, education, religion, family composition, lifestyle, ethnicity and political affiliation), but in other countries such a correlation is weak or almost non-existent. If in countries with low cycling levels—such as the United States, Canada, Great Britain and Australia—men, youngsters and students are strongly overrepresented among wheelers, while also in France and Belgium more men than women pedal, countries with large (Netherlands and Denmark) or moderate (the rest of Scandinavia, Germany, Austria and Switzerland) cycling volumes show a stronger correspondence between the demographic traits of bicyclists and those of the general population. In some countries, young metropolitan professionals are overrepresented among wheelers, while in general ethnic groups with non-Western roots seem underrepresented. Researchers have found no straightforward correlations between bicycle use and such demographic variables as education, wealth, income, family composition, religion, and car ownership; such correlations are usually weak and vary between countries.[10] However, there are indications that

*Transportation Research Record* 2010 (2007); Sammy Zahran et al., "Cycling and Walking: Explaining the Spatial Distribution of Healthy Modes of Transportation in the United States", *Transportation Research Part D: Transport and Environment* 13, no. 7 (2008); Xinyu (Jason) Cao, Patricia L. Mokhtarian, and Susan L. Handy, "Examining the Impacts of Residential Self-Selection on Travel Behaviour: A Focus on Empirical Findings", *Transport Reviews* 29, no. 3 (2009); Reid Ewing and Robert Cervero, "Travel and the Built Environment: A Meta-Analysis", *Journal of the American Planning Association* 76, no. 3 (2010); Meghan Winters et al., "Built Environment Influences on Healthy Transportation Choices: Bicycling versus Driving", *Journal of Urban Health* 87, no. 6 (2010); Joachim Scheiner, "Interrelations between Travel Mode Choice and Trip Distance: Trends in Germany 1976–2002", *Journal of Transport Geography* 18, no. 1 (2010); Grégory Vandenbulcke et al., "Cycle Commuting in Belgium: Spatial Determinants and 'Re-Cycling' Strategies", *Transportation Research Part A: Policy and Practice* 45, no. 2 (2011); Kevin J. Krizek, "Cycling, Urban Form and Cities: What Do We Know and How Should We Respond?" in *Cycling and Sustainability*, ed. Parkin; Maria Börjesson and Jonas Eliasson, "The Benefits of Cycling: Viewing Cyclists as Travellers Rather than Non-Motorists", in *Cycling and Sustainability*, ed. Parkin, 254, 256, 264, 266.

10 Ipek N. Sener, Naveen Eluru, and Chandra R. Bhat, "An Analysis of Bicyclists and Bicycling Characteristics: Who, Why, and How Much Are They Bicycling", *Transportation Research Record* 2134 (2009); Gregory B. Rodgers, "The Characteristics and Use Patterns of Bicyclists in the United States", *Journal of Safety Research* 25, no. 2 (1994), 86–88; Michael R. Baltes, "Factors Influencing Nondiscretionary Work Trips by Bicycle Determined from 1990 U.S. Census Metropolitan Statistical Area Data", *Transportation Research Record* 1538 (1996): 96–101; William E. Moritz, "A Survey of North American Bicycle Commuters: Design and Aggregate Results", *Transport Research Record* 1578 (1997): 98; Ministry of Transport, Public Works and Water Management, *Eindrapport*, 45; Colin G. Pooley and Jean Turnbull, "Modal

lifestyle, social environment, and status sensitivity or egalitarianism are relevant. The difference between the Netherlands and Belgium in this respect is striking indeed. In Belgium, people with lower education and lower income levels are overrepresented among bicyclists. For Belgians, it seems, pedalling in daily commuting is much more strongly linked to social status than for the

---

Choice and Modal Change: The Journey to Work in Britain since 1890", *Journal of Transport Geography* 8, no. 1 (2000): 14–15, 19; City of Copenhagen, Denmark, *Cycle Policy 2002–2012* (Copenhagen: The City of Copenhagen, Roads & Parks Department, 2002); Rietveld and Daniel, "Determinants"; Ruud Ververs and Arnold Ziegelaar, *Verklaringsmodel voor fietsgebruik gemeenten. Eindrapport* (Leiden: Research voor Beleid, 2006); John Parkin, Tim Ryley, and Tim Jones, "Barriers to Cycling: An Exploration of Quantitative Analyses", in *Cycling and Society*, ed. Dave Horton, Paul Rosen, and Peter Cox (Aldershot: Ashgate, 2007); Winters et al., "Utilitarian Bicycling"; Joachim Scheiner and Christian Holz-Rau, "Travel Mode Choice: Affected by Objective or Subjective Determinants?" *Transportation* 34, no. 4 (2007): 487–511; Joachim Scheiner, "Mobility Biographies: Elements of a Biographical Theory of Travel Demand", *Erdkunde* 61, no. 2 (2007); Bas de Geus et al., "Psychosocial and Environmental Factors Associated with Cycling for Transport among a Working Population", *Health Education Research* 23, no. 4 (2007); Van Hout, *Literature Search*; John Parkin, Mark Wardman, and Matthew Page, "Estimation of the Determinants of Bicycle Mode Share for the Journey to Work Using Census Data", *Transportation* 35, no. 1 (2008); Transport for London, *Cycling in London: Final Report* (London: Transport for London, 2008) 1, 8–11, 41, 45; Michael Smart, "US Immigrants and Bicycling: Two-Wheeled in Autopia", *Transport Policy* 17, no. 3 (2010); Catherine Emond, Wei Tang, and Susan Handy, "Explaining Gender Difference in Bicycling Behavior", *Transportation Research Record* 2125 (2009); Sener, Eluru, and Bhat, "An Analysis"; Eva Heinen, Bert van Wee, and Kees Maat, "Commuting by Bicycle: An Overview of the Literature", *Transport Reviews* 30, no. 1 (2010); Rik Verhoeven and Pieter M. Schrijnen, "Allochtonen onderweg: fietsgebruik onder immigranten" (presentation, Colloquium Vervoersplanologisch Speurwerk, Roermond, The Netherlands, 25–26 November 2010); Peter Pelzer, "Fietsmulticulturalisme", AGORA—*Magazine voor sociaalruimtelijke vraagstukken* 26, no. 4 (2010); Yan Xing, Susan L. Handy, and Patricia L. Mokhtarian, "Factors Associated with Proportions and Miles of Bicycling for Transportation and Recreation in Six Small US Cities", *Transportation Research Part D: Transport and Environment* 15, no. 2 (2010); Frank Goetzke and Tilmann Rave, "Bicycle Use in Germany: Explaining Differences between Municipalities with Social Network Effects", *Urban Studies* 48, no. 2 (2011); Pucher, Buehler, and Seinen, "Bicycle Renaissance in North America?", 454–58; John Pucher et al., "Walking and Cycling in the United States: Evidence from the National Household Travel Surveys", *American Journal of Public Health* 101, no. 1 (2011): 312–13; Shannon L. Sahlqvist and Kristiann C. Heesch, "Characteristics of Utility Cyclists in Queensland, Australia: An Examination of the Associations between Individual, Social and Environmental Factors and Utility Cycling", *Journal of Physical Activity and Health* 9, no. 6 (2011); Vandenbulcke et al., "Cycle Commuting"; Jennifer Bonham and Anne Wilson, "Women Cycling through the Life Course: An Australian Case Study", in *Cycling and Sustainability*, ed. Parkin.

Dutch. At the same time, some researchers have noted that the education levels of American, British, Australian and Danish bicyclists are above average.[11]

Generally speaking over the last decade, bicycling has gained more status as part of the trendy lifestyle choices ("cycle chic") of a metropolitan and well-educated "creative class" in many parts of the Western world. The notion of a recent bicycle-renaissance appears to be related to gentrification and neoliberal consumer capitalism as well as to a privileged middle-class perspective on pedalling. The increasing popularity of riding expensive, stylish and technically advanced bicycles among "yuppies" in some large European and American cities and the associated middle-class cycling activism may be at odds with the interests and attitudes of lower class citizens. Whereas the first group can afford to live close to their jobs in expensive uptown neighbourhoods and does not view driving as relevant for their social status, the lower class (including deprived ethnic groups), which is increasingly shunted to the cheaper peripheries, cannot and probably does not want to reduce its dependence on motorized traffic—at least if they do not cycle out of sheer necessity because they cannot afford motorized transport.[12]

Against this background, in the United States the construction of new urban cycling facilities sometimes evokes aversion among lower class, black urban residents who feel that they are elbowed out of their neighbourhoods. Therefore it is questionable whether bicycle policies always serve the interests

11 De Geus et al., "Psychosocial and Environmental Factors"; Peter Pelzer, "Bicycling as a Way of Life: A Comparative Case Study of Bicycle Culture in Portland and Amsterdam" (master's thesis, University of Amsterdam, 2010), 31, 61, 91; Giselinde Kuipers, "De fiets van Hare Majesteit: Over nationale habitus en sociologische vergelijking", *Sociologie* 6, no. 3 (2011); Vandenbulcke et al., "Cycle Commuting", 121; Rodgers, "The Characteristics", 86; Shafizadeh and Niemeier, "Bicycle Journey-to-Work", 84; William Moritz, "Adult Bicyclists in the United States: Characteristics and Riding Experience in 1996", *Transportation Research Record* 1636 (1998); Pucher, Komanoff, and Schimeck, "Bicycling Renaissance in North America?", 629; Winters et al., "Utilitarian Bicycling"; Sener, Eluru, and Bhat, "An Analysis"; Pucher, Buehler, and Seinen, "Bicycle Renaissance", 455; City of Copenhagen, *Cycle Policy*, 9; Sahlqvist and Heesch, "Characteristics of Utility Cyclists".

12 Richard Florida, *The Rise of the Creative Class: And How It's Transforming Work, Leisure, Community, and Everyday Life* (New York: Basic Books, 2002); Carstensen and Ebert, "Cycling Cultures", 47–50; Börjesson and Eliasson, "The Benefits of Cycling", 258; Dave Horton and John Parkin, "Conclusion: Towards a Revolution in Cycling", in *Cycling and Sustainability*, ed. Parkin; Giselinde Kuipers, "The Rise and Decline of National Habitus: Dutch Cycling Culture and the Shaping of National Similarity", *European Journal of Social Theory* 16, no. 1 (2012); Carlton Reid, *Roads Were Not Built for Cars: How Cyclists Were the First to Push for Good Roads & Became the Pioneers of Motoring* (Washington: Island Press, 2015), 259–60; Peter Cox, "Ideas in Motion: Cycling: Image and Imagery in the Cultural Turn: Review Essay", *Transfers: Interdisciplinary Journal of Mobility Studies* 2, no. 1 (2012); Smethurst, *The Bicycle*, 57–65, 145–46, 153–55.

of the financially weak who are pushed back to the outskirts of cities without such facilities and without adequate public transport. Anyway, this reversal of the post-war image of the two-wheeler as a poor man's vehicle qualifies the idea that it is by definition an egalitarian and emancipatory means of transport.[13] Whereas in the lower class and non-Western ethnic perspective driving still bestows social status and cycling is second-rate, younger parts of the privileged urban middle-class, who experience pedalling as convenient and enjoyable, feel that car-ownership is no longer relevant for their social standing.

## 4 Infrastructural Facilities

Traffic systems, infrastructures and other material provisions have received much attention in policy-oriented bicycle research, in particular because they can be purposively (re)shaped by means of planning and engineering. This category includes traffic rules and speeds for motorized and pedalling traffic; whether cyclists are segregated or not from cars on the one hand and from pedestrians on the other; the availability of (secure) parking space and its cost for cars and/or two-wheelers; the presence of bicycle ways and lanes, car restricted zones, marked routes and networks, separate bridges, viaducts, tunnels, traffic lights, repair shops, changing rooms and showers in the workplace, and storage capacity for bicycles at home. Additional elements in this category are bicycle rental facilities, adjusting public transportation to pedalling, and the costs and taxation of various modes of transport.

Existing traffic systems and infrastructures offer more or less possibilities to add specific modifications and amenities for cyclists. Many policymakers and bicycle researchers assume that the construction of facilities that make cycling efficient, comfortable, pleasant and safe will result in increasing numbers of people opting for the bicycle in daily commuting. Some of them display an unshakable optimism about the possibilities of promoting pedalling through infrastructural and traffic measures. Typically for many American bicycle experts, a leading professor of urban planning, John Pucher, strongly believes that "bicycling can be increased even under quite unfavourable circumstances,

13 Timothy A. Gibson, "The Rise and Fall of Adrian Fenty, Mayor-Triathlete: Cycling, Gentrification and Class Politics in Washington, DC", *Leisure Studies* 34, no. 2 (2015); John Stehlin, "Regulating Inclusion: Spatial Form, Social Process, and the Normalization of Cycling Practices in the USA", *Mobilities* 9, no. 1 (2014); Karel Martens, "Role of the Bicycle in the Limitation of Transport Poverty in the Netherlands", *Transport Research Record* 2387 (2013).

provided the right public policies are implemented".[14] He claims that the high cycling levels in the Netherlands, Denmark and parts of Germany as well as in some cities in other countries, are largely caused by policies and the wide availability and good quality of infrastructural facilities. A similar approach would be the solution for countries with little bicycle traffic.

In order to find out whether infrastructural adjustments and facilities indeed have encouraged (utilitarian) cycling, researchers have investigated to what extent and by whom they are used. Some of them have established a correlation between improved bicycle routes and networks and an increased modal share of the two-wheeler, but only under an array of specific conditions. Cycling paths and lanes should provide direct and continuous connections and they should be part of a larger network, which should be located not too far from a cyclist's point of departure and destination. The road conditions should be good for pedalling and the routes should avoid steep climbs, and the number of traffic lights and busy intersections with car traffic should be kept to a minimum. Furthermore, the stimulating effect of facilities on bicycle levels seems not to be the same among all user groups. It is stronger among relatively inexperienced cyclists, the elderly and women than among experienced and sporty riders, including many younger men. The first group pedals prudently and prioritizes the (assumed) safety of segregated facilities, while the latter group, characterized by a more assertive driving style, prefers to take roads with motorized traffic if that saves them travel time.[15] Also, more as a

---

14 John Pucher, "Bicycling Boom in Germany: A Revival Engineered by Public Policy", *Transportation Quarterly* 51, no. 4 (1997): 44. See also John Pucher, "Urban Travel Behavior as the Outcome of Public Policy: The Example of Modal-Split in Western Europe and North America", *Journal of the American Planning Association* 54, no. 4 (1988); John Pucher and Lewis Dijkstra, "Making Walking and Cycling Safer: Lessons from Europe", *Transportation Quarterly* 54, no. 3 (2000); John Pucher and Lewis Dijkstra, "Promoting Safe Walking and Cycling to Improve Public Health: Lessons from the Netherlands and Germany", *American Journal of Public Health* 93, no. 9 (2003); Pucher and Buehler, "Making Cycling Irresistible"; John Pucher and Ralph Buehler, "At the Frontiers of Cycling: Policy Innovations in the Netherlands", *World Transport Policy* 1313 (2008); Pucher, Buehler, and Seinen, "Bicycling Renaissance", 464, 471; Pucher and Buehler, *City Cycling*; cf. Marcia D. Lowe, *The Bicycle: Vehicle for a Small Planet*, Worldwatch Paper 30 (Washington, DC: Worldwatch Institute, 1989), 6, 10, 31, 35, 39–40; John Parkin and Glen Koorey, "Network Planning and Infrastructure Design", in *Cycling and Sustainability*, ed. Parkin; Horton and Parkin, "Conclusion"; Meghan Winters et al., "Motivators and Deterrents of Bicycling: Comparing Influences on Decisions to Ride", *Transportation* 38, no. 1 (2011).

15 Cathy L. Antonakos, "Environmental and Travel Preferences of Cyclists", *Transport Research Record* 1438 (1994); Robert B. Noland and Howard Kunreuther, "Short-Run and Long-Run Policies for Increasing Bicycle Transportation for Daily Commuter Trips", *Transport Policy* 2, no. 1 (1995); Shafizadeh and Niemeier, "Bicycle Journey-to-Work";

general rule, Dutch and Danish findings have revealed that a clear causal link between infrastructural policies and an increase in cycling can only be demonstrated if pull measures such as installing cycling networks are combined with push measures such as constraining traffic regulations for motoring and a substantial rise of parking rates for cars in town centres.[16]

---

Monique A. Stinson and Chandra R. Bhat, "Commuter Bicyclist Route Choice: Analysis Using a Stated Preference Survey", *Transportation Research Record* 1828 (2003); Terri Pikora et al., "Developing a Framework for Assessment of the Environmental Determinants of Walking and Cycling", *Social Science and Medicine* 56, no. 8 (2003); Janet E. Dickinson et al., "Employer Travel Plans, Cycling and Gender: Will Travel Plan Measures Improve the Outlook for Cycling to Work in the UK?", *Transportation Research Part D: Transport and Environment* 8, no. 1 (2003); Monique A. Stinson and Chandra R. Bhat, "An Analysis of the Frequency of Bicycle Commuting Using an Internet-Based Survey", *Transportation Research Record* 1878 (2004); Kevin J. Krizek, Pamela J. Johnson, and Nebiyou Tilahun, "Gender Differences in Bicycling Behavior and Facility Preferences", in *Research on Women's Issues in Transportation: Report of a Conference, Volume 2: Technical Papers*, Transportation Research Board Conference Proceedings 35 (Chicago: Transportation Research Board of the National Academies, 2004); Rietveld and Daniel, "Determinants"; Hunt and Abraham, "Influences"; Marie-José Olde Kalter, *Vaker op de fiets? Effecten van overheidsmaatregelen* (The Hague: Kennisinstituut voor Mobiliteitsbeleid, 2007); Nebiyou Y. Tilahun, David M. Levinson, and Kevin J. Krizek, "Trails, Lanes, or Traffic: Valuing Bicycle Facilities with an Adaptive Stated Preference Survey", *Transportation Research Part A: Policy and Practice* 41, no. 4 (2007); Deborah Cohen et al., "Impact of New Bicycle Path on Physical Activity", *Preventive Medicine* 46 (2008); Inger Mary Bernhoft and Gitte Carstensen, "Preferences and Behaviour of Pedestrians and Cyclists by Age and Gender", *Transportation Research Part F: Traffic Psychology and Behavior* 11, no. 2 (2008); Jan Garrard, Geoffrey Rose, and Sing Kai Lo, "Promoting Transportation Cycling for Women: The Role of Bicycle Infrastructure", *Preventive Medicine* 46 (2008); Gulsa Akar and Kelly J. Clifton, "The Influence of Individual Perceptions and Bicycle Infrastructure on the Decision to Bike", *Transportation Research Record* 2140 (2009); Emond, Tang, and Handy, "Explaining Gender Difference"; Jennifer Dill, "Bicycling for Transportation and Health: The Role of Infrastructure", *Journal of Public Health Policy* 30, no. 1 (2009); Heinen, Van Wee, and Maat, "Commuting"; John Pucher, Jennifer Dill, and Susan Handy, "Infrastructure, Programs, and Politics to Increase Bicycling: An International Review", *Preventive Medicine* 50 (2010); Gianluca Menghini et al., "Route Choice of Cyclists in Zurich", *Transportation Research Part A: Policy and Practice* 44, no. 9 (2010); Jacob Larsen and Ahmed El-Geneidy, "A Travel Behavior Analysis of Urban Cycling Facilities in Montréal, Canada", *Transportation Research Part D: Transport and Environment* 16, no. 2 (2011); Bonham and Wilson, "Women Cycling"; Krizek, "Cycling".

16 Ministry of Transport, Public Works and Water Management, *Eindrapport*, 34, 40–41; Ververs and Ziegelaar, *Verklaringsmodel*; I.J.M. Hendriksen et al., *Beleidsadvies. Stimuleren van fietsen naar het werk*, TNO-rapport (The Hague: Ministry of Health, Welfare and Sport, 2010); Thomas Krag, "Bicycle Promotion Strategies in Denmark", Polish Seminars on Bicycle Promotion, accessed 28 December 2017, http://www.Friefugle.dk/poland/promotion_tk_en.html; Olde Kalter, *Vaker op de fiets?*; Kevin J. Krizek, Ann Forsyth, and Laura

Other studies have further questioned the assumption that infrastructural policies bring about an increase in bicycle traffic. It is difficult to determine the precise impact of facilities on bicycle use: finding a correlation between the cycling volumes and the presence of cycling routes and other amenities is not the same as proving that the construction of facilities causes an increase of pedalling. There may be a tendency in bicycle research to underestimate or even overlook the impact of existing wheeling levels on other relevant factors. Instead of infrastructure triggering an upsurge of cycling levels, policies aimed at building and improving facilities can also be a result of existing bicycle practices or the preceding rise of the two-wheeler's modal split, which may have been advanced by other factors. A growing number of cyclists may entail an increasing need and demand for adapting the traffic system and built environment and a greater willingness of governments to meet such pressure, in particular if it is articulated by well-informed and vocal bicycle activists and lobbyists. Self-selection should also be taken into account: individuals who are motivated to pedal, may prefer to settle in a bicycle-friendly neighbourhood or area. In this light it is difficult to determine the extent to which bicycle use is influenced by the available infrastructure or the composition of the population, individual preferences, lifestyle, perceptions, attitudes and habits.[17] An American study even concludes that there is no clear evidence for a correlation between infrastructure and cycling levels, and that demographic factors are far more relevant. The authors assert "that people who cycle do so irrespective of a supportive transportation infrastructure. Such commonly accepted

---

Baum, *Walking and Cycling International Literature Review: Final Report* (Melbourne: Department of Transport, State of Victoria, 2009).

17 Chris Banister, "Planning for Cycling", *Planning Practice & Research* 5, no. 1 (1990); Baltes, "Factors"; Arthur C. Nelson and David Allen, "If You Build Them, Commuters Will Use Them: The Association between Bicycle Facilities and Bicycle Commuting", *Transportation Research Record* 1578 (1997); Chris Gardiner and Rosalie Hill, "Cycling on the Journey to Work: Analysis of Socioeconomic Variables from the UK 1991 Population Census Samples of Anonymised Records", *Planning Practice & Research* 12, no. 3 (1997); Jennifer Dill and Theresa Carr, "Bicycle Commuting and Facilities in Major U.S. Cities: If You Build Them, Commuters Will Use Them—Another Look", *Transportation Research Record* 1828 (2003); Gary Barnes and Kevin J. Krizek, "Estimating Bicycling Demand", *Transportation Research Record* 1939 (2005); Parkin, Wardman, and Page, "Estimation"; Krizek, Forsyth, and Baum, *Walking and Cycling*; Kevin J. Krizek, Susan Handy, and Ann Forsyth, "Explaining Changes in Walking and Bicycling Behavior: Challenges for Transportation Research", *Environment and Planning B: Urban Analytics and City Science* 36, no. 4 (2009); Heinen, Van Wee, and Maat, "Commuting"; Krizek, "Cycling"; Xing, Handy, and Mokhtarian, "Factors"; Simon Fraser and Karen Lock, "Cycling for Transport and Public Health: A Systematic Review of the Effect of the Environment on Cycling", *European Journal of Public Health* 21, no. 6 (2010).

route-related correlates of cycling as bicycle lanes, traffic conditions, and street connectivity ... remain insignificant".[18]

Some longitudinal studies, which compare cycling levels before and after installing new bicycle facilities in several American cities, show that their construction did not result in a substantial growth of (utilitarian) cycling. To be true, modest increases (especially of recreational cycling) were sometimes realized, but these seem to have occurred only in neighbourhoods where the bicycle's modal share was above average already before the new facilities were put in. Conversely, in the suburban outskirts, where bicycle levels were lowest to begin with, little change was accomplished, if the number of wheelers did not in fact go down.[19] German and British studies comparing the widely divergent levels of bicycle use among commuters and school children in several cities demonstrate that no direct causal link can be established between, on the one hand, cycling volumes and whether or not pedalling is a matter of course, and, on the other hand, the existing traffic infrastructure and current cycling policies.[20] The results of a British survey-study suggest that the construction and improvement of cycling facilities hardly brought about an increase in pedalling commuters.[21]

Overall, in countries with low average pedalling levels, such as the United States, Canada, Australia and Great Britain, the implementation of bicycle policies—the construction of infrastructural facilities and also the launching of promotion campaigns—have failed to generate substantial increases of utilitarian cycling, apart from a few local exceptions. The number of Brits, Americans, Canadians and Australians who cycle on a daily or regular basis does not exceed 2 per cent of the population, and the past fifteen years have even witnessed a decline in utilitarian two-wheeler traffic, despite the implementation of cycling policies.[22] In these countries such policies do not find

---

18 Moudon et al. "Cycling and the Built Environment", 259.

19 Kevin J. Krizek, Gary Barnes, and Kristin Thompson, "Analyzing the Effect of Bicycle Facilities on Commute Mode Share over Time", *Journal of Urban Planning and Development* 135 no. 2 (2009); Fay Cleaveland and Frank Douma, "The Impact of Bicycling Facilities on Commute Mode Share" (presentation, Transportation Research Board 88th Annual Meeting, Washington, DC, 11–15 January 2009).

20 Marcus Jones, "Promoting Cycling in the UK: Problems Experienced by the Practitioners", *World Transport Policy & Practice* 7, no. 3 (2001); Goetzke and Rave, "Bicycle Use".

21 Simon Kingham, Janet Dickinson, and Scott Copsey, "Travelling to Work: Will People Move Out of Their Cars?" *Transport Policy* 8, no. 2 (2001).

22 Karel Martens, "The Bicycle as a Feedering Mode: Experiences from Three European Countries", *Transportation Research Part D: Transport and Environment* 9, no. 4 (2004), 284; Parkin, Wardman, and Page, "Estimation", 2; Malcolm J. Wardlaw, "Assessing the Actual Risks Faced by Cyclists", *Tec* December (2002): 352; Transport for London, *Cycling,*

fertile ground in an established and widespread daily practice, and they tend to be at odds with the wider environmental and infrastructural planning, as well as with comprehensive transport policies. Inasmuch as cycling facilities have been put in, these are patchy and mainly geared to recreation and sports; for practical commuting purposes they hardly prove effective. Continuous cycle routes and networks for everyday utilitarian mobility are few and far between, whereby many people still view and experience cycling as stressful and dangerous.[23] Thus cycling remains limited to the minority of the extraordinarily motivated.

Bicycle policies seem to be more fruitful in countries where cycling levels are high already and riding a bicycle is a well-established and time-honoured practice, but their results have to be put in perspective. German studies suggest that bicycle traffic saw its largest growth before city governments, in the 1980s, and the Federal Government, from 2002 on, introduced policies to promote cycling. Other developments appear to have advanced pedalling: greater environmental awareness, increasing traffic congestion, rising fuel prices, citizens' initiatives, local activism and urban expansion, which entailed that the distances covered in daily traffic increased and people who used to walk changed to pedalling.[24] Similarly, the relation between, on the one hand, the internationally renowned Dutch policies and, on the other, the increasing cycling volumes since the mid-1970s, followed by a slight decline between the mid-1980s and mid-1990s and the stabilization of the modal split at around 27 per cent, is far from unambiguous. After some cities had started to build bicycle routes and networks and to ban cars from town-centres, and the central government began to subsidize the building of bicycle ways next to main

46–47; John Pucher and Ralph Buehler, "Cycling Trends and Policies in Canadian Cities", *World Transport Policy and Practice* 11, no. 1 (2005): 43, 45; Pucher and Buehler, "Why Canadians Cycle", 278; Austroads Incorporated, *The Australian National*, 8–9; Australian Bicycle Council and Austroads, *Gearing up*, 14; Pucher, Komanoff, and Schimeck, "Bicycling Renaissance", 3; Cleaveland and Douma, "The Impact", 2; Krizek, Handy, and Forsyth, "Explaining", 735–36; Pucher, Buehler, and Seinen, "Bicycle Renaissance", 452; Pucher et al., "Walking", 310–12; John Pucher and Ralph Buehler, "Walking and Cycling for Healthy Cities", *Built Environment* 36, no. 4 (2010); Rachel Aldred, "Governing Transport from Welfare State to Hollow State: The Case of Cycling in the UK", *Transport Policy* 23 (2012); Anna Goodman, "Walking, Cycling and Driving to Work in the English and Welsh 2011 Census: Trends, Socio-Economic Patterning and Relevance to Travel Behaviour in General". *PLoS* ONE 8, no. 8: e7190 (2013), accessed 28 December 2017, http://journals.plos.org/plosone/article?id=10.1371/journal.pone.0071790.

23 Krizek, Forsyth, and Baum, *Walking and Cycling*, 37, 40; cf. Horton and Parkin, "Conclusion".

24 Heath Maddox, "Another Look at Germany's Bicycle Boom: Implications for Local Transportation Policy & Planning Strategy in the USA", *World Transport Policy & Practice* 7, no. 3 (2001).

roads, in the second half of the 1970s and early 1980s the total number of pedalled kilometres by the Dutch rose by 35 per cent. However, the introduction of bicycle policies on a national scale in 1990, when the *Bicycle Master Plan* was launched in order to expand and improve the existing cycling infrastructure, was not followed by a further substantial increase in bicycle use. Whereas over the past decade bicycle traffic has increased considerably in Dutch inner cities, partly as a consequence of measures to discourage car use there, at the same time the two-wheeler lost ground to the car and public transport in rural and suburban areas. In the 1980s and 1990s, the number of kilometres travelled by car continued to grow sharply and its modal share rose from almost 46 to 49 per cent between 1980 and the mid-1990s. The net result of Dutch cycling policy during the last two decades is that the bicycle's modal split has remained at a similar level. Without cycling policy and car-restricted urban zones it would probably have decreased.[25]

The same is true of Denmark where the bicycle's modal share has dropped slightly since the late 1980s, while that of car and public transport continued to rise.[26] Apparently it is difficult to achieve a further growth of cycling even

25 Anthony G. Welleman, "Why a Bicycle Policy in the Netherlands?" In *The Bicycle: Global Perspectives: Papers Presented at the Conference vélo mondiale Pro Bike–Velo-city, Montreal, Canada, 13–17 September 1992*, ed. R. Boivin and J.-F. Pronovost (Québec: Vélo Québec, 1992); Peter F. Peters, *De haast van Albertine. Reizen in de technologische cultuur. Naar een theorie van passages* (Amsterdam: Uitgeverij De Balie, 2003), 198–209; Ministry of Transport, Public Works and Water Management, *Eindrapport*, 42–44; Ministry of Transport, Public Works and Water Management, *The Dutch Bicycle Master Plan*, 38, 83–84, 87–88; Joop Atsma, *Fietsen in Nederland ... een tandje erbij. Initiatiefnota met voorstellen voor actief fietsbeleid in Nederland* (The Hague: CDA Tweede Kamerfractie, 2008), 9–10; cf. Pucher and Buehler, "Walking and Cycling"; Ministry of Transport, Public Works and Water Management, *Eindrapport*, 47, 49; Jeroen Buis and Roefof Wittink, *The Economic Significance of Cycling: A Study to Illustrate the Costs and Benefits of Cycling Policy* (The Hague: VNG uitgeverij, 2000), 13; Kees van Goeverden and Tom Godefrooij, "Ontwikkeling van het fietsbeleid en—gebruik in Nederland", Bijdrage aan het Colloquium Vervoersplanologisch Speurwerk 25 en 26 november 2010, accessed 28 December 2017, https://www.cvs-congres.nl/cvspdfdocs/cvs10_062.pdf; Advisory Council for Transport, Public Works and Water Management, The Netherlands, *Wie ik ben en waar ik ga. Advies over de effecten van veranderingen in demografie en leefstijlen op mobiliteit* (The Hague: Advisory Council for Transport, Public Works and Water Management, 2010); Goeverden and Godefrooij, "Ontwikkeling", 2; Yusak Susilo and Kees Maat, "The Influence of Built Environment to the Trends in Commuting Journeys in the Netherlands", *Transportation* 34, no. 5 (2007); Fraser and Lock, "Cycling for Transport", 742.

26 Thomas Krag, "Cycling, Environment, Exercise and Health", in *Cost Benefit Analysis of Cycling*, ed. Gunnar Lind (Copenhagen: Tema Nord, Nordic Council, 2005), 64; Lowe, *The Bicycle*, 37; Pucher and Buehler, "Walking and Cycling"; Till Koglin, "Vélomobility—A Critical Analysis of Planning and Space" (dissertation, Lund University, 2013), 159, 171;

in Holland and Denmark, perhaps for the very reason that the two-wheeler's modal share in daily traffic is already so large compared to other countries.[27] Cycling policies may have exhausted their potential: the Dutch and Danes who still do not bicycle will probably, for several reasons, never do so. Danish and Dutch bicycle policies, which are a shining example for many bicycle advocates, researchers and urban planners in other countries, apparently have not so much brought about a substantial rise of the bicycle's modal share. The result is rather a continuance and facilitation of the existing cycle traffic at the same steady level, while also having made cycling more convenient, safe and enjoyable for the fairly large numbers of people who used to pedal anyway.[28]

## 5 Attitudes, Perceptions and Habits

Policy-oriented bicycle research used to be dominated by traffic engineers and mobility experts, who focused on technological and infrastructural problems and solutions, and showed little interest in the experiences of cyclists and the social meanings of pedalling. More recently, however, a growing number of scholars have begun to criticize the one-sided emphasis on the "hard" material conditions of cycling, particularly infrastructural facilities, which would go at the expense of "soft" interventions, like information, education, promotion and marketing, aimed at improving the image and status of the two-wheeler. These scholars have drawn attention to the individual motivation as

Thomas A. Sick Nielsen, Hans Skov-Petersen, and Trine A. Carstensen, "Urban Planning Practices for Bikeable Cities—The Case of Copenhagen", *Urban Research & Practice* 6, no. 1 (2013): 111–12.

27 On the exemplary international role of Dutch cycling policies, see Ministry of Transport, Public Works and Water Management, *Cities Make Room for Cyclists* (The Hague: Ministry of Transport, Public Works and Water Management, 1995); Ministry of Transport, Public Works and Water Management, *Cycling in the Netherlands* (The Hague: Ministry of Transport, Public Works and Water Management, 2006); Jan Ploeger, "Designing for Cycling: The New Dutch Design Manual", in *The Greening of Urban Transport: Planning for Walking and Cycling in Western Cities*, ed. Rodney Tolley (Chichester: John Wiley & Sons, 1997).

28 Ministry of Transport, Public Works and Water Management, *Eindrapport*, 41–44, 50–54; Ministry of Transport, Public Works and Water Management, *The Dutch Bicycle Master Plan*, 38, 83–84; Krizek, Forsyth, and Baum, *Walking and Cycling*, 37; see also Ruth Oldenziel and Adri Albert de la Bruhèze, "Contested Spaces: Bicycle Lanes in Urban Europe, 1900–1995", *Transfers* 1, no. 2 (2011). For critical accounts of Dutch cycling policies, see Ida H.J. Sabelis, "Diversity in Cycle Policies", in *Cycling Cultures*, ed. Peter Cox (Chester: University of Chester Press, 2015); Angela van der Kloof, "Lessons Learned through Training Immigrant Women in the Netherlands to Cycle", in *Cycling Cultures*, ed. Cox.

to whether or not to use the bicycle for transportation.[29] Two perspectives can be distinguished in their studies. The first assumes that the choice for a mode of mobility is based on a rational-instrumental assessment by individuals of its costs and benefits in the light of their circumstances and available options. The bicycle's usefulness in daily commuting is central in this perspective. The second perspective centres on the influence of so-called affective motives (norms and values, beliefs, perceptions, attitudes and habits) that are largely shaped by the social environment and the wider culture. This perspective, which dovetails with more general pleas for a cultural turn in transport and mobility studies,[30] stresses that cycling experiences are moulded in various ways and that they cannot be reduced to economic and other utilitarian considerations.[31]

---

29 Mark Wardman, Richard Hatfield, and Michael Page, "The UK National Cycling Strategy: Can Improved Facilities Meet the Targets?" *Transport Policy* 4, no. 2 (1997); Christopher Porter, John Suhrbier, and William L. Schwartz, "Forecasting Bicycle and Pedestrian Travel: State of the Practice and Research Needs", *Transportation Research Record* 1674 (1999); Jones, "Promoting Cycling"; James Harrison, "Planning for More Cycling: The York Experience Bucks the Trend", *World Transport Policy & Practice* 7, no. 3 (2001); Paul Rosen, *How Can Research into Cycling Help Implement the National Cycling Strategy? Review of Cycling Research Findings and Needs*, Report of Whitehall Summer Placement in the Department for Transport, CLT3 and CLT4 (York: University of York, Science and Technology Studies Unit, 2003); Jillian Anable and Birgitta Gatersleben, "All Work and No Play? The Role of Instrumental and Affective Factors in Work and Leisure Journeys by Different Travel Modes", *Transportation Research Part A: Policy and Practice* 39, nos. 2–3 (2005); Barnes and Krizek, "Estimating"; De Geus et al., "Psychosocial and Environmental Factors"; Transport for London, *Cycling*; Krizek, Handy, and Forsyth, "Explaining"; Heinen, Van Wee, and Maat, "Commuting". See also Veronique Van Acker, Bert van Wee, and Frank Witlox, "When Transport Geography Meets Social Psychology: Toward a Conceptual Model of Travel Behaviour", *Transport Reviews* 30, no. 2 (2010); S. Bamberg, "Understanding and Promoting Bicycle Use: Insights from Psychological Research", in Parkin, *Cycling and Sustainability*, Transport and Sustainability, Volume 1, ed. John Parkin; Winters et al., "Motivators and Deterrents"; Georgina S.A. Trapp et al., "On Your Bike! A Cross-Partal Study of the Individual, Social and Environmental Correlates of Cycling to School", *International Journal of Behavioral Nutrition and Physical Activity* 8, no. 123 (2011); Robert J. Schneider, "Theory of Routine Mode Choice Decisions: An Operational Framework to Increase Sustainable Transportation", *Transport Policy* 25 (2013); Tim Jones, "Getting the British Back on Bicycles: The Effects of Urban Traffic-Free Paths on Everyday Cycling", *Transport Policy* 20 (2012).

30 See, e.g., Colin Divall and George Revill, "Cultures of Transport: Representation, Practice and Technology", *Journal of Transport History* 26, no. 1 (2005); Mimi Sheller and John Urry, "The New Mobilities Paradigm", *Environment and Planning A: Economy and Space* (2006); Justin Spinney, "Cycling the City: Movement, Meaning and Method", *Geography Compass* 3, no. 2 (2009); Phillip Vannini, "Mobile Cultures: From the Sociology of Transportation to the Study of Mobilities", *Sociology Compass* 4, no. 2 (2010).

31 Anable and Gatersleben, "All Work"; De Geus et al., "Psychosocial and Environmental Factors"; Heinen, Van Wee, and Maat, "Commuting"; Goetzke and Rave, "Bicycle Use". See also Bas Verplanken et al., "Attitude versus General Habit: Antecedents of Travel Mode

Some English studies indicate that deciding whether or not to cycle is closely linked to the perception of benefits (fun, fitness or health, low costs, flexibility and relatively fast on short distances) and disadvantages (slow on long distances, too much physical effort and sweating, too much climbing, exposed to speedy motorized traffic and bad weather, and loss of social status). People who never pedal predominantly perceive insurmountable obstacles and policies will not change this perception. Interventions only lead to behavioural change if people already have considered the possibility of riding a bicycle, have cycling experience or have a positive view on it, thus reducing objections such as discomfort and safety-risks.[32] These studies also make clear that cycling experience among regular and motivated cyclists is linked to positive feel-

Choice", *Journal of Applied Social Psychology* 2, no .44 (1994); Bas Verplanken, Henk Aarts and Ad van Knippenberg, "Habit, Information Acquisition, and the Process of Making Travel Mode Choices", *European Journal of Social Psychology* 27, no. 5 (1997); Sebastian Bamberg, Izek Ajzen, and Peter Schmidt, "Choice of Travel Mode in the Theory of Planned Behavior: The Roles of Past Behavior, Habit, and Reasoned Action", *Basic and Applied Social Psychology* 25, no. 3 (2003); Sylvia Titze, Willibald J. Stronegger, Susanne Janschitz, and Pekka Oja, "Association of Built-Environment, Social-Environment and Personal Factors with Bicycling as a Mode of Transportation among Austrian City Dwellers", *Preventive Medicine* 47, no. 3 (2008); Bamberg, "Understanding and Promoting"; Michelle Daley and Chris Rissel, "Perspectives and Images of Cycling as a Barrier or Facilitator of Cycling", *Transport Policy* 18, no. 1 (2011); Luuk H. Engbers and Ingrid J.M. Hendriksen, "Characteristics of a Population of Commuter Cyclists in the Netherlands: Perceived Barriers and Facilitators in the Personal, Social and Physical Environment", *International Journal of Behavioral Nutrition and Physical Activity* 7, no. 1 (2011); Eva Heinen, Kees Maat, and Bert van Wee, "The Role of Attitudes toward Characteristics of Bicycle Commuting on the Choice to Cycle to Work over Various Distances", *Transportation Research Part D: Transport and Environment* 16, no. 2 (2011); Eva Heinen, Kees Maat, and Bert van Wee, "The Effect of Work-Related Factors on the Bicycle Commute Mode Choice in the Netherlands", *Transportation* 40, no. 1 (2013); Colin G. Pooley et al., "Policies for Promoting Walking and Cycling in England: A View from the Street", *Transport Policy* 27 (2013); Liang Ma, "The Objective versus the Perceived Environment: What Matters for Active Travel" (dissertation, Portland State University, 2014); Liang Ma and Jennifer Dill, "Associations between the Objective and Perceived Built Environment and Bicycling for Transportation, *Journal of Transport and Health* 2, no. 2 (2015); Rachel Aldred and Katrina Jungnickel, "Why Culture Matters for Transport Policy: The Case of Cycling in the UK", *Journal of Transport Geography* 34 (2014); Devon Willis, Kevin Manaugh, and Ahmed El-Geneidy, "Cycling under Influence: Summarizing the Influence of Attitudes, Habits, Social Environments and Perceptions on Cycling for Transportation", *International Journal of Sustainable Transportation* 9, no. 8 (2015); Lake Sagaris, "Lessons from 40 Years of Planning for Cycle-Inclusion: Reflections from Santiago, Chile", *Natural Resources Forum* 39, no. 1 (2015); Caroline E. Scheepers, "Opportunities to Stimulate Active Transport" (dissertation, Free University Amsterdam, 2016).

32 Anable and Gatersleben, "All Work"; Birgitte Gatersleben and Katherine M. Appleton, "Contemplating Cycling to Work: Attitudes and Perceptions in Different Stages of Change", *Transportation Research Part A: Policy and Practice* 41, no. 4 (2007).

ings: relaxation and fun, a sense of independence, freedom, self-confidence, self-control and flexibility, a pleasant sensory stimulation because of the exercise and being in the open air, the intense perception of and interaction with the environment, and pleasant childhood memories. On the other hand, many Brits who do not pedal can only see cyclists as reckless daredevils and accident-prone persons, who are responsible for trouble and dangerous situations in traffic. They also regard cycling as unaesthetic and uncomfortable, as "hot and sweaty" and associate it with athletic (or not so athletic middle-aged) men wearing helmets and Lycra outfits.[33] Anxieties about bodily performance and appearance impede many women and non-western immigrants to pedal.[34]

More in general, research of motivation makes clear that the decision whether or not to cycle is usually not taken exclusively on the basis of a calculation of costs and benefits and explicit views, but that it is also inspired by more intuitive perceptions, experiences and valuations. For this reason, several researchers have put the difference between instrumental and affective motivations into perspective. They argue that instrumental choices can only be understood in the context of affective motivations. In daily practice apparent objective cost and benefit assessments are usually imbued with subjective perceptions of advantages and disadvantages. Such perceptions are embedded in habits, routines, experiences and attitudes. When it comes to a cost and benefit assessment, for example with regard to the investment of time, the physical effort, the health effects, the (in)convenience, the (in)efficiency, the (lack of) safety and the financial costs or yields of cycling, judgments vary considerably between regular cyclists and people who hardly or never pedal. The last group identifies far more drawbacks—riding a bicycle is believed to be uncomfortable, too strenuous, too dangerous, too slow, or too individualistic; the weather and the roads or bicycle ways can be bad; no luggage and other passengers can be transported; it is difficult to communicate with traveling companions—than the first group and also evaluates the environmental conditions for bicycling more negatively.[35] Promotion campaigns aimed at

---

33 Transport for London, *Cycling*, 22, 29; Anable and Gatersleben "All Work"; Birgitta Gatersleben and David Uzzell, "Affective Appraisals of the Daily Commute: Comparing Perceptions of Drivers, Cyclists, Walkers, and Users of Public Transport", *Environment and Behavior* 39, no. 5 (2007); cf. Bonham and Wilson, "Women Cycling"; Rachel Aldred, "The Role of Advocacy and Activism", in *Cycling and Sustainability*, ed. Parkin, 97; Horton and Parkin, "Conclusion", 309; see also Marius C. Claudy and Mark Peterson, "Understanding the Underutilization of Urban Bicycle Commuting: A Behavioral Reasoning Perspective", *Journal of Public Policy & Marketing* 33, no. 2 (2014).

34 Rachel Aldred, "Incompetent or Too Competent? Negotiating Everyday Cycling Identities in a Motor Dominated Society", *Mobilities* 8, no. 2 (2013).

35 Noland and Kunreuther, "Short-Run and Long-Run", 75–76; Peter Gordon and Harry W. Richardson, "Bicycling in the United States: A Fringe Mode?" *Transportation Quarterly*

boosting the bicycle's image have a similar effect as building facilities: they mainly attract people who already cycle and who do not have to be convinced of the practical usefulness and fun of pedalling, while their impact on people who rarely or never ride a bicycle is limited or none.[36]

The positive or negative evaluation of cycling and the associated perception of its benefits and difficulties by individuals are also determined by the attitudes in their social environment and whether cycling is part of the learned pattern of daily habits, whether or not one has grown up with bicycles as a common means of transportation. Through a mutual influencing and enhancement of experiences with and perceptions of cycling, a self-fulfilling prophecy appears to be working here. A positive or negative image of cycling implies that either its pros or cons are stressed. These perceptions determine whether or not one chooses the bicycle for commuting and whether or not one develops cycling experience. And experience, in its turn, determines perception again. Opting or not opting for pedalling is embedded in an accumulation of corroborating and reinforcing meanings, perceptions, experiences and behaviours.[37]

## 6 The Relevance of History and National Culture

All in all, the available research offers no conclusive evidence that cycling increases substantially as a result of infrastructural planning and promotional activities. That is not to say that such policies are futile and would have to be discarded. At least they may counterbalance several social, economic and technological dynamics that all over the western world structurally impede pedalling: spatial up-scaling and increasing mobility over greater distances furthering car-driving and the use of public transport; the continuing (neoliberal)

52, no 1 (1998); Stinson and Bhat, "An Analysis", 124, 126, 128; Anable and Gatersleben, "All Work"; Barnes and Krizek, "Estimating"; Krag, "Cycling", 64–68; Gatersleben and Appleton, "Contemplating"; Gatersleben and Uzzell, "Affective Appraisals"; De Geus et al., "Psychosocial and Environmental Factors"; Krizek, Forsyth, and Baum, *Walking and Cycling*, 23–26; Akar and Clifton, "The Influence"; Heinen, Van Wee, and Maat, "Commuting"; Bonham and Wilson, "Women Cycling"; Horton and Parkin, "Conclusion"; Claudy and Peterson, "Understanding the Underutilization"; Victoria Morckel and Kathryn Terzano, "The Influence of Travel Attitudes, Commute Mode Choice, and Perceived Neighborhood Characteristics on Physical Activity", *Journal of Physical Activity and Health* 11, no. 1 (2014); Ma, "The Objective versus the Perceived"; Ma and Dill, "Associations".

36 Transport for London, *Cycling*, 40.

37 Pelzer, "Bicycling", 14, 99–100; Stinson and Bhat, "An Analysis", 128; Barnes and Krizek, "Estimating", 9; Van Acker, Van Wee, and Witlox, "When Transport Geography Meets"; Bamberg, "Understanding and Promoting".

prevalence of efficiency, speed and economic values; the regular priority of motorized transport in traffic policies; growing prosperity and car ownership; and the proportional increase of the ageing population and non-Western ethnic minorities.[38]

Be that as it may, the diverse findings and partly uncertain conclusions of the authors in the field of social-scientific and traffic engineering studies of bicycling, tacitly rather than explicitly, call into question the basic premise of cycling policies: the belief that cycling can be advanced directly and in the short term through targeted technological and social interventions. To a large extent pedalling levels seem to be determined by factors which are not amenable to rational decision-making and planning: geographical, climatological and environmental conditions, demographic characteristics, socially and culturally determined attitudes, experiences, habits and perceptions, and the popular image and social status of the bicycle. It is not entirely clear how these factors influence bicycle use, what their relative weight is, how they interact, and how they affect the outcomes of bicycle policies. Moreover, what is lacking in policy-oriented bicycle research is the consideration that most of the relevant determinants—land use patterns, the built environment, and traffic infrastructures, attitudes and motivations, meanings and perceptions, and habits and routines—have taken shape and evolved in long-term, path-dependent developments[39]—and also, largely, in the context of the modern nation state.

There are good reasons to question the assumption that the travel behaviour of people can be changed in the short term through targeted policy measures. In their historical research on the development of British commuter traffic from the late nineteenth century, Colin G. Pooley and Jean Turnbull demonstrate that historical shifts in mobility patterns can be identified—for example, before the Second World War most people walked and cycled to work while after 1960 car-driving became dominant—but that within different periods individual travel behaviour showed a large degree of rigidity: few people switched to another means of transport. Their conclusion is that the individual's choice for a particular mode of mobility is largely determined by habits and routines, many of which, in turn, go back to prevailing social practices and cultural values.[40] Such findings suggest that the historical dimension of

---

38 Sudhir C. Rajan, "Automobility and the Liberal Disposition", *Sociological Review* 54, no. 1 (2006); Oliver Schwedes, "The Field of Transport Policy: An Initial Approach", *German Policy Studies* 7, no. 2 (2011); Aldred, "Governing Transport".

39 Cf. Pelzer, "Bicycling".

40 Pooley and Turnbull, "Modal Choice", 15, 23; Colin G. Pooley, Jean Turnbull, and Mags Adams, *A Mobile Century? Changes in Everyday Mobility in Britain in the Twentieth Century* (Aldershot: Ashgate, 2005); cf. Scheiner, "Mobility Biographies".

bicycling is relevant and they question the assumption that policies can bring about substantial changes in pedalling behaviour, at least in the short term.

Strikingly, policy-oriented researchers have not taken notice of the many historical works on bicycling published in the past three decades—at least I did not find any references to such studies in their papers. Some of them refer in passing to the possible impact of history and culture, in particular if their surveys fail to establish correlations between wheeling levels and other factors, while at the same time they play down that influence.[41] Typical is the assertion of John Pucher and his co-author Ralph Buehler that "policies appear to be far more important than history and culture in explaining ... cycling trends".[42] Comparing American and European cycling levels, they claim that "[t]he much higher levels of cycling in Europe are not simply historical artefacts or culturally determined".[43] Their way of reasoning suggests that policies can be made and implemented apart from historical and cultural contexts. Apparently, they do not consider that the more or less successful cycling policies and the extensive bicycle infrastructures in countries such as the Netherlands and Denmark could only be realized because of the bicycle-minded culture which had emerged and had been upheld in these nations since the early twentieth century.

Although more and more policy-oriented studies—more often implicitly than explicitly—suggest that the degree to which bicycle use can be substantially increased through policies depends on social-cultural contexts, only a few social-scientific researchers clearly acknowledge that historical factors may be highly relevant and deserve more serious attention.[44] Considering research into the relation between policies and infrastructures on the one hand and the volume of pedalling traffic on the other, the American bicycle scholars Gary Barnes and Kevin Krizek, for example, have pointed out that local variations in cycling levels across different American regions and cities cannot be reasonably explained by differences in policies and infrastructures. "Unmeasured factors, perhaps cultural or historical", they write, "appear to play an extremely large role in determining the level of cycling in an area". In their conclusion

41 See, e.g., the passing reference to "culture, custom and habit" in Pucher, Dill, and Handy, "Infrastructure, Programs, and Politics", 121.

42 Pucher and Buehler, "Walking and Cycling", 408.

43 Pucher and Buehler, "Why Canadians Cycle", 277; cf. Pucher and Buehler, "Making Cycling Irresistible".

44 See Barnes and Krizek, "Estimating", 45, 50; Krizek, Handy, and Forsyth, "Explaining", 725, 737; Parkin, Riley, and Jones, "Barriers"; David Skinner and Paul Rosen, "Hell is Other Cyclists: Rethinking Transport and Identity", in *Cycling and Society*, ed. Horton, Rosen, and Cox, 83–96; Sagaris, "Lessons"; Scheepers, *Opportunities*.

they assert: "It seems that local or even 'subcultural' attitudes and perhaps history play a very substantial role in the perception of bicycling as an appealing or even 'normal' thing for an adult to do". They add that "soft factors such as culture and attitudes" should be researched "in some systematic way", without indicating, however, how this should be done.[45] Together with Susan Handy and Ann Forsyth, Krizek also suggests that the disregard for history is related to bicycle researchers' strong and optimistic belief in planning and design and their one-sided and possibly biased focus on the practical effects of their studies. Their work is, according to these authors, "fraught with practical challenges as well as political ones: expectations are high, interventions are modest, and effects may be unclear", while planners and policymakers "have a responsibility to understand the limitations of the available evidence and not misuse that evidence in making the case for bicycle and pedestrian interventions".[46]

However, for Krizek and his co-authors this appears to be no reason to fundamentally question the basic approach and purpose of policy-oriented bicycle research and to take up his earlier suggestion that historical and cultural analysis should be included. On the contrary, in an evaluative survey he and his co-authors Ann Forsyth and Daniel Rodríquez call for more research along the established lines on the basis of more refined data collection and analysis, more sophisticated social-scientific theories and models as well as more precise quantitative methods in order to increase the usefulness of such work for policymaking.[47] In my view, the relevance of such an appeal and the implied belief in procedural rationality is disputable, and perhaps even counterproductive, because it may undermine the very societal (and also scholarly) bearing of such research.[48] Apart from the fact that history and culture are beyond planning and design, one of the main reasons that these "soft" factors

45 Barnes and Krizek, "Estimating", 45, 50.

46 Krizek, Handy, and Forsyth, "Explaining", 725, 737; cf. D. Ogilvie et al., "Promoting Walking and Cycling as an Alternative to Using Cars: Systematic Review", *British Medical Journal* 329, no. 7469 (2004); Krizek, "Cycling", 123, 125–26; Krizek, Forsyth, and Baum, *Walking and Cycling*; Ann Forsyth and Kevin J. Krizek, "Promoting Walking and Bicycling: Assessing the Evidence to Assist Planners", *Built Environment* 36, no. 4 (2010); Lin Yang et al., "Interventions to Promote Cycling: Systematic Review", *British Medical Journal* 341, c5293 (2010); Parkin and Koorey, "Network Planning".

47 Ann Forsyth, Kevin Krizek and Daniel Rodríguez, "Non-Motorized Travel Research and Contemporary Planning Initiatives", *Progress in Planning* 71, no. 4 (2009).

48 Having read dozens of research reports of quantitative bicycle research, I cannot escape the impression that many are full of truisms and that their conclusions are often trivial. See, e.g., Kevin J. Krizek and Rio W. Roland, "What Is at the End of the Road? Understanding Discontinuities of On-Street Bicycle Lanes in Urban Settings", *Transportation Research Part D: Transport and Environment* 10, no. 1 (2005).

appear as residual categories and remain invisible in social-scientific bicycle research, is precisely that they cannot be probed on the basis of the prevailing quantitative methods. Since bicycle-researchers do not question their basic assumption that bicycling can be planned and designed, and do not put in perspective their narrow approach, they continue to disregard the cultural and historical dimension, and therefore also the tenacity and persistence of pedalling patterns.

As a first move towards bridging the gap between historical and policy-oriented bicycle studies, as well as a broader and also international-comparative perspective, I would suggest that, generally, three partly contrasting bicycle cultures can be distinguished in the western world on the basis of different volumes and purposes of bicycle use; different meanings, images and perceptions of pedalling; different patterns of engrained cycling behaviour (*habitus*[49]); different characteristics of cyclists and their motivations; and differences in the nature of cycling policies and activism. There is a marked contrast between the bicycle culture in the Netherlands and Denmark, which provides a prominent role for the two-wheeler in daily transport, and is historically rooted in its image of a "democratic horse" and "civilizing tool", and the English-speaking countries and to a certain extent Germany, in which the bicycle has a marginalized or exclusive position, as either the poor man's humble utensil or as an alternative and trendy vehicle. The third bicycle culture can be found in France, Italy and Belgium, where the popularity of cycling centred on sports and (professional) racing: the bicycle was (and is) especially glorified as a record-breaker, while pedalling for utilitarian purposes, with the exception of the Flemish part of Belgium since the 1970s, has declined to rather low levels.

These cycling cultures have taken shape in specific historical trajectories and in the context of modern nation states. In the following parts I will sketch these trajectories and contexts on the basis of existing studies[50] with a focus

49 The concept of 'national habitus' was coined by the historical sociologist Norbert Elias in order to refer to culturally and socially shaped patterns of behaviour that are self-evident on a national scale. See Kuipers, "The Rise and Decline".

50 In the field of bicycle history (and, to a certain extent, also sociology), the focus has shifted from the technological development of bicycles to the social, cultural, and political dimensions of pedalling. For a historiographical overview, see Manuel Stoffers and Harry Oosterhuis, "Ons populairste vervoermiddel. De Nederlandse fietshistoriografie in internationaal perspectief", *Bijdragen en Mededelingen betreffende de Geschiedenis der Nederlanden/The Low Countries Historical Review* 124, no. 3 (2009): 390–418; Manuel Stoffers, Harry Oosterhuis, and Peter Cox, "Bicycle History as Transport History: The Cultural Turn", in *Mobility in History: Themes in Transport: T2M Yearbook 2011*, ed. Gijs Mom, Peter Norton, Georgina Clarsen, and Gordon Pirie (Neuchâtel: Alphil, 2010). For

on three phases in the bicycle's history from its introduction in the late nineteenth century onwards resulting in growing diversity in cycling patterns between countries. First, when bicycles made their entry into society in the late nineteenth century, specific meanings and values were attached to pedalling, and particular wheeling practices were highlighted and promoted. Second, in the first half of the twentieth century the two-wheeler established itself as a means of transport for the masses and at the same time bicycle practices and experiences were increasingly affected by growing motorized traffic. Third, after the Second World War, the nationally diverse cycling patterns that had evolved in the previous period consolidated in restraining or enhancing vicious circles. The various relevant factors—cycling volumes and practices; meanings, perceptions and public images; attitudes and habits; land use, urban design and traffic infrastructures; government policies and bicycle lobbying and activism—mutually supported and strengthened each other in either an inhibiting or stimulating way and hardened in positive and negative spirals.

## 7 Modernity and Nationalism

The introduction of bicycles in late nineteenth-century society—the "velocipede mania" in the 1860s which was followed in the 1890s by a "bicycle boom" in many parts of the western world—was generally caught up in praise of modernity. The new vehicle was strongly associated with scientific and technological innovation, social progress and individual liberation, in particular among the liberal and urban middle-class citizens who had sufficient means and time to afford and ride it. The two-wheeled "freedom machine" enabled flexible mobility at an unprecedented speed, and it thus involved not only a new experience of time and space, but also self-autonomy and a widening of one's mental horizon.[51] Although the two-wheeler was introduced in postal

---

sociological contributions, see Horton, Rosen, and Cox, eds., *Cycling and Society*; Cox, *Cycling Cultures*.

51 John Pinkerton, "Who Put the Working Man on a Bicycle?" *Cycle History 8: Proceedings of the 8th International Cycling History Conferences*, ed. Nicholas Oddy and Rob van der Plas (San Francisco: Van der Plas Publications, 1998), 101; Glen Norcliffe, *The Ride to Modernity: The Bicycle in Canada, 1869–1900* (Toronto, Buffalo, and London: University of Toronto Press, 2001); Joachim Radkau, "Das Fahrrad in den Technikvisionen der Jahrhundertwende, oder: das Erlebnis in der Technikgeschichte", *Wege zur Fahrradgeschichte*, ed. Volker Briese, Wilhelm Matthies, and Gerd Renda (Bielefeld: BVA Bielefelder Verlag, 1995); Nadine Besse, ed., *The Velocipede, Object of Modernity 1860–1870/Le velocipede, objet de modernité 1860–1870* (Saint-Étienne: Musée d'Art et d'Industrie, 2008); Joachim Krausse, "Versuch auf's Fahrrad zu kommen. Zur Technik und Ästhetik der Velo-Evolution", in

services, police and fire departments, and the army before 1900,[52] the first civilian cyclists did not so much use it for utilitarian rather than for sporting and leisure purposes. Riding the velocipede and the high-wheeler, which was risky and required agility, was largely restricted to athletic young men.[53] For women and older men, pedalling became feasible only after the more practical and comfortable "safety bicycle" came onto the market. Their use of the new vehicle was above all recreational: touring in the countryside and enjoying nature, which provided townsfolk with a counterbalance to the supposedly harmful and unhealthy sides of industrial society.[54] Cycling, an activity that combined physical exercise and mental respite, reflected and fostered a growing anxiety about individual as well as collective health. Riding on two wheels was a way to take part in modernity's dynamism, while at the same time, by keeping balance and mastering the machine and experiencing inner tranquillity, to be in control of its disruptive restlessness.

In order to defend their interests against authorities who impeded their freedom of movement as well as against other users of roads such as coachmen and pedestrians, from the 1870s onwards, upper- and middle-class bicycle

---

*Absolut modern sein: Culture technique in Frankreich 1889–1937*, ed. Hans Joachim Neyer (Berlin: Elefanten Press, Staatliche Kunsthalle, 1986); Peter Hinrichs, Ingo Kolboom, and Hans Joachim Neyer, "Zwischen Fahrrad und Fliessband. Culture technique in Frankreich zwischen Belle Epoque and Front Populaire", in *Absolut modern sein*, ed. Neyer; Titia Berlage, ed., *De fiets* (Rotterdam: Museum Boymans-Van Beuningen, 1977); Anne-Katrin Ebert, "Zwischen 'Radreiten' und 'Kraftmaschine'. Der bürgerliche Radsport am Ende des 19. Jahrhundert", *Werkstatt Geschichte* 44 (2007); Duncan R. Jamieson, "Bicycle Touring in the Late Nineteenth Century", in *Cycle History 12: Preceedings of the 12th Interntational Cycling History Conference*, ed. Andrew Ritchie and Rob van der Plas (San Francisco: Cycle Publishing, 2002); Lars Amenda, "Mit dem Fahrrad um die Welt—Radfernreisen vor hundert Jahren", in *Das Fahrrad: Kultur, Technik, Mobilität*, ed. Mario Bäumer, 112–16 (Hamburg: Junius, 2014); Jim Fitzpatrick, *Wheeling Matilda: The Story of Australian Cycling* (Kilcoy: Star Hill Studio, 2013); Smethurst, *The Bicycle*, 32–39, 67–104.

52 Jim Fitzpatrick, *The Bicycle in Wartime* (Washington: Brassey's, 1998); G.D. Cornelissen de Beer, "Invoering en gebruik van het rijwiel bij de Europese legers in de 19e eeuw", *Armamentaria. Jaarboek Legermuseum* 19 (1984–1985): 60–87.

53 Nick Clayton, "The Quest for Safety: What Took So Long?" in *Cycle History 8*, ed. Oddy and van der Plas; see also Wiebe E. Bijker, *Of Bicycles, Bakelites, and Bulbs: Toward a Theory of Sociotechnical Change* (Cambridge, MA: MIT Press, 1995), 19–100.

54 Gary A. Tobin, "The Bicycle Boom of the 1890s: The Development of Private Transportation and the Birth of the Modern Tourist", *Journal of Popular Culture* 7, no. 4 (1974): 838–49; Richard Holt, "The Bicycle, The Bourgeoisie and the Discovery of Rural France, 1880–1914", *British Journal of Sports History* 2, no. 2 (1984): 127–39; Catherine Bertho-Lavenir, *La Roue et le stylo. Comment nous sommes devenus touristes* (Paris: Odile Jacob, 1999); Christopher Thompson, "Bicycling, Class and the Politics of Leisure in Belle Epoque France", in *Histories of Leisure*, ed. Rudy Koshar, 131–46 (Oxford and New York: Berg, 2002).

hobbyists organized themselves into local clubs as well as national associations. Such organizations, which were part of civil society, associated pedalling with particular social and political values.[55] They lobbied for the improvement of traffic infrastructure and the civil right of riding on public roads, while at the same time they pressed wheelers to behave as civilized and self-disciplined traffic participants by learning the proper art of cycling. The individual freedom afforded by the new vehicle should be balanced by decency and responsible citizenship. The meaning which these organizations bestowed on bicycling reflected not only the values of bourgeois respectability and liberalism, but also nationalist ideals. Lobbying for more and better roads and other traffic facilities served the cause of connecting the nation. Their promotion of bicycle tourism as a way to discover native landscapes, the unspoiled, "traditional" countryside and national heritage, as well as to bridge the distance between town and countryside and between different regions radiated national pride. Bicycle shows and parades also became part of nationalist celebrations. Since cycling clubs sought official recognition by government authorities, it was not unusual that their members paraded in uniforms and rode in formation—the similarity with horse-riding including military cavalry was obvious—in order to present themselves as patriotic citizens. Pointing out the bicycle's military potential was also part of the scheme to win the approval of the authorities.

55 Roman Sandgruber, "Cyclisation und Zivilisation. Fahrradkultur um 1900", in *Glücklich ist, wer vergisst ...? Das andere Wien um 1900*, ed Hubert Ch. Ehalt, Gernot Heiss, and Hannes Stekl (Vienna, Cologne, Graz: Hermann Böhlaus, 1986); Les Bowerman, "Clubs: Their Part in the Study of Cycles and Cycling History", *Cycle History 5: Proceedings of the 5th International Cycling History Conference*, ed. Rob van der Plas (San Francisco: Bicycle Books, 1995); Catherine Bertho-Lavenir, "Normes de comportement et contrôle de l'espace. Le Touring Club de Belgique avant 1914", *Le Mouvement Social* 178 (1997); Alex Poyer, *Les premiers temps des vélo-clubs. Apparition et diffusion du cyclisme associative français entre 1867 et 1914* (Paris: L'Harmattan, 2003); Thomas Burr, "National Cycling Organizations in Britain, France, and the United States, 1875–1905", in *Cycle History 18: Proceedings of the 18th International Cycling History Conference*, ed. Rob van der Plas (San Francisco: Van Der Plas Publications, 2009); Glen Norcliffe, "Associations, Modernity and the Insider-Citizens of a Victorian Highwheel Bicycle Club", *Journal of Historical Sociology* 19, no. 2 (2006); Raymond Henry, "Origins and Brief History of the Fédération Française de Cyclotourisme", in *Cycle History 19: Proceedings of the 19th International Cycle History Conference*, ed. Anne Henry (St. Etienne: Musée d'Arts et d'Industries, 2010); Norbert Stellner, *Radfahrervereine in der bayerischen Provinz. Raum Mühldorf/Altötting 1882–1994* (Regensburg: EditionVulpis, 2000); Mikko Kylliäinen, Cycling towards Civil Society: Estonian Cycling History in the 19th Century", in *Cycle History 21: Proceedings of the 21st International Cycling History Conference*, ed. Andrew Ritchie (Birmingham: Cycling History Publishing, 2012); Reid, *Roads Were Not Built*, 123–33, 143–58, 173–82.

To be sure, such patriotic manifestations were partly a rhetorical means to gain general approval. In some countries, however, pedalling was instilled with national values in a more fundamental and lasting way. In the early twentieth century, the Netherlands and Denmark came to be regarded by their own populations (as well as others) as cycling nations par excellence, while in France, Belgium and Italy cycle racing became a source of national pride. In the English-speaking countries, Germany and most other western nations, on the other hand, the two-wheeler was not linked to national distinctiveness, although Britain set the tone in the organization of cycle clubs and pedalling as amateur sports, and, together with France, was also leading in bicycle engineering and production, while the invention of the bicycle was claimed by Germans as well as Brits and French.[56]

## 8 Democratization and Status Decline

In the late nineteenth century, bicycles were expensive luxury items and therefore restricted to the upper and middle classes. In the first three decades of the twentieth century, ever more efficient mass production of safety bicycles entailed falling prices and their widespread adoption in daily traffic and for other utilitarian purposes.[57] The two-wheeler enabled a longer distance be-

56 See Jacques Seray, *Deux Roues. La véritable histoire du vélo* (Rodez: Éditions de Rouergue, 1988) 19, 26, 46–47, 83, 112; Tony Hadland and Hans-Erhard Lessing, *Bicycle Design: An Illustrated History* (Cambridge, MA: MIT Press, 2014), 45–53; Smethurst, *The Bicycle*, 53–57.

57 Ross D. Petty, "Peddling the Bicycle in the 1890s: Mass Marketing Shifts into High Gear", *Journal of Macromarketing* 15, no. 1 (1995); Roger Lloyd-Jones and Myrddin John Lewis, *Raleigh and the British Bicycle Industry: An Economic and Business History, 1870–1960* (Burlington, VT: Ashgate, 2000); Bruce Epperson, "Failed Colossus: Strategic Error at the Pope Manufacturing Company, 1878–1900", *Technology and Culture* 41, no. 2 (2000); Bruce Epperson, "After Pope: The Pope Manufacturing Company and the American Bicycle Industry, 1899–1990", in *Cycle History 10: Proceedings of the 10th International Cycling History Conference*, ed. Rob van der Plas (San Francisco: Van der Plas Publications, 2000); Bruce Epperson, "Chasing the 'isms': Fordism, Taylorism, Popeism, and the Search for Meaning in the History of the American Bicycle Industry", in *Cycle History 20: Proceedings of the 20th International Cycling History Conference*, ed. Gary Sanderson (Chesham: John Pinkerton Memorial Publishing Fund, 2010); Paul Rosen, *Framing Production: Technology, Culture, and Change in the British Bicycle Industry* (Cambridge, MA: MIT Press, 2002); Thomas Burr, "Building Community, Legitimating Consumption: Creating the US Bicycle Market, 1876–1884", *Socio-Economic Review* 4, no. 3 (2006); Tony Hadland, *Raleigh: Past and Presence of an Iconic Bicycle Brand* (San Francisco: Cycle Publishing, 2011); Robert J. Turpin, "'Our Best Bet Is the Boy': Bicycle Marketing Schemes and American Culture after World War I", in *Cycle History 22: Proceedings of the 22nd International Cycling History Conference*, ed. Andrew Ritchie (San Francisco: Cycling History, 2012); Carlo Mari, "Putting the Italians

tween home and work, and thus contributed to suburbanization. Traders, shopkeepers and artisans used it to transport goods or offer their services. In the countryside, it advanced the opening up of isolated settlements: schooling and dating opportunities broadened, distant relatives and friends as well as new consumption options came within reach, and participation in social and club life on a regional and even national scale was facilitated. In some countries, the bicycle was employed to bridge long distances in sparsely populated and barren areas for economic purposes, for example in Sweden's northern forest regions and Australia's Western territories during the big gold rush.[58]

The interbellum period saw the onset of national differences in bicycle use and its public image, which have left their mark to this day. These variations evolved from (1) the diverging effects of growing motoring traffic; (2) the ensuing traffic policies implemented by governments; (3) the association of class and status distinctions with car-driving versus bicycle-riding, (4) the responses by bicycle organizations to these developments, and (5) their varying positions vis-à-vis professional cycle racing.

In several countries, the upsurge of utilitarian cycling among the lower middle and working class incited a social status decline of the two-wheeler. In Germany and Britain for example, where class and status distinctions were marked, the aristocracy and bourgeoisie more and more turned their back on the vehicle and exchanged it for the motorcycle and the car in order to distinguish themselves from the pedalling masses. Although the volume of cycle traffic was greater than ever between the First World War and the mid-1950s, the bicycle's aura as an icon of modernity was eclipsed by the automobile. The changing image of the two-wheeler from innovative to outmoded, was at odds with the growing practical use of the vehicle in the first half of the twentieth century.[59] Even so, membership of middle-class cycling associations

---

on Bicycles: Marketing at Bianchi, 1885–1955", *Journal of Historical Research in Marketing* 7, no. 1 (2015).

58 Tiina Männistö-Funk, "The Prime, Decline, and Recalling of Rural Cycling: Bicycle Practices in 1920s' and 1930s' Finland Remembered in 1971–1972", *Transfers* 2, no. 2 (2012); Anna-Maria Rautio and Lars Östlund, "'Starvation Strings' and the Public Good: Development of a Swedish Bike Trail Network in the Early Twentieth Century", *The Journal Of Transport History* 33, no. 1 (2012); Fitzpatrick, *Wheeling Matilda*, 16–33; Georgina Clarsen, "Pedaling Power: Bicycles, Subjectivities and Landscapes in a Settler Colonial Society", *Mobilities* 10, no. 5 (2015).

59 Karl Hodges, "Did the Emergence of the Automobile End the Bicycle Boom?", in *Cycle History 4: Proceedings of the 4th International Cycle History Conference*, ed. Rob van der Plas (San Francisco: Bicycle Books, 1995); Adri A. Albert de la Bruhèze and Frank C.A. Veraart, "Fietsen en verkeersbeleid. Het fietsgebruik in negen West-Europese steden in de twintigste eeuw", NEHA-*jaarboek voor economische, bedrijfs- en techniekgeschiedenis* 62

declined, while at the same time they barred or failed to draw the lower classes. German, British, and Italian workers established their own organizations, in which the bicycle was put into action for the socialist cause.[60] As a consequence, the cycle lobby became hampered by organizational and ideological fragmentation. In this dynamic, the two-wheeler was more and more regarded as a "humble utensil", as the typical lower class transport mode.[61] In the United States, where the car became a mass product and an affordable means of transportation for the common man between the two world wars, the bicycle was socially marginalized even earlier and more rapidly than in European countries. Driving a car became part of the American dream and the two-wheeler was viewed as the vehicle for losers and eccentrics or for those with no status to lose such as youngsters and students. Already in the interwar period American wheeling levels were much lower than European ones.[62]

At the same time the influence of British, American and German pressure groups of cyclists, which in the preceding decades had lobbied successfully for improving the traffic infrastructure, dwindled when driving started to grow after the First World War. Not only did governments intensify their interference in traffic, experts gained more influence in policies at the expense

(1999); Ruth Oldenziel, Martin Emanuel, Adri A. Albert de la Bruhèze, and Frank Veraart, eds., *Cycling Cities: The European Experience* (Eindhoven: Foundation for the History of Technology, 2016); Daniel Hart London, "Keeping a Respectable Distance: The Rise and Fall of the Bicycle as an Instrument of Gentility", in *Cycle History 20*, ed. Sanderson; Nicolas Lefevre, "Popularité du Cyclisme et Cyclisme Populaire. Pour en Fin avec le Mythe et le Misérablilisme", In *Cycle History 23: Proceedings of the 23rd International Cycling History Conference*, ed. Andrew Ritchie (San Francisco: Cycling History, 2013); Stoffers and Ebert, "New Directions".

60 Ralf Beduhn and Jens Klocksin, eds., *Rad-Kultur-Bewegung. 100 Jahre rund ums Rad: Rad- und Kraftfahrerbund Solidarität. Illustrierte Geschichte 1896–1996* (Essen: Klartext, 1995); Rüdiger Rabenstein, *Radsport und Gesellschaft. Ihre sozialgeschichtlichen Zusammenhänge in der Zeit von 1867 bis 1914* (Hildesheim, München, and Zürich: Weidmann, 1991); Denis Pye, *Fellowship Is Life: The Story of the Clarion Cycling Club* (Bolton: Clarion, 2004); Stefano Pivato, "The Bicycle as a Political Symbol: Italy, 1885–1955", *International Journal of the History of Sport* 7, no. 2 (1990) 172–87; Mari, "Putting the Italians on Bicycles", 134.

61 Fitzpatrick, *Wheeling Matilda*, 88.

62 Paul Rubenson, "Missing Link: The Case for Bicycle Transportation in the United States in the Early 20th Century", in *Cycle History 16: Proceedings of the 16th International Cycling History Conference*, ed. Andrew Ritchie (San Francisco: Cycling History, 2006); Evan Friss, "The Path Not Taken: The Rise of America's Cycle Paths and the Fall of Urban Cycling", in *Cycle History 20*, ed. Sanderson; David V. Herlihy, *Bicycle: The History* (New Haven, CT: Yale University Press, 2004), 298–305, 328; Pryor Dodge, *The Bicycle* (Paris and New York: Flammarion, 1996), 180; John Forester, "American Cycling History from the 1940s; As I Remember It", John Forester's Home Page, accessed 28 December 2017, http://www.johnforester.com/Articles/Social/My%20History.htm; Reid, *Roads Were Not Built*, 249.

of laymen such as cycle lobbyists. The modernist, forward-looking traffic and urban planning creed, in which upscaling and efficiency was central, prioritized the facilitation of motorized traffic and the construction of public transport networks. Policymakers, urban planners and traffic engineers viewed motoring in terms of progress and economic growth. Bicycle transportation was disparaged as out-dated, slow, inefficient and unsafe, as an impediment to a smooth and speedy circulation of traffic.[63] In the English-speaking countries, and to a lesser extent also in Germany, this approach forced bicyclists, although still ever-present on public roads, on the defensive already before the Second World War. Lower-class cyclists who used their means of transport for utilitarian purposes and out of sheer necessity, did not have a voice in traffic policies.[64] Policymakers and traffic experts largely excluded cycling from their (middle-class and future-oriented) frame of reference and thus made it invisible as a useful and convenient mode of transport. Moreover, because in general cyclists, unlike (middle-class) motorists, did not pay taxes for road use, their associations were no match for the much stronger car lobby, in particular

63 Burkhard Horn, "Vom Niedergang eines Massenverkehrsmittels. Zur Geschichte der städtischen Radverkehrsplannung" (thesis, Gesamthochschule Kassel, 1990); Burkhard Horn, "Geschichte der städtischen Radverkehrsplanung", in *Handbuch der kommunalen Verkehrsplanung*, ed. Tilman Bracher et al., Ch. 2.1.1.2. (Berlin: VDE Verlag, 2002); Nicholas Oddy, "Cycling's Dark Age? The Period 1900–1920 in Cycling History", in *Cycle History 15: Proceedings of the 15th International Cycling History Conference*, ed. Rob van der Plas (San Francisco: Van der Plas and Cycle, 2005); Albert de la Bruhèze and Veraart, "Fietsen en verkeersbeleid"; Jeffrey Brown, Eric A. Morris, and Brian D. Taylor, "Planning for Cars in Cities: Planners, Engineers, and Freeways in the 20th Century", *Journal of the American Planning Association* 75, no. 2 (2009); Jennifer Bonham and Peter Cox, "The Disruptive Traveller? A Foucauldian Analysis of Cycleways", *Road & Transport Research* 19, no. 2 (2010); Oldenziel and Albert de la Bruhèze, "Contested Spaces"; Martin Emanuel, "Understanding Conditions for Bicycle Traffic through Historical Inquiry: The Case of Stockholm", *Journal of the Institute of Urban Transport of India*, December (2010); Martin Emanuel, "Constructing the Cyclist: Ideology and Representations in Urban Traffic Planning in Stockholm, 1930–1970", *Journal Of Transport History* 33, no. 1 (2012); Aldred, "Governing Transport"; Koglin, *Vélomobility*; Till Koglin and Tom Rye, "The Marginalisation of Bicycling in Modernist Urban Transport Planning", *Journal of Transport & Health* 1, no. 4 (2014); Till Koglin, "Vélomobility and the Politics of Transport Planning", *GeoJournal* 80, no. 4 (2014); Till Koglin, "Organisation Does Matter—Planning for Cycling in Stockholm and Copenhagen", *Transport Policy* 39 (2015): 55–62.

64 More in general, they were less outspoken about their cycling experiences than the late nineteenth-century upper and middle-class pioneers for whom the vehicle was novel and special, and they have left behind far less historical sources in which their voice can be heard. See, e.g., Männistö-Funk, "The Prime, Decline, and Recalling"; also Tiina Männistö-Funk, "The Crossroads of Technology and Tradition: Vernacular Bicycles in Rural Finland, 1880–1910", *Technology and Culture* 52, no. 4 (2011).

if it was supported by the powerful automobile industries. If the economic crisis of the 1930s and the hardships during and after World War II impelled the massive utilization of the two-wheeler, from the 1950s on growing prosperity fostered car-ownership and driving as well as mobility on motorcycles, scooters and mopeds, while post-war traffic policies further cleared the way for the ascendency of motoring on the roads. Functional differentiation of the built environment, urban sprawl and the upscaling of land use patterns entailed an increase of the number and distances of daily trips. Making room for moving and parked cars was the tenet of modernist urban design, which could be implemented in cities that had been destroyed in the Second World War.[65]

In several countries, such as Germany and to a lesser extent also Britain and the United States, bicycle paths were planned and sometimes built before the Second World War, but mostly only locally and not systematically. Moreover, they were generally poorly constructed, too narrow, incomplete, and not direct and continuous. Officially cycling tracks served the safety and convenience of cyclists, but the main purpose, backed up by governmental authorities, planners, police and motoring organizations, was to keep wheelers away from highways and serve the facilitation and speeding up of motorized traffic.[66] In the Anglo-Saxon world, the construction of cycling infrastructure was half-hearted, ironically in part because bicycle rights advocates did not support it. They divined that separated facilities, even if they were incomplete, implied the suggestion that cyclists did not belong on regular roads and should be banned from them. British cycle associations insisted that wheelers should behave and be treated as drivers of a vehicle with the full right, like motorists, to move on public roads. This argument in favour of so-called vehicular cycling is still current in Britain and the United States. Although vehicular cycling was controversial among bicycle lobbyists and activists and many of them advocated segregated facilities, policymakers adopted it because this approach took the

65 Nicholas Oddy, "The Flaneur on Wheels?" in *Cycling and Society*, ed. Horton, Rosen, and Cox; David L. Patton, "Aspects of a Historical Geography of Technology: A Study of Cycling, 1919–1939", in *Cycle History 5*, ed. van der Plas; Pooley and Turnbull, "Modal Choice"; Pooley, Turnbull, and Adams, *A Mobile Century*; Peter Cox, "'A Denial of Our Boasted Civilisation': Cyclists' Views on Conflicts over Road Use in Britain, 1926–1935", *Transfers* 2, no. 3 (2012); Reid, *Roads Were Not Built*.

66 Volker Briese, "From Cycling Lanes to Compulsory Bike Path: Bicycle Path Construction in Germany, 1897–1940", in *Cycle History 5*, ed. van der Plas; Bonham and Cox, "The Disruptive Traveller"; Laura Golbuff and Rachel Aldred, *Cycling Policy in the UK: A Thematic and Historical Overview* (London: University of East London, 2011); Anne-Katrin Ebert, "When Cycling Gets Political: Building Cycling Paths in Germany and the Netherlands, 1910–40", *Journal of Transport History* 33, no. 1 (2012); Reid, *Roads Were Not Built*, 159–72, 251–58; Rautio and Östlund, "'Starvation Strings'"; Fitzpatrick, *Wheeling Matilda*, 16–33.

least effort and was cheapest.[67] As the volume and speed of motorized traffic on the roads kept increasing, however, vehicular cycling came with the unintended consequence of a growing preoccupation with the danger of cycling. This caused more and more people to abandon pedalling for commuting; only a minority of strongly motivated and experienced cyclists was not going to be put off by its real or alleged risks.

At the same time cycling suffered from a further loss of social status, now also among the working classes: only those without a driver's licence or who could not afford a car (the lowest income groups, youngsters, students and women) merely pedalled out of sheer necessity.[68] On the basis of his own experience as a devoted wheeler in California during the 1940s and 1950s, the American engineer and bicycle activist John Forester relates that he faced a social stigma. He was impeded in his professional career and his wife felt embarrassed because the neighbours saw him commuting to work on his bicycle. Only certain groups, according to Forester, could use the bicycle in daily commuter traffic without losing standing. "The small community of cyclists that had always existed consisted largely of persons who could resist the social convention that despised cycling: those with no status to lose (working class, students) and those whose status was proof against derision (doctors, professors), and those who chose not to obey convention".[69]

---

67 John Forester, "Ideas in Motion: The Bicycle Transportation Controversy", *Transportation Quarterly* 55, no. 2 (2001); John Forester, "Objective and Psychological Explanations for Differences in the Bicycling Programs of Different Nations", in *The Bicycle: Global Perspectives: Papers Presented at the Conference vélo mondiale Pro Bike–Velo-city, Montreal, Canada, 13–17 September 1992*, ed. Robert Boivin and Jean-François Pronovost (Québec: Vélo Québec, 1992); Gordon and Richardson, "Bicycling"; John Forester, "Bicycling, Transportation and the Problem of Evil", American Dream Coalition, accessed 28 December 2017, http://www.americandreamcoalition.org/safety/CTEvil.pdf; John Forester, "Fight for Your Right to Cycle Properly!", John Forester's Home Page, accessed 28 December 2017, http://www.johnforester.com/; John Pucher, "Cycling Safety on Bikeways vs. Roads", *Transportation* Quarterly 55, no. 4 (2001); John Pucher and Ralph Buehler, "Cycling for a Few or For Everyone: The Importance of Social Justice in Cycling Policy", *World Transport Policy & Practice* 15, no. 1 (2009); Bruce D. Epperson, "The Great Schism: Federal Bicycle Safety Regulation and the Unraveling of American Bicycle Planning", *Transportation Law Journal* 37, no. 2 (2010).

68 Thomas Fläschner, "Out of Date, Out of Mind: Public Awareness of the Bicycle during the 1950s in Germany's Saarland State", in *Cycle History 13: Proceedings of the 13th International Cycling History Conference*, edited by Nick Clayton and Andrew Ritchie (San Francisco: Cycle Publishing/Van der Plas Publications, 2003); Thomas Fläschner, "Stahlroß auf dem Aussterbe-Etat. Zur Geschichte des Fahrrads und seiner Verdrängung in den 50er Jahren", *Eckstein. Journal für Geschichte* 9 (2000).

69 Forester, *American Cycling History*, 9.

As a consequence of the marginalization of pedalling, up to now infrastructural cycling policies have received little support among tax-paying citizens. Two American transportation experts, who refute pleas for more bicycle facilities and pushing back motoring, probably voice the views of a large segment of the American population when they argue that such policies do not make sense in the United States because a substantial rise of pedalling levels is unlikely to come about. They consider such policy as undesirable because its benefits would not outweigh its costs, and it would violate the interests of most Americans: "we strongly object to the use of disincentives to driving as an inducement to bicycling. [...] It is bad policy to damage severely the welfare of 99 per cent (or even 98 per cent, assuming a consequent doubling of bicycling's current modal share) to benefit the 1 per cent (or, eventually, 2 per cent) who bicycle".[70] This demonstrates that in countries with low cycling levels, not only the United States, but also Britain, Canada and Australia, democratic legitimacy is a thorny issue that troubles cycling policies: substantial returns on investments depend on consistent long-term planning, whereas politicians and their voters may expect relatively quick results. If substantial results fail to occur in the short run, the political will and legitimacy to continue investing in bicycle infrastructure may be undermined. Against this background, the analysis of costs and benefits has gained in importance in bicycle research, the more so under neoliberal governance.[71]

## 9 The Dutchness and Danishness of Pedalling

The twentieth-century development of bicycling in the Netherlands and Denmark differed significantly from that in other countries. Already around

---

70 Gordon and Richardson, "Bicycling", 11.

71 P. Hopkinson and Mark Wardman, "Evaluating the Demand for New Cycle Facilities", *Transport Policy* 3, no. 4 (1996); Buis and Wittink, *The Economic Significance*; Glen F. Koorey, "Why a Cycling Strategy on Its Own Will NOT Increase Cycling" (presentation, New Zealand Cycling Conference, North Shore, 10–11 October 2003); Parkin and Koorey, "Network Planning", 156–57; cf. Gunnar Lind, ed., *Cost Benefit Analysis of Cycling* (Copenhagen: Tema Nord, Nordic Council, 2005); Kevin J. Krizek et al., "Analysing the Benefits and Costs of Bicycle Facilities via Online Guidelines", *Planning, Practice & Research* 22, no. 2 (2007); Krizek, "Cycling"; Börjesson and Eliasson, "The Benefits of Cycling"; Horton and Parkin, "Conclusion"; Nick Cavill et al., "Economic Analyses of Transport Infrastructure and Policies Including Health Effects Related to Cycling and Walking: A Systematic Review", *Transport Policy* 15, no. 5 (2008); Kjartan Saelensminde, "Cost-Benefit Analyses of Walking and Cycling Track Networks Taking into Account Insecurity, Health Effects and External Costs of Motorized Traffic", *Transportation Research Part A: Policy and Practice* 38, no. 8 (2004); Aldred, "Governing Transport".

the First World War, the Netherlands was seen as a cycling country par excellence by Dutch and foreigners alike, and in Denmark the bicycle became a national token from the 1920s on. The two-wheeler's establishment as a widely shared means of transportation and its lasting popularity in both countries was not so much, or at least not only, related to favourable material conditions (minor or no differences in elevation, relatively short distances, high levels of urbanization and compact historical towns) and the cycling volume in itself, which hardly differed from their neighbouring countries until the 1950s. What set the Netherlands and Denmark apart from the rest of the Western world, was rather the socio-political meaning attached to pedalling and its public image.[72]

The large and influential National Dutch Wheelers' Association, founded in 1883, steadily promoted the bicycle as a widely accessible and convenient means of transportation. In public expressions and events, the liberal and national-minded bourgeois citizens who directed the association promoted wheeling in terms of supposedly longstanding Dutch qualities and certain civil virtues, such as independence, self-control, soberness, modesty and stability. The Dutchness of pedalling was underlined by comparing it with ice-skating. On the other hand, the bourgeois cycling-vanguard mostly considered (commercial) cycle racing as vulgar and indecent, as contrary to the image of wheelers as respectable and responsible road users. This view made itself felt in government policies: road cycling races became rare as a consequence of prohibitions in a traffic law adopted in 1905. Touring, on the other hand, and also practical cycling were actively promoted. When the bicycle came within reach of the popular masses, the Dutch Wheeler's Association advocated it as an egalitarian means of transportation—"the democratic horse" as the editor of the organization's periodical phrased it—that would bring progress for all

72 Anne-Katrin Ebert, *Radelnde Nationen. Die Geschichte des Fahrrads in Deutschland und den Niederlanden bis 1940* (Frankfurt am Main: Campus, 2010); Anne-Katrin Ebert, "Cycling towards the Nation: The Use of the Bicycle in Germany and the Netherlands, 1880–1940", *European Review of History* 11, no. 3 (2004); Anne-Katrin Ebert. "Het 'paard der democratie'. Fatsoenlijk fietsen in Nederland, 1900–1920", in *Fatsoenlijk vertier. Deugdzame ontspanning voor arbeiders na 1870*, ed. Christianne Smit, 209–37 (Amsterdam: Bert Bakker, 2008); Frank C.A. Veraart, "Geschiedenis van de fiets in Nederland 1870–1940. Van sportmiddel naar massavervoermiddel" (thesis, Technische Universiteit Eindhoven, 1995); Kaspar Hanenbergh and Michiel Röben, eds., *Ons stalen ros. Nederland wordt een land van fietsers 1820 tot 1920* (Utrecht: Uitgeverij Ons Stalen Ros, 2015); Carstensen and Ebert, "Cycling Cultures"; Marie Kastrup, "Hverdagens beskedne demokrati. Analyser af cyklen som symbol pa danskhed" (thesis, University of Copenhagen, 2007); Walther Knudsen, Thomas Krag, and Dansk Cyklistforbund, *På cykel i 100 år. Dansk Cyklistforbund 1905–2005* (Copenhagen: Dansk Cyklistforbund, 2005); Koglin, *Vélomobility*; Koglin, "Vélomobility and the Politics".

ranks.[73] Its diffusion among the working classes was viewed as a way to integrate them into the nation.

The technical design of the two-wheeler that became standard in the Netherlands, reflected the idea of the vehicle as a practical and civilizing tool. The solid and sturdy Dutch wheeler (*Hollandrad* as Germans still refer to it), was not meant to break speed records, to work oneself, dressed in sports outfits and leaning forward, into a sweat, or to incite sensual or aesthetic feelings, but to ride neatly upright, in regular and decent clothes and in an unhurried pace. Equipped with luggage carrier, chain guard, dress-guards and lighting, and usually painted in black, it was tailored to everyday use as well as civil standards of propriety. With pictures of cyclists against the backgrounds of typical Dutch landscapes, historical towns, windmills, the Dutch flag and folk in traditional costumes, the marketing of the Dutch bicycle industry stressed the embeddedness of wheeling in national culture.[74] The Dutch image of the two-wheeler differed from its aura in other countries—in France, Belgium and Italy in particular—where bicycle models were often geared to sports and racing. Associations with lightness and flying as well as with fashion and eroticism—if cycling women were depicted—were also prominent in advertising the vehicle, but such attributes were in general lacking in the Netherlands.

Developments in Denmark largely were similar to Dutch patterns. The Danish Bicycle Club and Cycling Federation, founded in 1881 and 1905 respectively, dissociated themselves from cycle racing and promoted the construction of bicycle ways and touring, which was marked as a national pastime. Unlike the German, British and American bicycle organizations, which were divided along class-lines and over different bicycle activities (sports versus utilitarian use and touring, as well as the English ideal of gentleman-amateur sports versus professional and commercial racing), the Dutch and Danish associations spoke with one voice and could claim to represent the common interests of all cyclists in the country. Contrary to German and British workers, the Dutch and Danish labour movements did not develop a distinct socialist vision on bicycling and the attitudes among their constituencies were largely in line with

73 Ebert. "Het 'paard der democratie'", 236; on the topic of cycling women, see also Anne-Katrin Ebert, "Liberating Technologies? Of Bicycles, Balance and the 'New Woman' in the 1890s", ICON: *Journal of the International Committee for the History of Technology* 16 (2010).

74 Ebert, "Het 'paard der democratie'", 223–24; Manuel Stoffers, "Exoot wordt icoon: de inburgering van de fiets. Van gewild buitenlands product naar nationaal vervoermiddel", in *Ons stalen ros*, ed. Hanenbergh and Röben; Sue-Yen Tjong Tjin Tai, "Hoe werd de fiets een Nederlands product? Hollandse degelijkheid vormt de fiets voor dagelijks gebruik", in *Ons stalen ros*, ed. Hanenbergh and Röben.

the national and civil values (practicality, modesty, simplicity, soberness, level-headedness, and diligence) which the middle-class cycling lobby disseminated. In both countries the definition of the two-wheeler as a practical transport mode for all citizens and as a civilizing and assimilating utensil prevailed. In this way it was moulded into a vehicle of national identity, while at the same time cycle racing did not incite national pride, as happened in France, Italy and Belgium. The bicycle's enduring omnipresence and popularity and the riding style in the Netherlands and Denmark—characterized by a Dutch sociologist as "distinction through simplicity"—was and is related to the fairly egalitarian social ethos and distaste for showing off and status distinctions.[75] This is underlined by the fact that (in particular female) members of the Dutch royal family and cabinet ministers regularly appeared (and still appear) in public on a bicycle. In Denmark, images of pedalling women played an important role in linking the bicycle with liberal attitudes, emancipation and equality, which have been presented as national virtues. In particular during the First and Second World Wars, cycling was connected to qualities that supposedly contrasted the Dutch and Danish nations with everything that characterized the militaristic belligerent nations, and especially German authoritarianism and hyper-masculinity.

In the Netherlands and Denmark, the growth of motoring—which in both countries was slower than in other parts of the Western world, partly because there was no large automobile industry and cars were more heavily taxed than in car-producing countries—led to a decrease in bicycle use, but to a much lesser extent than elsewhere and neither did it entail a social devaluation of the vehicle.[76] Again, the approach of the Dutch and Danish cycling associations and government policies played a major part. Whereas in other countries bicycle traffic and motoring were increasingly considered as mutually exclusive and conflicting, in the Netherlands and Denmark their complementary nature and shared needs (good roads, signposting, traffic safety, and largely separated facilities) were underlined—which reflected that most drivers were also accustomed to pedalling. Therefore the Dutch and Danish cycling organizations, which promoted bicycle ways, succeeded in influencing government policies more effectively than the marginalized bicycle interest groups elsewhere. From the early twentieth century onwards, bicycle ways were constructed in these countries, at first mostly for leisure touring, but increasingly also for utilitarian purposes, and over the decades the bicycle infrastructure

75 Kuipers, "The Rise and Decline", 24.

76 Vincent van der Vinne, *De trage verbreiding van de auto in Nederland 1896–1939* (Amsterdam 2007); Koglin, *Vélomobility*.

steadily expanded. While the Dutch Wheelers' Association played an initiating role, the government used the revenues of the—generally resented—bicycle tax levied between 1924 and 1941, for funding cycling facilities. The growing contrast between, on the one hand, the Netherlands and Denmark and, on the other, the English-speaking world and, to a lesser extent, Germany, can be partly explained on the basis of the long-term effects of the infrastructural facilities put in decades earlier, which in their turn had been advanced by the previously established image of the bicycle as a widely shared means of transportation as well as a vehicle for national distinctiveness.[77]

While, from the 1950s onwards, daily bicycle-use strongly diminished in most western countries, it remained relatively high in the Netherlands and Denmark. Whereas in the late 1930s the level of bicycle traffic was even higher in Denmark, in the late 1940s and the 1950s the Netherlands developed and maintained the highest bicycle density in the world.[78] Although the Dutch and Danish governments did not promote cycling actively until the 1970s and motoring swelled rapidly from around 1960, bicycle traffic was not hampered to the same extent as elsewhere and it continued to be more visible. The leftist, "green" bicycle activism that arose in the 1970s, affirmed the self-evident view of pedalling as a sensible means of mobility. The more radical lobbyists, just like the established ones earlier on, became involved in policy-making by local and national governments.

## 10 Contrasting National Cycling Cultures

The twentieth-century trends in, on the one hand, the Netherlands and Denmark and, on the other, the English-speaking world and, partly, also Germany, have resulted in contrasting bicycle cultures. In the first two countries the benefits of cycling are self-evident: it is largely part of people's natural daily routine from an early age. The bicycle is used first of all for practical purposes; its role in leisure and sports, although also significant, is secondary. The demographic characteristics of wheelers largely represent those of the

77 Welleman, "Why a Bicycle Policy"; Albert de la Bruhèze and Veraart, "Fietsen en verkeersbeleid"; Oldenziel and Albert de la Bruhèze, "Contested Spaces".

78 Peter Eloy Staal, *Automobilisme in Nederland. Een geschiedenis van gebruik, misbruik en nut* (Zutphen: Walburg Pers, 2003), 115; Adri A. Albert de la Bruhèze and Frank C.A. Veraart, *Fietsverkeer in praktijk en beleid in de twintigste eeuw. Overeenkomsten en verschillen in fietsgebruik in Amsterdam, Eindhoven, Enschede, Zuidoost-Limburg, Antwerpen, Manchester, Kopenhagen, Hannover en Basel* (Den Haag: Ministry of Transport, Public Works and Water Management/Stichting Historie der Techniek, 1999), 50.

population as a whole and cycling is hardly associated with a particular alternative lifestyle or political viewpoint. Government and bicycle interest groups cooperate and cycling policies are hardly disputed. Apart from non-European immigrants, the Dutch and the Danes hardly need to be convinced of the usefulness and benefits of bicycle transport. The emphasis is on the improvement of the general infrastructural preconditions, and facilitation of the already high levels of bicycle traffic, especially for the benefit of the flow of *all* traffic. Since many Dutch and Danes both drive and pedal, cyclists and motorists are not pitted against each other to the same extent as in car-dominated countries. Wheelers enjoy a high level of security in traffic and bicycle-riding is not regarded as particularly dangerous.[79]

In the United States, Britain, Canada and Australia, where the bicycle was pushed out by the car to a much greater extent than in most European countries, and cycling policies, if they exist at all, are contested and do not elicit broad support, the two-wheeler is rather used for leisure and exercise than in daily commuting. For many people pedalling is a typical childhood and youth experience at best and bicycles are often regarded as toys rather than useful vehicles. Younger men are strongly overrepresented among wheelers, while women and the elderly are underrepresented. In the public perception of utilitarian cycling, negative valuations as abnormal, eccentric, inferior, unsafe, uncomfortable and (too) strenuous abound. Also, pedalling is associated with either poverty and low social status (although nowadays well-educated people are overrepresented) or, on the other hand, an exclusive "Lycra-and-helmet, sporty-and-skilled" and fashionable yuppie practice, whereas some wheelers, in particular bicycle couriers, are labelled as "kamikaze riders".[80] Many mo-

79 Ministry of Transport, Public Works and Water Management, *The Dutch Bicycle Master Plan*; Ministry of Transport, Public Works and Water Management, *Cycling in the Netherlands*; Berlage, *De fiets*; Welleman, "Why a Bicycle Policy"; Ministry of Transport, Public Works and Water Management, *Eindrapport*, 11, 27, 41; Huwer, "Let's Bike", 43; Buis and Wittink, *The Economic Significance*, 38; Pete Jordan, *In the City of Bikes: The Story of the Amsterdam Cyclist* (New York: Harper Collins, 2013); Florentine M.W. Bax, "Bicycle Use in the Netherlands versus the United States" (thesis, Rijksuniversiteit Groningen, 2004): Pelzer, "Bicycling"; City of Copenhagen, *Cycle Policy*; Krag, "Cycling"; Krag, "Bicycle Promotion"; Koglin, *Vélomobility*; Koglin, "Vélomobility and the Politics".

80 Peter L. Jacobsen, Francesca Racioppi, and Harry Rutter, "Who Owns the Roads? How Motorized Traffic Discourages Walking and Bicycling", *Injury Prevention* 15, no. 6 (2009); Chris Rissel et al., "Representations of Cycling in Metropolitan Newspapers: Changes over Time and Differences between Sydney and Melbourne, Australia", BMC *Public Health* 10, no. 371 (2010); Elliot Fishman, Simon Washington, and Narelle L. Haworth, "Understanding the Fear of Bicycle Riding in Australia", *Journal of the Australasian College of Road Safety* 23, no. 3 (2012); Horton and Parkin, "Conclusion", 310; Aldred, "The Role of Advocacy", 97; Dave Horton, "Fear", in *Cycling and Society*, ed. Horton, Rosen, and Cox;

torists view wheelers as a nuisance and as incompetent and dangerous road users. Reporting on his experience as a commuter cyclist in Birmingham (Britain), geographer Phil Jones points out how he "was being loaded with a whole series of labels: 'fit', 'healthy', 'eco-friendly', 'sustainable' (and also 'mad', 'crazy', 'reckless'). No longer myself, I was constructed as a 'cyclist'."[81] Both in the public perception and in their self-image Anglo-Saxon wheelers are "a breed apart" and this impedes the "normalization" of cycling.[82] Many British and American bicyclists, women even more than men, seem to struggle with the image of being either a "hard-core" cyclist and "'too much' of a cyclist" or a "bad" (that is unskilled and irresponsible) cyclist.[83]

In the English-speaking world, and also in Germany, the minority of regular and determined cyclists not only share a strong sensitivity for the partly bicycle-hostile and unsafe traffic conditions, but also pronounced motives and great appreciation of, and identification with, their vehicle.[84] In general, in

---

see also Rachel Aldred, "On the Outside: Constructing Cycling Citizenship", *Social and Cultural Geography* 11, no. 1 (2010); Aldred, "Incompetent or Too Competent?"; Gibson, "The Rise and Fall of Adrian Fenty"; Stehlin, "Regulating Inclusion"; Pucher, Komanoff, and Schimeck, "Bicycling Renaissance", 13; Ben Fincham, "Bicycle Messengers and the Road to Freedom", in *Against Automobility*, ed. Steffan Böhm, Campbell Jones, Chris Land, and Matthew Peterson (Malden: Blackwell, 2006); Aldred, "The Role of Advocacy", 91, 97; Horton and Parkin, "Conclusion", 309–10.

81 Phil Jones, "Performing the City: A Body and a Bicycle Take on Birmingham, UK", *Social & Cultural Geography* 6, no. 6 (2005): 814; cf. L. Basford et al., *Drivers' Perceptions of Cyclists* (London: Department of Transport, and Crowthorne: Transport Research Limited, 2002), 29; Simon Christmas et al., *Cycling, Safety and Sharing the Road: Qualitative Research with Cyclists and Other Road Users* (London: Department for Transport, 2010).

82 Transport for London, *Cycling*, 22, 29; Aldred, "The Role of Advocacy", 97; Horton and Parkin, "Conclusion", 310; Birgitta Gatersleben and Hebba Haddad, "Who Is the Typical Bicyclist?" *Transportation Research Part F: Traffic Psychology and Behaviour* 13, no. 1 (2010), 47; Rebecca Steinbach et al., "Cycling and the City: A Case Study of How Gendered, Ethnic and Class Identities Can Shape Healthy Transport Choices", *Social Science & Medicine* 72, no. 7 (2011); Peter Cox, "Cycling Cultures and Social Theory", in *Cycling Cultures*, ed. Cox, 14–42; Dave Horton and Tim Jones, "Rhetoric and Reality: Understanding the English Cycling Situation", in *Cycling Cultures*, ed. Cox, 63–77; Rachel Aldred, "Who are Londoners on Bikes and What Do They Want? Negotiating Identity and Issue Definition in a 'Pop-up' Cycle Campaign." *Journal of Transport Geography* 30 (2013); Pooley et al., "Policies for Promoting"; Aldred and Jungnickel, "Why Culture Matters"; Jennifer Bonham and Marily Johnson eds., *Cycling Futures* (Adelaide: University of Adelaide Press, 2015).

83 Aldred, "Incompetent or Too Competent?"; see also Aldred, "On the Outside"; Deborah McCarthy, "'I'm a Normal Person': An Examination of How Utilitarian Cyclists in Charleston South Carolina Use an Insider/Outsider Framework to Make Sense of Risks", *Urban Studies* 48, no. 7 (2011).

84 Rosen, "Up the Vélorution"; Les Lumsdon and Rodney Tolley, "The National Cycle Strategy in the UK: To What Extent Have Local Authorities Adopted Its Model Strategy Spproach?"

these countries a much more explicit reflection on the significance and experience of pedalling is present than in the Netherlands and Denmark, where it is an entrenched routine without there being a need for political-ideological arguments about its benefits.[85] In Britain and America, the striving for being treated as equal traffic participants and for wheelers' safety is (and apparently still needs to be) articulated in political terms such as the civil right of having free access to mobility regardless of the means of transportation. In the United States in particular, a militant and politicized cycling movement and subculture have developed, in which the glorification of two-wheelers is intrinsically linked to fundamental criticism of the dominance of motoring and the interrelated urban planning, economic prerequisites and lifestyle.[86] In a similar vein, many German wheelers seem to distinguish themselves from the majority of the population by their conscious lifestyle, "green" political affiliation and critical attitude towards the dominant car-oriented traffic culture and policies. Until the 1970s, the use-pattern and popular image of the two-wheeler in Germany resembled those in the English-speaking world. Since the 1970s, however, German pedalling levels have risen considerably above those of Britain, the United States, Canada and Australia, and over the last three

*Journal of Transport Geography* 9, no. 4 (2001): 293; Pucher, Komanoff, and Schimeck, "Bicycling Renaissance", 3–4; Sener, Eluru, and Bhat, "An Analysis"; Pucher and Buehler, "Making Cycling", 4–5; Aldred, "The Role of Advocacy"; Australian Bicycle Council and Austroads, *Gearing up*, 15–16; John Pucher, Jan Garrard, and Stephen Greaves, "Cycling Down Under: A Comparative Analysis of Bicycling Trends and Policies in Sydney and Melbourne", *Journal of Transport Geography* 19, no. 2 (2011); Daley and Rissel, "Perspectives and Images"; Horton and Parkin, "Conclusion", 310–11; Bonham and Wilson, "Women and Cycling".

85 Paradoxically, this may also explain why there has been more (popular as well as academic) interest in cycle history in Great Britain, the United States, Canada, Germany, and France than in the Netherlands and Denmark. In the last two countries, pedalling seems to be so obvious and so uncontested that it does not give cause to substantial reflection or foster a deeply felt need to pursue its history. See Harry Oosterhuis and Manuel Stoffers, "A Strong Presence, but a Weak History: The Bicycle in Dutch Historiography", *Cycle History* 21, ed. Ritchie; Manuel Stoffers, "Cycling as Heritage: Representing the History of Cycling in the Netherlands", *Journal of Transport History* 33, no. 1 (2012).

86 Zachary Mooradian Furness, "'Put the Fun between Your Legs!' The Politics and Counterculture of the Bicycle" (dissertation, University of Pittsburgh, 2005); Zack Furness, "Critical Mass, Urban Space and Vélomobility", *Mobilities* 2, no. 2 (2007); Theodore J. Buehler, "Fifty Years of Bicycle Policy in Davis, CA" (Davis: University of California, 2007), htttp://www.davishistoryresearch.org/3-authors/BuehlerThesisFinalDraft.pdf; J. Harry Wray, *Pedal Power: The Quiet Rise of the Bicycle in American Public Life* (Boulder and London: Paradigm Publishers, 2008); Jeff Mapes, *Pedaling Revolution: How Cyclists Are Changing American Cities* (Corvallis: Oregon State University Press, 2009). On the history of British cycling policies, see Golbuff and Aldred, *Cycling Policy in the UK*.

decades in several parts of the country the cycling culture has shifted into the direction of the Dutch and Danish one. At the same time, however, the car, supported by an all-powerful industrial and automobile lobby, has preserved its absolute priority in traffic policies, while the bicycle infrastructure and its upkeep, which are geared to touring rather than daily commuting, lags far behind that in Denmark and the Netherlands.

Apart from being fragmentary, inconsistent and controversial, British, American, Australian, and, to a lesser extent, German bicycle policies do not elicit broad public interest or support. In particular, with respect to discussions about the (lack of) safety of bicycle riding, policymakers as well as cycling activists have different views on the need for separate bicycle facilities. Some of them oppose the segregation of bicycle and motorized traffic and argue that cyclists should be enabled to share the regular roads with other vehicles and that they as well as motorists should adapt their traffic behaviour and driving skills to this situation.[87] Furthermore, the highlighting of safety issues, including the promotion of bicycle helmets, has had an adverse effect: the image of the bicycle as a risky vehicle and cyclists as extremely vulnerable road users has again and again been confirmed and even strengthened.[88] In countries with low cycling levels, it is difficult to change this image of the bicycle—and the reality underpinning it. Bicycle studies strongly suggest that there is a statistical correlation—called safety in numbers—between the (lack of) safety of pedalling and the volume of cycling traffic. The statistical risk of an accident is relatively smaller or larger as more or less cyclists use the roads.[89] As long as cycling levels are low and motorized transportation is dominant, the

---

87 Gordon and Richardson, "Bicycling", 11; Forester, "Ideas in Motion"; Forester, "Objective and Psychological"; John Forester, "The Place of Bicycle Transportation in Modern Industrialized Societies", American Dream Coalition, accessed 28 December 2017, http://americandreamcoalition.org/safety/forester.pdf; Martin Wachs, "Creating Political Pressure for Cycling", *Transportation Quarterly* 52, no. 1 (1998); Susan Blickstein and Susa Hanson, "Critical Mass: Forging a Politics of Sustainable Mobility in the Information Age", *Transportation* 28, no. 4 (2001); Susan G. Blickstein, "Automobility and the Politics of Bicycling in New York City", *International Journal of Urban and Regional Research* 34, no. 4 (2010) 886–905; Furness, "Put the Fun"; Furness, "Critical Mass"; Chris Carlsson, ed., *Critical Mass: Bicycling's Defiant Celebration* (Oakland: AK Press, 2002); Wray, *Pedal Power*; Mapes, *Pedaling Revolution*; Pelzer, "Bicycling"; Bax, "Bicycle Use".

88 Warren Salomon, "How Australian Cyclists Got Their New Helmet Laws", in *The Bicycle*, ed. Boivin and Pronovost; Wardlaw, "Three Lessons"; Rosen, *How Can Research*; Horton, "Fear".

89 Peter L. Jacobsen, "Safety in Numbers: More Walkers and Bicyclists, Safer Walking and Bicycling", *Injury Prevention* 9, no. 3 (2009); Conor C.O. Reynolds et al., "The impact of Transportation Infrastructure on Bicycling Injuries and Crashes: A Review of the Literature", (*Environmental Health* 8, no. 47 (2009), https://ehjournal.biomedcentral.com/

risk of an accident seems large and pedalling is (perceived as) dangerous. And as long as cyclists are depicted as potential victims, most people will shy away from heavy traffic and only a few will use the bicycle in daily traffic or let their children pedal. Cycling among children and young people has considerably declined in recent decades, resulting in fewer people who in fact get used to ride a bicycle in traffic and more people who do not develop any bicycle skills at all. Lack of cycle experience and the self-perpetuating association of biking with danger discourage people later in life from using a bicycle as a means of transport. Not only real risks, but also the related image of pedalling as dangerous, strengthen the sense of insecurity and hamper bicycle promotion policies in the English-speaking world.[90]

Apart from the marked contrast between on the one hand the Netherlands and Denmark and, on the other, the English-speaking countries, and, to a large extent, Germany, a third national bicycle culture can be distinguished, that of France, Italy, Belgium, and, to a lesser extent, Spain.[91] The nationalist

---

articles/10.1186/1476-069X-8-47; Wardlaw, "Three Lessons"; Wardlaw, "Assessing"; Buis and Wittink, *The Economic Significance*, 38.

90 Pucher, Komanoff, and Schimeck, "Bicycling Renaissance"; Wardman, Hatfield, and Page, "The UK National Cycling Strategy"; Transport for London, *Cycling*, 17–18, 20–23, 29, 45, 50; Krizek, Forsyth, and Baum, *Walking and Cycling*, 6, 23, 25–26; Wardlaw, "Assessing", 353; Stinson and Bhat, "An Analysis", 124; Noland and Kunreuther, "Short-Run and Long-Run"; Parkin, Wardman, and Page, "Estimation"; Akar and Clifton, "The Influence"; Wardlaw, "Three Lessons"; Horton, "Fear"; Bonham and Cox, "The Disruptive Traveller"; Oldenziel and Albert de la Bruhèze, "Contested Spaces".

91 Richard Holt, *Sport and Society in Modern France* (London and Oxford 1981); Pivato, "The Bicycle as a Political Symbol"; Stefano Pivato, "Italian Cycling and the Creation of a Catholic Hero: The Bartali Myth", *International Journal of the History of Sport* 13, no. 1 (1996); R.J.B. Bosworth, "The Touring Club Italiano and the Nationalization of the Italian Bourgeoisie", *European History Quarterly* 27, no. 3 (1997); Georges Vigarello, "Le Tour de France", in *Les lieux de mémoire*, vol. 3, ed. Pierre Nora (Paris: Gallimard, 1992); Philippe Gaboriau, *Le Tour de France et le vélo. Histoire sociale d'une épopée contemporaine* (Paris: L'Harmattan, 1995); Daniele Marchesini, *L'Italia del Giro d'Italia* (Bologna: Il Mulino, 1996); Hugh Dauncey and Geoff Hare, eds., *The Tour de France: A Century of Sporting Structures, Meanings and Values* (London: Routledge, 2003); Christopher S. Thompson, *The Tour de France: A Cultural History* (Berkeley, CA: University of California Press, 2006); Anthony Cardoza, "Making Italians? Cycling and National Identity in Italy: 1900–1950", *Journal of Modern Italian Studies* 15, no. 3 (2010); Bernat López, "The Failed Vuelta Ciclista a Espana of 1913 and the Launching of the Volta a Catalunya (1911–1913): Centre versus Periphery in the Struggle for the Governance of Cycling in Early Twentieth-Century Spain", *Sport in History* 30, no. 4 (2010); Bernat López, "Sport, Media, Politics and Nationalism on the Eve of the Spanish Civil War: The First Vuelta Ciclista a Espana", *The International Journal of the History of Sport* 27, no. 4 (2010); Hugh Dauncey, *French Cycling; A Social and Cultural History* (Liverpool: Liverpool University Press, 2012); Stijn Knuts, "Converging and Competing Courses of Identity Construction: Shaping and Imagining Society through

dimension of cycling in these countries did not concern, as in the Netherlands and Denmark, daily commuting, which since the 1960s fell to similar low levels as in many other parts of the western world, but was connected to sports and (professional) racing. Around 1900 there was considerable social resistance against competitive professional racing in the Netherlands, Denmark, Britain and Germany because most bicycle organizations in these countries not only followed the English sports ideal of the gentleman-amateur, but also prioritized touring and, later on, utilitarian bicycling as well. Leading French, Belgian (Flemish as well as Walloon), Italian and Spanish cycling associations, on the other hand, embraced and promoted bicycle racing, which entailed the commercial involvement of the (sports) media and the bicycle industry. Together with bourgeois lobbyists, bicycle manufacturers and newspapers organized and sponsored races in order to attract customers and subscribers. In the many local, national and international seasonal races, which replaced or were embedded in more traditional community entertainment and which attracted mass audiences, the achievements and sporting virtues of native racing heroes were celebrated, widely publicized and associated with the nation's vitality. Annual highlights such as the long-distance and staged road races *Tour de France* (from 1903 onwards), *Ronde van België* (Tour of Belgium, from 1906 onwards) and *Ronde van Vlaanderen* (Flanders, from 1913 on), *Giro d'Italia* (from 1909 onwards), and the Spanish *Vuelta Ciclista* (from 1935 on) became national events and grew into cherished traditions. Since the racers crossed the entire country, the extensive media reports covered its geographic contours, spurring the spectators along the roads and the reading audience to identify with the nation. In Belgium, bicycle racing was also instrumental in the emancipation struggle of the (lower-class) Flemish population against the dominant Francophone upper classes.

Most of the professional racers originated from the working class; for them a cycling-career offered an attractive opportunity to reap local, national or even international fame, make money and climb the social ladder. Although socialists and communists criticized the commercialization of professional pedalling and accused the bourgeois organizers of exploiting working-class racers for capitalist purposes, the sport enjoyed broad popularity among the lower as well as the middle classes. With the exception of the Flemish bicycle culture

---

Cycling and Bicycle Racing in Belgium before World War Two" (dissertation, KU Leuven, 2014). Cf. Rabenstein, *Radsport*; Rüdiger Rabenstein, "The Vienna–Berlin Race of 1893 and Its Influence on the Cycling Movement in Germany and Austria", *Cycle History 4*, ed. van der Plas; Andrew Ritchie, *Quest for Speed: A History of Early Bicycle Racing, 1868–1903* (San Francisco: Cycle Publishing/Van der Plas Publications, 2011).

and volumes, which have tended to shift towards Dutch and Danish patterns, the racing nations show rather low utilitarian pedalling levels while their infrastructural cycling facilities and policies (with local exceptions, such as many Northern Italian towns) are lagging behind those of North-Western and Central European countries. The overrepresentation of men among cyclists and the strong presence of racing bicycles in these countries signal a continuing strong association of cycling with sports, although the popularity of touring is growing and rent-bicycles have been introduced in several tourist cities.[92]

## 11 Conclusion: The Relevance of History for Bicycle Policies

Natural, spatial and demographic factors cannot adequately explain the large international differences in cycling levels. Also, the often-assumed causal link between infrastructural design and promotional activities on the one hand and pedalling volumes on the other has not been confirmed empirically. Bicycle studies do not provide conclusive evidence that "hard" infrastructural policies bring about an increase in cycling. An inverse relation cannot be ruled out: the building of facilities and their use may be the consequence of a preceding surge in cycling, caused by other factors and advancing the demand for cycle provisions. In that case infrastructures principally serve the needs of already accustomed wheelers—an effect which is in itself not without merit, although much less spectacular and visible than facilities causing an upsurge of cycling. In a similar vein, "soft" policies, such as education, promotion and marketing, primarily affect people who already pedal and believe in its benefits, whereas those who never or seldom mount a bicycle, are barely reached, let alone convinced, so that among them changes in perception and behaviour are not realized. Policies during the last two decades have largely failed to generate significant increases in utilitarian cycling in countries with low to average pedalling levels, whereas in countries with relatively high cycling volumes,

92 Philippe Gaboriau, "Les trois ages du vélo en France", *Vingtième Siècle. Revue d'histoire* 29 (1991); Giuseppe Di Donna, "Bicycle World in Italy", in *The Bicycle*, ed. Boivin and Pronovost; Jean-René Carré, "La situation de la bicyclette en France", in *The Bicycle*, ed. Boivin and Pronovost; Peter Cox, "Strategies Promoting Cycle Tourism in Belgium: Practices and Implications", *Tourism Planning & Development* 9, no. 1 (2012); Dauncey, *French Cycling*; Francis Papon, "The Evolution of Bicycle Mobility in France", *Cycle History* 22, ed. Ritchie; Francis Papon, "Historiographical Needs in the Study of Bicycling Mobility in France", *Mobility in History* 5, no. 1 (2014); Ester Anaya and Santiago Gorostiza, "The Historiography of Cycling Mobility in Spain in the Twentieth Century", *Mobility in History* 5, no. 1 (2014); Santiago Gorostiza and Ester Anaya, "Under the Shade of Bahamontes: Bicycle Decline in Spain between 1950 and 1975", *Cycle History 23*, ed. Ritchie.

such policies have contributed to a consolidation of its existing, relatively high level, rather than to further growth.

The widely diverging national cycling volumes and their large degree of permanence are rooted in diverse long-term national trajectories which have shaped built environments and traffic infrastructures as well as the collective meanings attributed to cycling and interrelated attitudes, perceptions, experiences and habits. These factors are largely immune to direct and short-term planning and, in particular, to policies which are based on the assumption that travel behaviour is motivated by rational decision-making. Taken together, the factors that are relevant for cycling levels appear to be trapped in either an inhibiting or stimulating vicious circle, in dynamics from which it is difficult to break out. In countries where land use patterns, urban planning and traffic infrastructure are not conductive to cycling, and the two-wheeler is not broadly regarded as an obvious means of transport, few people use it. As long as cycling continues to be a fringe mode of an exceptionally motivated and skilled minority, motoring will dominate traffic, the idea will prevail that pedalling is deviant, inferior, uncomfortable and dangerous, and there will be a lack of sufficient social pressure, democratic support and willingness among policy-makers for changing the traffic infrastructure and the image of the bicycle. Although governments in the English-speaking world have made efforts to promote bicycling, overall, apart from some modest results on the local level, the outcomes have been disappointing because such efforts are not structurally embedded in policies and lack continuity. In Denmark and the Netherlands, and perhaps increasingly in some other countries, such as Germany and Flanders, on the other hand, the enduring high or increased bicycle volumes and the familiarity of the majority (or a substantial part) of the population with bicycle-riding guarantee broad or adequate support for bicycle policies. The steady and structural development and upkeep of cycling facilities warrants that pedalling remains attractive and a matter of course. In these countries, bicycle policies have contributed less to a significant growth of bicycling than to a consolidation of its existing level.

Policymakers and social-scientific bicycle researchers have largely disregarded the persistent influence of history and (national) culture on current cycling levels and patterns. Both determining factors, which are largely invisible in policy-oriented bicycle research, put limits on what policies can realize in the short run. As a corrective to the over-optimistic belief in rational planning and in order to develop more realistic and effective policies, it may be advisable for policymakers and bicycle researchers to consider the historical and national specific interrelations between natural and built environments, traffic infrastructures, meanings and perceptions, and habits and attitudes with regard to cycling. Adopting historical knowledge and an international comparative angle may temper unrealistic expectations among bicycle researchers and policymakers, and help them to attune policies to what is feasible and what is not within existing bicycle cultures. Also, it may be wise to shift the focus in bicycle policies from rational planning to nudging strategies in order to influence through more subtle, socio-psychological and cultural means the engrained habits and attitudes that play such a crucial, but not always clearly visible motivational role in traffic behaviour and mobility patterns. Finally, efforts to promote bicycling can only be successful in an enduring way if politicians and other policymakers have the courage to defy powerful car lobbies and to introduce structural measures that discourage and curb motoring.

# PART 2

## *Political and Economic Shaping of the Bicycle*

CHAPTER 4

# Waves of Cycling: Policies of Cycling, Mobility, and Urban Planning in Stockholm since 1970

*Martin Emanuel*

## 1 Introduction

This chapter outlines the variations in cycling, policy attention to cycling, and bicycle planning in Stockholm in the period from 1970 to 2014. It also assesses the integration of bicycle policy and planning into wider city traffic and urban planning policies and developments.[1]

When contemplating Stockholm from the point of view of mobility, few people will probably think of cycling. Those familiar with Sir Peter Hall's and others' comparative scholarship on urban planning and development will be more likely to think of the capital of Sweden as a public transportation-oriented city, while others will know Stockholm as one of a few cities that have successfully (now 10 years ago) implemented a congestion charging scheme in the city centre.[2] Even so, like in many other European cities, cycling levels in Stockholm, as measured in traffic counts at the inner city cordon, have boomed during the last decade. From its low point of below one per cent of all traffic in the mid-1970s, the number of cyclists, most of them commuters, has over the decades accumulated to constitute around 10 per cent of traffic in its entirety.[3]

Indeed, cyclists are no longer invisible, or made invisible, in the streets of Stockholm. The situation is far removed from the post-war period, when cyclists all over Europe (as in the case of Dublin studied by the historian, Erika

1 This article is partly based on the Stockholm chapter of a comparative work on the cycling histories of 14 European cities: Martin Emanuel, "Stockholm: Where Public Transit Eclipses Cycling", in *Cycling Cities: The European Experience: Hundred Years of Policy and Practice*, ed. Ruth Oldenziel, et al. (Eindhoven: Foundation of the History of Technology, 2016).

2 Peter Hall, *Cities in civilization: Culture, innovation, and urban order* (London: Weidenfeld & Nicolson, 1998); Robert Cervero, *The transit metropolis: A global enquiry* (Washington, D.C.: Island Press, 1998); Anders Gullberg and Karolina Isaksson, eds., *Congestion taxes in city traffic: Lessons learnt from the Stockholm trial* (Lund: Nordic Academic Press, 2009).

3 Travel surveys of all trips made by Stockholmers (not just commuting to and from the city centre) show stable cycling levels of around five per cent since the 1980s.

© KONINKLIJKE BRILL NV, LEIDEN, 2019 | DOI:10.1163/9789004289970_005

Hanna) were "rendered largely invisible in the ways that the city was quantified and planned"—notwithstanding the fact that they made up a considerable part of urban traffic. This did not only lead to the marginalization of cyclists, but also left historians with often limited and scattered archival evidence.[4]

Since the 1970s, cycling has become more visible in municipal archives, not least because of the emergence of bicycle policy and planning. In response to the concerns of cyclists and bicycle activists regarding the difficult situation they faced in European cities—which, during the preceding decades, had been transformed to cater for urban automobility—politicians had traffic departments develop bicycle plans.[5] During its now more than 40-year long history, bicycle planning has become professionalized and wider in scope. From an initially quite narrow focus on infrastructure provision, bicycle planning today includes traffic management and "soft measures", such as campaigning and marketing techniques, in an attempt to encourage more and safer cycling.[6]

The Stockholm case reveals that the post-1970 rise in cycling was not necessarily linear or even continuous. As we will see in this chapter, the waves of cycling have been paralleled by similar waves of bicycle policy, which depends, amongst other things, on economic conditions and political turns. Today, cycling policy and planning are arguably experiencing tailwinds. It remains unclear, however, whether present-day cycling initiatives are stand-alone initiatives and possibly fads, or if they are well-integrated in wider takes on traffic and urban planning and thus stand a better chance of prevailing if the political and public enthusiasm for cycling erodes. To express the matter in another way, while bicycle policies today render cyclists visible, this chapter asks: are such policies integrated well enough in "the larger scenario" to make them relevant? The chapter thus traces the position of cycling in relation to a

---

4 Erika Hanna, "Seeing like a cyclist: Visibility and mobility in modern Dublin, c. 1930–1980", *Urban History* 42, no. 2 (2015); Martin Emanuel, "Constructing the Cyclist: Ideology and Representations in Urban Traffic Planning in Stockholm, 1930–70", *Journal of Transport History* 33, no. 1 (2011).

5 Ruth Oldenziel et al., eds., *Cycling Cities: The European Experience: Hundred Years of Policy and Practice* (Eindhoven: Foundation of the History of Technology, 2016); Hugh McClintock, "Post-war planning and provision for the bicycle", in *The bicycle and city traffic: Principles and practice*, ed. Hugh McClintock (London: Belhaven, 1992).

6 For state-of-the-art contributions in English of different generations of literature on cycling policy and planning, see e.g. Mike Hudson, *The bicycle planning book* (London: Open books, 1978); Hugh McClintock, *The bicycle and city traffic: Principles and practice* (London: Belhaven, 1992); *Planning for cycling: Principles, practice and solutions for urban planners* (Boca Raton, Fla.: CRC Press, 2002); John R. Pucher and Ralph Buehler, *City cycling* (Cambridge, Mass.: MIT Press, 2012).

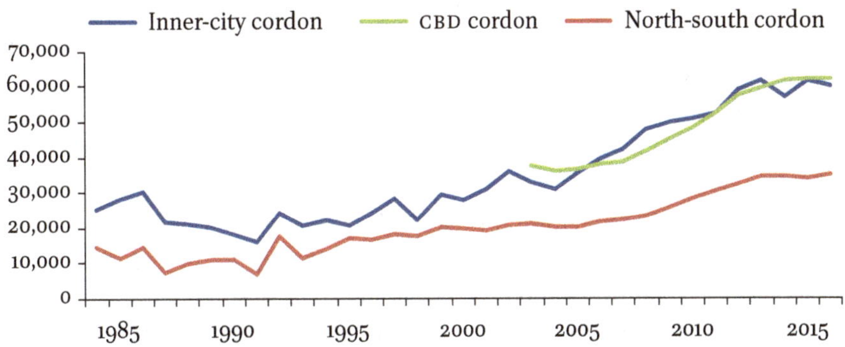


GRAPH 4.1 Number of daily cyclists in high-season (May) to and from the city centre of Stockholm (blue line), to and from the central business district (green line), and passing the bridges separating the north and the south part of the city (red line)
SOURCE: DATA PROVIDED BY THE CITY OF STOCKHOLM, AVAILABLE AT HTTP://MILJOBAROMETERN.STOCKHOLM.SE/TRAFIK/

number of important traffic projects and trajectories in urban development in Stockholm—traffic zoning, road building schemes and congestion charging, as well as regional enlargement and urban densification—with the purpose of assessing the level of integration of cycling policy and planning in the overall policy and planning initiatives.

## 2 Cycling and Traffic Zoning

As in other European cities, cycling in Stockholm declined until the mid-1970s, then stabilized and increased slightly, although starting from a low level. Largely due to the radical modernization of the capital of Sweden compared to other cities, marginalization of cyclists was greater than elsewhere.[7] According to traffic counts undertaken by the Traffic Department in October every year, cycling to and from the city centre reached its lowest level in 1974—at this point bicycle traffic constituted less than one per cent of all traffic—but then more than trebled until 1982.[8] This minor revival of cycling in European cities is commonly framed as a consequence of the 1973–74 oil crisis. Indeed, the economic recession of the 1970s and the trebling of gasoline prices in Sweden during this decade surely contributed to the decrease in car traffic in Stockholm

7 Emanuel, "Constructing the Cyclist".
8 Martin Bergman, *Effektivare cykeltrafik: Planeringsmässiga förutsättningar för ökad cykelanvändning* (Stockholm: Royal Institute of Technology, 1994), 18.

(a decline of 5 per cent over the period of 1968–82). The car recession may also have had an indirect effect to the increasing bicycle levels in 1974–82. But the changing state of the market needs to be balanced with a reorientation of local traffic policy in Stockholm during the 1970s. This policy was intended to curb automobility, and it had more to do with urban environmental protection than with fuel-related geopolitics.[9]

In the late 1960s and early 1970s, residents everywhere in urban Europe began to question urban automobility and technocratic planning. Leaving differences in their overall political programs aside, anti-car sentiments were shared by urban planning intellectuals, the New Left, and a great part of the overall counter-culture of the time.[10] In Stockholm, urban environmental groups took to the streets, partly in the form of bicycle rallies (see illustration 4.1). An epic struggle ensued: the so-called "battle of the elms". The activist group, Alternative City (*Alternativ stad*)—the Stockholm branch of the Friends of the Earth founded in 1969—had a common cause with groups committed to conserving the historic city. In 1971, this coalition successfully defeated the plans of the authorities of cutting down a number of ancient and beloved elms to make way for a new subway exit.[11]

In response to this grassroots movement, policy makers embraced a new traffic philosophy. In the 1960s, Stockholm's urban planners had drafted plans to rebuild parts of the inner city following the same hierarchical principles as in the suburbs, with the double aim of achieving safety for cyclists and pedestrians, as well as uninterrupted flows for motorists. The new context of the 1970s compelled them to develop a traffic policy that avoided major interventions in the urban fabric: traffic zoning (in Swedish: *trafiksanering*). They now aimed to limit and regulate, rather than to encourage, car traffic, while at the same time facilitating other modes of transport. They also wanted to lead through traffic around rather than through residential neighbourhoods, providing pedestrians and cyclists with calmer and safer environments.[12]

Similar attempts to come to terms with urban automobility proliferated all over Europe, but there were differences. The zoning schemes in Stockholm and other Swedish cities were inspired by the British Buchanan Report, "Traffic

9 Martin Emanuel, *Trafikslag på undantag: Cykeltrafiken i Stockholm 1930–1980* (Stockholm: Stockholmia, 2012), 281–82.

10 See e.g. Zach Furness, *One Less Car: Bicycling and the Politics of Automobility* (Philadelphia, PA: Temple University Press, 2010), 52–69; Dave Horton, "Environmentalism and the Bicycle", *Environmental Politics* 15, no. 1 (2006).

11 Emanuel, *Trafikslag på undantag*, 279–82.

12 Ibid., 311–13; Stig Holmstedt, *Ett halvsekel i Stockholmstrafiken: Politik, planing och utbyggnader* (Stockholm: Stockholmia, 2012), 155–56.

ILLUSTRATION 4.1 The bicycle was promoted by the urban environmental movement as complementary to public transport in curbing car dependence. It was also a symbol in the struggle against what critics understood as an overly car-oriented urban redevelopment, as accentuated in the case of this demonstration held in May 1970.
PHOTO: EIE HERLITZ

in Towns", from 1963, rather than the German "traffic calming" or the Dutch "*woonerf*" (literally, living yard). Contrary to the two latter models, which emphasised the co-existence of different modes of traffic, so that motorists would need to adjust to pedestrians and cyclists, Buchanan and the zoning schemes aimed at traffic differentiation, steering motorized traffic away from neighbourhoods to larger main streets.[13]

The traffic zoning programs, however, became battlefields from the start. Although there had been political consensus regarding the principles of the programs in 1970, there were fierce debates both within and outside City Council when the same principles boiled down to specific proposals. In 1973–74, a large "traffic consultation" was held as part of a new participatory manner to carry out traffic planning.[14] The initially radical proposals were progressively eroded as they were recast within the municipality and included in official

13 *Ett halvsekel i Stockholmstrafiken*, 153–56. For more on the Buchanan report and its influece on British urban planning, see Simon Gunn, "The Buchanan Report, Environment and the Problem of Traffic in 1960s Britain," *Twentieth Century British History* 22 (2011), 521–542.

14 Emanuel, *Trafikslag på undantag*, 314–16; Holmstedt, *Ett halvsekel i Stockholmstrafiken*, 157–82; Tom Miller and Ralf Österberg, *Medborgarinflytande i kommunal planering: Försök*

planning documents. The Conservative Party opposed the zoning schemes and Ulf Adelsohn, the new commissioner of finance in 1976, resisted restrictions on urban automobility. The schemes also faced heavier public resistance in the late 1970s and the early 1980s. Many of the remedial measures in the final 1977 traffic plan were never realized, while many "hard" measures such as street closures were replaced by "softer" ones: bans on cars by means of signage, speed limits, and narrow streets parts.[15]

The nascent bicycle planning in Stockholm developed parallel to the traffic zoning policy. The engineers in Stockholm's Traffic Department were slow in responding to the new appreciation of cycling visible elsewhere in the early 1970s. It took several petitions from social democratic councillors before the Urban Planning Department deemed it necessary to draft a proposal for a network of bicycle routes in the city centre. From the start, it was clear that in the city centre, experts would have to abandon the cherished principle of separating cyclists from other traffic. The proposal thus included a variety of alternative solutions: combined pedestrian and cycling lanes; allowing cyclists access to public transit lanes; and mixing traffic in calmer streets. The traffic engineers singled out the zoning schemes as the best solution for realizing bicycle lanes or streets reserved for cyclists. In 1972, the social democratic traffic commissioner, Inge Hörlén, endorsed a proposal for a bicycle network as an input for the forthcoming traffic zoning programs.[16]

To bolster its in-house expertise, the municipal Traffic Department recruited the bicycle-minded engineer, Sven Ekman, in 1974. The department asked Ekman—who was spotted during the consultation process for the traffic zoning schemes—to elaborate a bicycle plan for Stockholm. His proposal was included in the city's 1975 overall traffic plan. His ideas also survived the revision of the plan under the new conservative majority that came to power in 1978. Ekman took issue with the lack of bicycle lanes along the city's main avenues. He argued at some length that bicycle lanes along the central city's main streets would be preferable on several grounds. These streets offered the flattest, quickest, and most navigable routes, while also reaching the most common destinations for cyclists. Whatever the official policy, Ekman argued,

---

*till utvärdering av trafiksamrådet i Stockholm 1973–1974* (Stockholm: Statens råd för byggnadsforskning, 1977).

15 Holmstedt, *Ett halvsekel i Stockholmstrafiken*, 184–203; Miller and Österberg, *Medborgarinflytande i kommunal planering: Försök till utvärdering av trafiksamrådet i Stockholm 1973–1974*, 24–25.

16 Stockholms kommunfullmäktiges handlingar (Documents of the Stockholm Municipal Council, hereafter KF), Utlåtande (Report) 166/1972. For details, see Emanuel, *Trafikslag på undantag*, 279–310.

cyclists would still use these streets. His political realism, however, kept him from actually proposing the construction of such bicycle lanes. Half the network consisted of calmer side streets with mixed traffic.[17]

No such inhibitions troubled a new activist group, the Bicycle Chain (*Cykelkedjan*), founded in the mid-1970s in the wake of the cycling demonstrations organized by Alternative City several years earlier. Despite its activist tactics, politicians in Stockholm urged the engineers to consider *Cykelkedjan*'s criticism: that the city's plan was rather unambitious and limited to new bicycle routes that were "easy to build, without impinging too much on cars". In 1979, *Cykelkedjan* presented its own alternative plan for a denser and more comprehensive bicycle network than the Traffic Department had proposed. For one thing, the activists projected their bicycle lanes in the main streets. The Traffic Department, while politely welcoming the alternative plan, argued that they had a more holistic (realistic) view on traffic issues than the opinions of a biased activist group. In practice, the extension of the bicycle network stagnated.[18]

The 1975/78 bicycle plan did bring some concreteness and funding to bicycle-related measures, but bicycle planning had low priority from the perspective of comprehensive planning. Although the traffic zoning schemes were promoted by city officials and politicians as the best opportunity to cater for cyclists in the city centre, the needs of cyclists had a subordinate position in the schemes. By pushing through-traffic from residential areas to the main streets, local streets would become more accommodating to pedestrians and cyclists. Despite having calmer environments within the zones, cyclists too were impacted by the street closings and one-way regulations that were primarily aimed at motorists. More importantly, however, the redistribution of car traffic made the situation worse for cyclists on the main streets they favoured.[19]

## 3 The Bicycle Renaissance Cut Short

A few years into the 1980s, public and political interest in cycling began to wane. The major political parties withdrew their support and cut their budgets

---

17 Emanuel, *Trafikslag på undantag*, 317–19; "Cykelplan 1978 för Stockholms kommun" (Stockholm: Gatukontoret, Stockholms stad, 1979), 15–17; telephone interview of Martin Emanuel with Sven Ekman 11.4.2005; interview of Martin Emanuel with Sven Ekman 29.2.2012.

18 *Trafikslag på undantag*, 320–22.

19 "Trafikplan 87. Diskussionsunderlag", (Stockholm: Gatukontoret, Stockholms kommun, 1987), 22; telephone interview of Martin Emanuel with Hans Genefors 30.7.2012.

directed at pro-cycling measures. When the Traffic Department was reorganized in 1985, there was no longer a specific person responsible for cycling.[20] Meanwhile, traffic counts in the 1980s showed that the number of cyclists entering the city centre dropped by a third.[21] The low interest in bicycle policy and planning during the 1980s may be framed as a result of, on the one hand, specific design choices in Stockholm around 1980, and on the other hand the bolstering economy of the decade and a limited political will to curb urban automobility.

After the implementation of some "easy" (non-conflicting) bicycle lanes in the second half of the 1970s, there remained only places where conflicts of interest were unavoidable. In this situation, the conservative traffic commissioner, Sture Palmgren, pursued what his traffic committee called "simpler and cheaper solutions": painting "bicycle lanes" on widened pavements—supposedly to accelerate the completion of a coherent network of bicycle lanes. While such solutions did not infringe on the space for cars and commercial traffic, they did create chronic conflicts between cyclists and pedestrians. These conflicts in turn had a detrimental effect on public support for cycling policies.[22] Considering the many competing demands for precious street space like bus lanes, car parking, and opportunities for loading and unloading, Palmgren argued that bicycle lanes were not a priority.[23] Instead, authorities subjected cyclists to campaigns to improve their behaviour in traffic.

In the meanwhile, automobility rose by a fourth, enthused by higher incomes, car-oriented work benefits (leased cars, free gasoline), and the urban sprawl. Efforts to curb automobility in the 1980s by investing in public transit and restrictive parking measures met with little success. Instead, proponents of new highways argued that only more roads could relieve Stockholm's city centre of automotive congestion; and indeed, in the mid-1980s engineers added a few new links to the highway network.[24] The re-introduction of relatively pro-car policies in Stockholm should be understood in the context of a generally

---

20 Sven Ekman, "Cykelplanering på 1980-talet i Stockholm", in *Ger mer cykeltrafik bättre städer? Rapport från ett seminarium anordnat i Linköping 22–23 augusti 1988 av Väg- och trafikinstitutet (VTI), Cykelfrämjandet och Föreningen för oskyddade trafikanter* (FOT) (Linköping: Statens väg- och transportforskningsinstitut, 1988), 54–59; Bergman, *Effektivare cykeltrafik*, 19; telephone interview of Martin Emanuel with Sven Ekman 11.4.2005.

21 "Cykelräkningar 2009" (Stockholm: Trafikkontoret, Stockholms stad, 2010).

22 "Trafikplan 87. Diskussionsunderlag", 25, 45.

23 KF Protokoll (Protocol) 7 October 1985, attachment 9.

24 Per Lagström, *Trafikutvecklingen i Stockholms stad 1945–1993* (Stockholm: Stockholm Konsult, 1994), 23–31; Holmstedt, *Ett halvsekel i Stockholmstrafiken*, 209–43; "Förslag till trafikprogram 93" (Stockholm: Stadsbyggnadskontoret, Stockholms stad, 1993),

less car-hostile, and also less environmentally-concerned, public debate. The recession of the late 1970s and early 1980s resulted in an understanding that automobility had reached its saturation point, which made the debates over the downsides of automobility abate.[25]

At the same time, cyclists in Stockholm lacked any strong spokespeople. As elsewhere, Stockholm's urban environmental movement lost its momentum in the 1980s.[26] The movement was, however, somewhat institutionalized in the local Stockholm Party. Following the local elections in 1979, the Stockholm Party had a potentially pivotal position between the right-wing and the left-wing parties. In its negotiations with the Social Democratic Party, the Stockholm Party wanted more funding for bicycle lanes, as well as—and this was unacceptable to the prospective partner—an explicit goal of reducing urban automobility by 50 per cent. Instead of concurring with these demands, the major parties (social democrats and conservatives) opted for collaboration to prevent the unpredictable newcomer from having any influence.[27] Moreover, Cykelkedjan disbanded in 1985. Although the more traditional Swedish Cyclists' Federation (*Cykelfrämjandet*) remained, the organization was frustrated by its small impact on municipal bicycle planning.[28] The Swedish Association of Cyclists (*Svenska Cykelsällskapet*), founded in 1980 following a schism within the federations' leadership, was somewhat more successful, although this was mostly in its claims for coherent bicycle routes in the suburbs.[29]

---

7–9; "Storstadstrafik 2: Bakgrundsmaterial", ed. Storstadstrafikkommittén (Stockholm: Allmänna förl., 1989), 16:5.

25 Emin Tengström, *Bilismen—i kris? En bok om bilen, människan, samhället och miljön* (Stockholm: Rabén & Sjögren, 1991), 121–28; Andrew Jamison, *The making of green knowledge: Environmental politics and cultural transformation* (Cambridge: Cambridge University Press, 2001), 90–93, 111–12.

26 Ulf Stahre, *Den alternativa staden: Stockholms stadsomvandling och byalagsrörelsen* (Stockholm: Stockholmia, 1999).

27 Sven Lilja, "Miljöpartier i Stockholmspolitiken 1966–2006", in *Makten i stadshuset: Stockholms lokalpolitik under 1900-talet*, ed. Anders Gullberg and Sven Lilja (Stockholm: Stockholmia, 2008); Börje Karlsson, "Stockholmspartiets uppgång och fall", in *Från almstrid till trängselskatt: De stora slagen om Stockholm*, ed. Börje Karlsson and Torbjörn Tenfält (Stockholm: Stockholmia, 2011).

28 Gatu- och fastighetskontorets arkiv (Traffic and Property Management Department Archive, held at the Traffic Department Archive, City of Stockholm; hereafter GFK), D.nr. 1989–321–2083:1–2. CF till Stockholms gatukontor, Trafikavdelningen, 20.10.1989; CF till Stockholms gatukontor, Trafikavdelningen, 6.12.1989; Lars Thörngren, "Stockholm som cykelstad", *Cykling* 1995:1, 19.

29 Interview with Nils-Göran Nilsson 28.1.2013. Nilsson worked with cycling-related matters in the Traffic Department during the 1980s and 1990s.

Increasing car traffic levels combined with renewed political attention to urban automobility made bicycle proponents within and outside city council worry about the safety of cyclists. The Traffic Department was straightforward in its responses to inquiries about improved conditions for cyclists: in a situation of stiff competition for space and resources, the current political leadership had a limited interest in bicycle planning.[30] In its draft traffic program from 1987, the department estimated that 40 per cent of the 1978 bicycle plan had been accomplished. The funds for extending it further were at this point small and would, according to the Traffic Department, not lead to the completion of the plan "in the foreseeable future".[31]

The minor "renaissance" of cycling of the 1970s was cut short in Stockholm. Whereas the overall development pattern of cycling in Western European cities is similar up until the 1980s, in that decade the pathways in different cities diverge. This is a complex issue, which did not always follow national lines, and which deserves more attention than it will receive here. In any case, during the 1980s, cycling moved off the mainstream policy agenda in many European countries, including Sweden, Great Britain and France.[32]

## 4 Cycling and the Dennis Agreement

Nearly a decade passed before policy makers and planners in Stockholm renewed their interest in cyclists and bicycle planning. The new stance towards cycling was not confined to Stockholm, but was discernible in many other European cities as well. It was becoming prominent also in national transport policies around Europe. The new discourse of sustainability began to take hold. Next to local and regional environmental problems connected to transport, which were taken more seriously than before, the issue of global warming and the publication of the Brundtland Report in 1987 made governments introduce new transport policies, in which concerns about health and the environment complemented the traditional transport agendas of efficiency, safety and equity.[33]

---

30 See e.g. documents in GFK 1990–321–1113 and 1991–321–1191.

31 "Trafikplan 87. Diskussionsunderlag", 45.

32 Oldenziel et al., *Cycling Cities* (in particular case studies of Stockholm, Manchester, and Lyon); Laura Golbuff and Rachel Aldred, "Cycling policy in the UK: A historical and thematic overview" (London: University of East London, 2011).

33 Emin Tengström, "Transport Sustainability in Denmark, Sweden and the Netherlands", in *Making urban transport sustainable*, ed. Nicholas Low and Brendan Gleeson (Basingstoke: Palgrave Macmillan, 2003), 134–37; Golbuff and Aldred, "Cycling policy in the UK", 10–14.

Large-scale urban infrastructure projects were, however, a more exciting prospect than bicycle planning at the outset of the 1990s. In September 1992, the major political parties reached an agreement (the Dennis Agreement, named after the negotiator appointed by the government, Bengt Dennis) on an extensive infrastructure package for the Stockholm region including roads, rails, and road pricing. Policy makers expected road pricing to fund the scheme rather than curb congestion. The agreement was clearly a grand political compromise rather than a clear-cut political choice.[34] Nevertheless, until the treaty broke down in 1997, it did have significance. The Urban Planning Department was commissioned to draw up a new traffic plan, which proposed the completion of a coherent bicycle network with only slight changes in comparison with the 1978 bicycle plan.[35] Meanwhile, the Traffic Department was instructed to come up with area-wise "traffic and street environment plans" based on the new conditions provided by the Dennis Agreement: bypasses were expected to reduce inner-city traffic by 30 per cent, while part of the income from the road pricing scheme would be earmarked for improving the urban environment.[36]

For cycling, this had two effects. Firstly, the expected traffic reduction, the proponents of the Dennis package argued, would leave space for bicycle lanes. Secondly, the funding for environmental improvements would make possible larger investments in the bicycle network—although the smaller political parties had their doubts about whether it would ever become reality. Working on the area-specific plans was allegedly an eye-opener to the officials regarding cyclists' precarious situation in the city. Moreover, in the 1990s, a new generation of traffic planners, hired to replace the old 1960s cohort that had reached retirement age, began to make its mark in Stockholm's Traffic Department. A few of the newcomers were dedicated cyclists who fuelled a new interest in bicycle planning.[37]

It is, however, politicians who stand out as the driving force in the resurgence of interest in cycling in Stockholm. A petition from the Left Party in 1993 triggered the revision of the bicycle plan of 1978. The leftist councillor, Iris Birath, argued that because a lot of resources were being spent on the accessibility

34 Holmstedt, *Ett halvsekel i Stockholmstrafiken*, 269–72. The Dennis agreement has been studied at length. See in particular Gullberg and Isaksson, *Congestion taxes in city traffic*.

35 "Förslag till trafikprogram 93", 32.

36 Holmstedt, *Ett halvsekel i Stockholmstrafiken*, 279–83.

37 Interview of Martin Emanuel with Staffan Ericsson 24.1.2013, Nils-Göran Nilsson 28.1.2013, and Krister Isaksson 28.1.2013. Ericsson was head of the division that drew up the area-specific plans. Isaksson was hired in 1999 with the special assignment to take care of cycling matters within the Traffic Department and the implementation of the recently adopted bicycle plan.

of car traffic, the city also had a responsibility to help cyclists.[38] The Traffic Department found the present bicycle plan well-suited as a basis for a continued extension of the network—although the funding was missing—but the traffic committee, led by the conservative, Carl-Erik Skårman, argued that the Dennis Agreement and the decrease in traffic in the city centre that it would bring, prompted a revision of the bicycle plan, giving special attention to the realization of coherent routes for cyclists throughout the city. [39]

The new red-green majority in office during the years 1995–98 increased the resources for extending the bicycle network, which at the end of the term had resulted in 46 km of new bicycle lanes, 35 km out of which were in the outer city.[40] The Traffic Committee put pressure on the Traffic Department to take measures in places where cyclists were particularly prone to accidents, thereby tying together existing links of the network. This also meant bicycle lanes in busy main streets—against the will of the traffic engineers, who were emphasising the scarcity of and stiff competition concerning space.[41]

To sum up, the politicians put pressure on civil servants in Stockholm to increase their attention upon cyclists during the 1990s. Cyclists were by no means at the centre of attention in the Dennis Agreement. But the traffic reduction that it was supposed to bring was important for a reassessment among politicians in the dominant parties on both left and right of the possibilities to provide cyclists with an infrastructure of their own in the city centre—and thus for bicycle planning being brought back to the policy table within the Traffic Department.

## 5 Spearheading Bicycle Planning

The manner, in which the local Stockholm Party used its pivotal position following the local election in 1998 proved crucial for the increased attention, ambition and conflicts over cycling towards the turn of the century. After a draft of the bicycle plan was circulated in 1997, a final version was approved by the traffic committee in early 1998. Compared to its predecessor, the 1998

38 KF Motion 51/1993.

39 KF Utlåtande 74/1994; Protokoll 16/5 1994, 14, Yttrande (Statement) 106/1994.

40 Bicycle funding increased from an annual 1–2 MSEK in the early 1990s to 7 MSEK in 1995 and 30 MSEK in 1997.

41 GFK 1995–322–929. GFK-tjut 2.2.1995, "Cykelplan 1995. Förslag till konkreta åtgärder"; GFN-protokoll 14.3.1995; GFK-tjut 23.5.1995, "Cykelplan 1995. Förslag till konkreta åtgärder, återremiss"; GFN-protokoll 13.6.1995; GFK-tjut 17.8.1995, "Cykelplan 1995. Förslag till konkreta åtgärder, återremiss".

plan paid attention to a broader spectrum of issues related to cycling, such as maintenance and bicycle parking, and also included a proposal for a public bicycle scheme. Still, it was essentially a plan for the building of a network of main bicycle lanes. The lion's share of the budget was allocated to the construction of bicycle lanes. The plan would be implemented over the following 5–15 years. The longer period was reserved for those segments (typically main streets in the city centre) where the realization of bicycle lanes depended on reductions in car traffic levels. The authors of the plan acknowledged that it was, of course, "technically possible" to implement the plan more quickly, but not without considerable negative effects on other modes of transport.[42]

Cyclists and environmental organizations expressed concerns about the plan's excessively long execution time, its dependence on the reduction of car traffic, and the absence of bicycle lanes in the main streets of the city centre.[43] The Stockholm Party agreed, and called the 1998 bicycle plan a "paper tiger". Balancing between the two traditional power blocks, the party embraced cycling as the most tangible issue to promote urban sustainability.[44] Following the local election, the party used its influence to put pressure on the city government for the improvement of the conditions for cyclists. Although their coalition partner, the Conservative Party, had only a few years earlier announced its resistance to bicycle lanes in the city centre, the budget for 1999 included an amendment to the bicycle plan. The amendment specified that bicycle lanes would be established within the mandate period to ten city centre streets, with painted bicycle tracks in another three locations.[45]

The original 1998 bicycle plan was quite straightforward. Its amendments, however—in particular a few on-street bicycle lanes, or bicycle tracks—caused serious conflicts and aroused media attention. The Chamber of Commerce and professional drivers were outraged because of the consequences the plan would have on sales and accessibility. The social democrats argued that the original plan, developed under their mandate, had a more balanced approach in relation to the different modes of traffic, while the Green Party and bicycle

42 Daniel Henricson and Cecilia Strömbom, "Cykelplan 1998 för Stockholms innerstad" (Stockholm: Gatu- och fastighetskontoret, Stockholms stad, 1998).

43 GFK 1997–322–2700:25. GFK-tjut 9.6.1998, "Cykelplan 1998 för Stockholms innerstad". See also Anders Orrenius, "Cykling: Utbyggt nät för cyklister dröjer", *Dagens Nyheter* 24.3.1998.

44 Svante Linusson, "De politiska förhandlingarna", in *1000 meter cykelfält som skakade Stockholm*, ed. Erik Beckman and Svante Linusson (Bromma: Forum08, 2009); interview of Martin Emanuel with Stella Fare 1.2.2013. See also Ander Orrenius, "Stockholmspartiet: 'Stockholm bör bli en stad för cyklister'", *Dagens Nyheter* 15.5.1998; Lena Hallerby, "Trafikfrågorna sp:s melodi", *Svenska Dagbladet*, 15.5.1998.

45 Ibid., 11–15; Lilja, "Miljöpartier i Stockholmspolitiken 1966–2006"; GFK 1996–322–3059:1. Carl-Erik Skårman (M) till GFN 5.11.1996.

experts within the Cycling Association preferred bicycle lanes to painted tracks for safety reasons.[46]

Despite all the criticism, the Stockholm Party successfully pushed for rapid implementation of the amendments. The angry electorate in the next elections summarily refuted its success; the party lost its place in the city council. The harsh conflicts that surrounded the implementation of the plan strongly indicate that this was nothing that would have happened by itself, that is, without the help and support of the Stockholm Party. According to Svante Linusson, who was responsible for cycling issues within the Stockholm Party, every inch of the bicycle infrastructure was heavily debated in the traffic committee and elsewhere.[47] During the following electoral term, the Conservative Party, now in minority, did not file a single cycling-related proposal to the Municipal Council. The Stockholm Party had been instrumental in ensuring the swift implementation of bicycle lanes and tracks on several of the main streets in Stockholm. Their efforts can be considered the first steps towards Stockholm becoming truly more accommodating towards cyclists.

## 6 Cycling in the Shadow of Congestion Charging

Cycling had been a fiercely debated issue around the turn of the century, but in the following years the heat of the debate largely dissipated as a series of new proposals were tabled and a new traffic congestion scheme was implemented in Stockholm.[48] Road tolls and congestion charging had been discussed in Stockholm with varying intensity since 1970. They had also been flagged in several serious proposals. Following the 2002 election they resur-

---

46 For details of the various cases, including the manifold critiques towards the amendments of the plan, see in particular the documents in GFK 1999–322–775. GFK-tjut 25.4.1999, "Cykelprogram 1999 för innerstaden, inriktningsbeslut"; GFN-protokoll 1.6.1999, § 43. Media coverage of the conflicts was extensive, see e.g. Pralen Melander, "Cykla i stan. Motstånd mot cykelbanor", *Dagens Nyheter* 5.5.1999; Peo Österholm, "Stockholms handelskammare: City riskerar utarmas", *Svenska Dagbladet* 8.10.1998; Niklas Svensson, "'Trafikkaoset måste lösas: Näringslivet och LO till attack mot Cederschiöld", *Expressen* 9.11.1998; Sven-Anders Eriksson, "'Politikerna är ute efter billiga poäng'", *Aftonbladet* 17.4.2000. The Stockholm chapter of the Swedish Trade Union Confederation and the Chamber of Commerce presented their joint criticism towards the amendments of the bicycle plan as emblematic of the incompetence of the new majority to face urban transport problems in a critical post-Dennis situation. See e.g. Ulf Stahre, *Den gröna staden: Stadsomvandling och stadsmiljörörelse i det nutida Stockholm* (Stockholm: Atlas, 2004), 55.

47 Interview with Stella Fare 1.2.2013; Linusson, "De politiska förhandlingarna".

48 Interview of Martin Emanuel with Per Bolund 9.3.2005 and Krister Isaksson 28.1.2013.

ILLUSTRATION 4.2 The amendments to the 1998 bicycle plan, and especially the bicycle tracks painted on the street in Hornsgatan and Fleminggatan, became the subject of fierce debate. The local Stockholm party had put pressure on the right-wing coalition to expand the city's agenda for cycle planning. In the subsequent elections (2002), social democratic mayoral candidate, Annika Billström, tried to capitalize on the motorists' resistance to the new bicycle tracks. In the publicity stunt featured here, Billström, paintbrush in hand, vows to banish the contested tracks put in place by her political opponents. Ultimately, Billström won the election, but the bicycle tracks remained intact.

PHOTO: JAN DÜSING, PUBLISHED IN *EXPRESSEN*, SEPTEMBER 17, 2002

faced as a policy instrument for curbing urban automobility. The small Green Party used its unique pivotal position to bring congestion charging into the negotiations with the Social Democratic Party both locally in Stockholm and on the national level. Despite immense opposition, political tensions and complex juridical considerations, a full-scale congestion charge trial was carried out between January and July 2006. It was followed by a referendum in the fall of 2006, which led to its introduction in 2007.[49]

The so-called Stockholm Congestion Tax Trial also involved public transport improvements and park-and-ride facilities. Improvements for cycling were initially not part of the package. In fact, many cycling advocates worried about the consequences of the local elections which had seen the Stockholm Party lose all its seats on the Municipal Council.[50] During the previous term, the Social Democratic Party had shown little interest in cycling. Prior to the election, their representative, Annika Billström, had even threatened to abolish the new bicycle tracks because of safety concerns (see illustration 4.2).[51] These threats were never realized. Convinced by the Traffic Department about the positive effects of the new bicycle facilities, the new majority of the traffic committee continued to extend the bicycle network on main streets, preferably in the form of bicycle lanes, not painted tracks.[52]

Investments in cycling facilities thus continued, although at a slower pace than before, as the Green Party put pressure on the social democrats to follow suit with the 1998 bicycle plan.[53] The Left Party, in turn, reinitiated its old pet project, the introduction of a public bicycle scheme, and succeeded in carrying it through within the term (1998–2002).[54]

Half-way through the term, the congestion charging trial made its imprint on bicycle planning, as the majority parties commissioned the Traffic Department to intensify its work to extend the bicycle network. In their agreement over the congestion charging trial, signed in April 2004, they emphasised the need to

49 For details see Gullberg and Isaksson, *Congestion taxes in city traffic*. See also Holmstedt, *Ett halvsekel i Stockholmstrafiken*, 360–62.

50 Henrik Eriksson, "Politisk punka", *Nya Cykeltidningen* 2002:3–4, 3.

51 Ola Wong, "Inga banor för Fare om s får bestämma", *Dagens Nyheter* 11.8.2001; "Cykelbanorna kan rivas upp nästa år", *Svenska Dagbladet* 27.9.2001; Torbjörn Tenfält, "Cykelbanorna riskerar att tas bort", *Dagens Nyheter* 10.6.2002.

52 GFK 2002–322–3242:1. GFK-tjut 28.8.2002, "Cykelbanor och cykelfält i Stockholms innerstad"; GFN-protokoll 1.11.2002, § 43.

53 From an annual peak of 40 MSEK during the preceding mandate period, funding was cut to 20–25 MSEK in 2004, while a mere 15 MSEK was planned to be spent on new bicycle roads in 2005. Anders Sundström, "Satsning på cykelbanor minskar kraftigt", *Dagens Nyheter* 27.1.2003; Linusson, "De politiska förhandlingarna", 11–15; interview of Martin Emanuel with Per Bolund 9.3.2005.

54 GFK 2005–322–00330. GFN-protokoll 2005–03–08, § 38.

investigate the possibilities of improving the capacity of the bicycle network, so that it would be able to house those commuters that would switch their cars for bicycles. Five months later, the department delivered a report containing the projects that could be initiated quickly if extra funding was provided.[55]

Furthermore, during this term new bicycle plans for the suburbs and the city centre were initiated and developed within the Traffic Department, and launched in 2005 and 2006, respectively. Like its predecessor from 1998, the core of the 2006 plan for the city centre was foremost a network of main bicycle lanes, although its scope was wider. Importantly, and in accordance with the recently launched Traffic Safety Program, the authors assumed that those streets where bicycle lanes were difficult to realize, and where cycling would thus take place in mixed traffic, would be subject to a speed limit of 30 km/h. If that would not go through, the authors suggested other measures that would in effect bring down the actual speed to similar levels.[56] The Traffic Safety Program did not have sufficient back-up in practice; speed limits have until recently been very difficult to change in Stockholm.[57]

The realization of the new bicycle plan also had a difficult start. In 2006, the next political shift to the right brought support for cycling to an abrupt end. The new conservative traffic commissioner, Mikael Söderlund, argued that the bicycle plan focussed single-mindedly on cycling, failing to address the accessibility of all modes of traffic. He opposed bicycle lanes on main streets and significantly reduced funding for cycling facilities. The Traffic Department advised against placing bicycle lanes on parallel streets, but in vain.[58]

The 2006 election was also a blow to the congestion charging scheme that had been promoted particularly by the Green Party. Not that the city never implemented the scheme—it did—but the new right-wing majority chose to use the revenues for road investments rather than improvements to public transit

55 GFK 2004–322–1864. Jan Valeskog (S), Ann-Marie Strömberg (V), Per Bolund (Mp) till GFN 18.5.2004; GFK-tjut, "Intensifierat arbete med cykelfält/cykelbanor", 2.9.2004; GFN-protokoll 21.9.2004, § 11. A year later a similar joint call about intensified extension of the bicycle network in the city centre was issued by the majority parties. GFK 2005–322–805. Karin Wanngård m.fl. (S), Åsa Romson (Mp), Ann-Marie Strömberg (V) till TRN 8.3.2005; GFK-tjut 21.6.2005, "Intensifiera utbyggnaden av cykelbanor i Stockholms innerstad"; TRN-protokoll 22.09.2005, § 25.

56 "Cykelplan 2006 för Stockholms innerstad" (Stockholm: Trafikkontoret, 2006), 15–16.

57 Krister Spolander, "Cykel i huvudstad", in *1000 meter cykelfält som skakade Stockholm*, ed. Erik Beckman and Svante Linusson (Bromma: Forum08, 2009), 66.

58 Ibid., 67; Trafikkontorets arkiv (Traffic Department Archive, City of Stockholm, hereafter T) D.nr. 2007–313–01112. TK-tjut 15.5.2007, "Planeringsprinciper för nya cykelbanor i Stockholm med konkreta exempel"; TRN-protokoll 28.8.2007, § 31. T 2008–313–04271. TK-tjut 9.12.2008, "Cykelplanering 2009. Lägesbeskrivning"; TRN-protokoll 20.1.2009.

as suggested in the original plan.[59] Shortly after, a new round of "negotiating planning" (planning based on political negotiation) was initiated. The result was a huge traffic investment package for the Stockholm region, including many of the road and rail projects that had been discarded in the past. In this sense, congestion charging, initially intended to curb urban automobility, had been skewed to enable road projects which would otherwise not have been realized.[60] And while the improved capacity of the cycling infrastructure was earlier considered a way of passage to some of those commuters that would give up their cars, the political shift in 2006 brought a drastic end to the support of cycling and bicycle planning.

## 7 Cycling and the Attractive, Accessible City

Following a period of little interest by the right-wing majority in the City of Stockholm, cycling policy resurfaced as an important issue in the 2010 local elections. While its influence on actual voting behaviour is difficult to assess, in the media cyclists were portrayed as a decisive voter group. The right-wing coalition remained in power but began to re-evaluate its position on cycling. Being goaded by the opposition, a broad-based and vigorous cycling community, and pro-cycling media, the coalition made cycling an essential part of their strategy to make the city "accessible" and "attractive".

Cycling as part and parcel of the attractive city is of course not solely a Stockholm phenomenon, nor entirely new. Bicycle policy and planning has become a way to strengthen the cultural prestige of a city.[61] As Trine Agervig Carstensen and Anne-Katrin Erbert have pointed out, bicycle has become the centre point of urban lifestyles that are considered important in creating "urban liveability". In contemporary cities, progressive cycling policies are expected to attract "high-income groups and groups specifically enthusiastic about wholesome urban lifestyles".[62]

The turnaround in the Conservative Party's attitude to cycling and cyclists was, partly, a reaction to initiatives by the opposition parties. As Ulla Hamilton

59 Holmstedt, *Ett halvsekel i Stockholmstrafiken*, 360–62.

60 Gullberg and Isaksson, *Congestion taxes in city traffic*; Tim Richardson, Karolina Isaksson, and Anders Gullberg, "Changing Frames of Mobility through Radical Policy Interventions? The Stockholm Congestion Tax", *International Planning Studies* 15, no. 1 (2010).

61 Oldenziel et al., *Cycling Cities*.

62 Trine Agerwig Carstensen and Anne-Katrin Ebert, "Cycling Cultures in Northern Europe: From 'Golden Age' to 'Renaissance'", in *Cycilng and Sustainability*, ed. John Parkin (Bingley: Emerald, 2012), 47–48.

replaced Mikael Söderlund as traffic commissioner in 2008, the social democrats presented their strategy for making Stockholm a "walking and cycling-friendly city", while the Green Party forwarded a 14-point program, which would turn Stockholm into "a living city to walk and bicycle in".[63] Other political parties followed suit and announced their own programs on how the City of Stockholm could better accommodate cyclists, all with a slightly different focus.[64]

Most political parties were eager to show how much they cared for the rapidly growing group of urban cyclists. In the spring and summer of 2010, daily papers even portrayed urban middle-class cyclists as a potentially decisive voter group in the coming local elections, while warning the Conservative Party about their failure to respond to the priorities of this potential group of supporters. To quote one author, the new cyclists were indeed "classical right-wing voters" in terms of income and education, but they had other goals in their lives, embracing sociability and sustainability, which were often better represented by the Green Party.[65] Indeed, surveys conducted within the city administration suggested that cyclists in Stockholm tended to be middle-aged people with high income and education, in comparison to the average Stockholm resident.[66]

Whether it was a response to such calls or not, shortly before the election, Hamilton publicly refuted her predecessor's resistance to bicycle lanes on main streets and announced that the Traffic Department had initiated the development of another new bicycle plan.[67] Launched in 2012, the new policy focused on commuting and coherent bicycle routes, with the ambitious aim of increasing commuter cycling from 10 to 15 per cent by 2030.[68] As the plan was approved by the Municipal Council, Hamilton situated it not only within

63 T 2009–314–986. Carin Jämtin (S) till kommunstyrelsen 4.5.2009, "Skrivelse angående gång- och cykelvänlig stad"; T 2009–314–1171. Emilia Hagberg (Mp) till kommunstyrelsen 26.5.2009, "Gör Stockholm till en levande stad att gå och cykla i".

64 "FP:s nya cykelprogram", http://lottaedholm.wordpress.com/2009/08/27/fps-nya-cykelprogram; "Stockholmsvänsterns Cykelplan 2010", http://sv.scribd.com/doc/43119601/Cykelplan; "96 åtgärder för en cykelvänlig region", Miljöpartiet de gröna i Stockholmregionen, 21.8.2010.

65 See e.g. Mattias Kamgren, "Cyklisterna kan avgöra valet. Hård kamp om väljare på två hjul". *Mitt i Östermalm* 14.9.2010; P.M. Nilsson, "Högertrafik på cykelbanorna. En ny urban medelklass växer fram", *Dagens Nyheter* 3.5.2010.

66 "Att cykla i Stockholm: Så tycker stockholmarna" (Stockholm: Utrednings- och statistikkontoret, Stockholms stad, 2006).

67 T 2010–313–01743. TK-tjut 4.6.2010, "Framtagande av nya cykelplan för Stockholm. Genomförandebeslut"; TRN-protokoll 14.6.2010, § 11.

68 "Cykelplan 2012" (Stockholm: Trafikkontoret, Stockholms stad, 2012).

an assemblage of other traffic-related planning documents, but also related it to the environmental and health policies of the city, the 2010 comprehensive plan, and, ultimately, the attractiveness of the city: "Accessibility in Stockholm and how well it works is and will be an important part of the attractiveness of Stockholm today and in the future.... Combined with the other modes of traffic ... the bicycle is an important part of an accessible and attractive Stockholm". In its review of the bicycle plan, the city executive office, responsible for strategic matters regarding cycling, wrote that it would use it as a point of departure for developing the "bicycle brand" of the city.[69]

The bicycle plan is part of a broader array of traffic policy documents, most prominently the Urban Mobility Strategy (*Framkomlighetsstategin*). For the first time, the city's overall strategy now prioritizes walking, cycling, and public transit over automobility as energy-efficient modes of transport making effective use of existing space. Policy makers also acknowledge that cycling facilities must sometimes have precedence over parking spaces and car lanes.[70] Hamilton was careful, however, to portray the conservative approach to traffic policy as "holistic" and responsible, balancing the needs of all modes of transit including utility traffic. Whereas the attractive city as promoted by the left-wing oppositional parties is a "green" or "liveable" city imbued with environmental and social connotations, the attractive city of the Conservative Party (and the Traffic Department) tends to refer to accessibility and efficiency. In view of Stockholm's elaborate transit system, Hamilton saw no point in comparing Stockholm to cycling cities like Copenhagen and Amsterdam.[71]

Per Ankersjö, leader of the Centre Party (the "green" conservative party) in Stockholm, begged to disagree. As commissioner for the urban environment, he profiled himself as "bicycle mayor" of Stockholm and secured ca. €100 million for bicycle-related investments in 2012. In justifying this budget, the city declared its intention to "compete with cities such as Copenhagen and Amsterdam when it comes to bicycle-friendliness and safety".[72]

The current (2014–2018) red-green majority in Stockholm, with a Green Party traffic commissioner, is not likely to scale down its ambitions for, and investments in, cycling—at least not anytime soon. A thriving local cycling community will help secure that. Indeed, in the last five-year period or so, the Cyclists' Federation has resurfaced as a serious player in traffic-related

69 KF Utlåtande 24/2013, Protokoll 18.2.2013, § 22.

70 Daniel Firth, "Framkomlighetsstrategin" (Stockholm: Stockholms stad, Trafikkontoret, 2012), 5, 17–18, 27.

71 Interview of Martin Emanuel with Ulla Hamilton 28.12.2013.

72 "Kommunstyrelsens förslag till budget 2012", www.stockholm.se/budget2012.

debates. Numerous bloggers and cycling aficionados closely monitor every step of the process. Stockholm's cyclists, many of whom are socio-economically strong bicycle commuters, no longer passively accept substandard cycling infrastructure or police campaigns that single out cyclists' traffic behaviour as the source of unsafety.[73]

## 8 The Bicycle City and the Car City?

Rocketing bicycle commuting levels in Stockholm during the last decade coincide with rapid population growth in the city. In the 1980s, a twenty-year long decline in the population of the City of Stockholm (and stagnation in the whole region) was broken, and the long tendency of suburban development further away from the city was succeeded by a process of densification of the city centre and just beyond. For the first time in over a hundred years, population growth in the city centre exceeded that of the suburbs.[74]

The "urban renaissance" may be described as the result of a resurgent interest in the city and what it has on offer. It is also the outcome of new urban planning ideals which appreciate the traditional city and its dense grid pattern. Further, this renaissance is the result of a transition from an industrial to a service society that has taken place since around 1970.[75] Today's urban planning policies in Stockholm strongly support re-urbanization. The latest development plan for the Stockholm region envisions densification of the city centre and a few regional centres. The 2010 comprehensive plan shares the ambition of its predecessor from 1999 to 'build the city inwards' and promotes densification of the city centre and the suburbs just beyond, as well as in a couple of suburban "focal points". The compact city that will supposedly be the result of this planning strategy is expected to bring about shorter (walkable, "bikeable") distances and a better basis for public transport to compete with the car.[76]

The headline of the 2010 comprehensive plan, the "Walking City", refers to a city in which one can safely walk between the different parts without

---

73 For an example, see the blog "Cyklistbloggen", run by two passionate commuter cyclists in Stockholm: http://www.cyklistbloggen.se.

74 Lars Nilsson, "The Return to the City: Twentieth Century Urban Development in Sweden", in *Reclaiming the City: Innovation, Culture, Experience*, ed. Marjaana Niemi and Ville Vuolanto (Helsingfors: Finnish Literature Society, 2003), 47.

75 Emanuel, *Trafikslag på undantag*, 351–52.

76 *Regional utvecklingsplan för Stockholmsregionen: RUFS 2010* (Stockholm: Regionplanenämnden, Stockholms läns landsting, 2010); *Promenadstaden: Översiktsplan för Stockholm* (Stockholm: Stadsbyggnadskontoret, 2010).

the barriers of traffic-related infrastructure or large green areas. Presenting the outline of the plan in 2008, Kristina Alvendal, the conservative urban planning commissioner, forwarded it as a way to "heal the wounds" of modernist planning. Ultimately, she challenged the primacy of the car: "The role of the car may be important, but it must be subordinated to the conditions of the city".[77] Regarding the bicycle, the plan expressed new attitudes towards its usefulness: "A fundamental starting point must be to treat the bicycle as a means of transportation, not for play and recreation".[78]

Cycling thus receives great attention as an important mode of mobility in both overall traffic policy documents and in comprehensive urban planning documents. Nonetheless, despite the political commitment to cycling, Stockholm's infrastructural plans cater to automobility; these plans have proven hard to beat. Despite the €100 million investment plan for cycling, more funds than ever are going to railroad and subway extensions and infrastructural projects related to motor vehicles: the budget of the €12 billion investment package defined in the 2007 transport infrastructure agreement for the Stockholm region (*Trafiksatsning Stockholm*) splits equally between road and rail projects. By way of contrast, the package leaves little more than 1 per cent for investments in cycling and walking.[79]

Cars and public transit continue to dominate regional traffic policy. Rising levels of bicycle commuting to the city centre are not the whole story. Counting all trips in the city (not only those to and from the city centre), the share of cycling has remained stable at around 5 per cent since the 1980s; in Greater Stockholm, cycling has decreased from 7 to 4 per cent—in a similar relation as re-urbanization mirrors continued suburbanization in the region as a whole.[80] Other larger cities, such as London and Manchester, along with more medium-sized ones in "bicycle nations", like Eindhoven, Amsterdam, and Copenhagen, exhibit a similar conflicting trend of inner city increase in cycling levels and outer city decline.[81]

The City of Stockholm is taking conscious steps towards reshaping the city into one which welcomes cyclists, not only by means of bicycle infrastructures,

77 Kristina Alvendal, "Stockholm blir en promenadstad", *Svenska Dagbladet* 15.10.2008.

78 *Promenadstaden: Översiktsplan för Stockholm*, 21.

79 Per Lundin, "Att tänka om staden med historia: En introduktion till Bilstaden", in Bilstaden: USA visade vägen, ed. Uno Åhrén/Per Lundin (Stockholm: Stockholmia, 1960, rev. ed. 2010); "Trafiksatsning Stockholm. FAQ", http://trafiksatsningstockholm.se/wordpress/wp-content/themes/trsthlm/illustrations/TSS_FAQ.docx.

80 Nilsson, "The Return to the City", 58–59. Modal split graphs from TRAFA, the Swedish government agency providing decision-makers with policy advice concerning traffic policy.

81 Golbuff and Aldred, "Cycling policy in the UK", 35; Oldenziel et al., *Cycling Cities*.

but also from the point of view of a larger urban planning perspective with both material (compact city, shorter distances) and cultural (the "good" city as one with spontaneity and small-scale mobility) components. Central parts of Stockholm appear to be in a sociotechnical process that may yield a "cycling city". But present-day understandings on where the bicycle belongs seem to be limited to the city centre and just beyond. The suburbs and the Stockholm region will—this appears to be the general understanding—remain the car city. Not to forget, of course, Stockholm's longstanding and outstanding system of public transit.

There are signs, however, that cycling is making its way into regional planning considerations. Regional policy makers are realizing the potential of bike-train-bike in the mobility chain as they struggle with overcrowded parking facilities at commuter hubs and the merging of city and regional cycling schemes—although evaluations suggest that much still needs to be done. Also, a regional bicycle plan focussing on commuting and high-speed cycling has recently been developed for the Stockholm region.[82] It remains to be seen whether these are the first steps towards mitigating the reflex reaction to disregard cycling when large-scale infrastructure and investments are brought to the negotiation table.

## 9 Conclusion

Cycling in Western cities has hardly experienced a smooth renaissance since the mid-1970s. As this detailed case study of Stockholm has shown, the wave-like development of cycling in the city has been dependent on several factors, including the state of the market, the intensity of environmental and sustainability debates, overall mobility and urban planning policies, party-politics and urban power-relations beyond traditional politics. This chapter has focused on the provision of dedicated bicycle infrastructure such as bicycle lanes, while downplaying the importance of other efforts to improve conditions for cyclists, such as the providing of bicycle parking facilities, safety-related measures in interparts, and traffic signals specifically targeted at cyclists. Bicycle infrastructures in city centres are, however, the measures that most clearly involve a redistribution of space and that, consequently, tend to be accompanied by the

82 Trafikverket Region Stockholm 2014; "Utvärdering av kombinationsresor med cykel och kollektivtrafik", Stockholms läns landsting, Trafikförvaltningen, 2014. "Regional cykelplan för Stockholms län 2014–2030", (Stockholm: Trafikverket Region Stockholm, 2014).

most intense conflicts. The processes behind them are thus good indicators of priorities and politics in urban mobility.

The first supportive phase of cycling in the 1970s was emblematic of prevailing criticisms of urban automobility. It is unlikely, however, that this criticism would have been channelled into bicycle planning without cyclists and activists putting pressure on politicians to act. The second supportive phase of cycling in Stockholm, the second half of the 1990s, was, even though it had broader resonance, driven by a small political party—and was thus also very conflict-laden and plagued by overwhelmingly negative media assessments of cycling. It also resulted in a more cautious bicycle policy during the following decade. In the third, now on-going, supportive phase of cycling in Stockholm, the embracing of cycling seems to be almost omnipresent. Although past representations of cycling as dangerous remain strong, cycling has broad support, since it, to put it bluntly, attracts "the right kind of people": the young and urban middle class, which is increasingly at the centre of attention of urban governance.

Intriguingly, both in the late 1990s and the early 2010s, it was the smaller, self-proclaimed "green" parties on the political centre-right which profiled themselves as bicycle-friendly and managed to push right-wing coalitions to initiate more ambitious bicycle planning. The surge in cycling policy in Stockholm has thus not been dependent on strong mayors, as in the cases of London (Ken Livingstone, Boris Johnson) and Bogota (Enrique Peñalosa), but on the dedication of the leaders of the smaller political parties which, in comparison to the larger parties, appeared to have more to win than to lose when working against urban automobility.[83]

This chapter has not attended specifically to how bicycle planning and policy have been refined or broadened in scope. It has shown, however, that the integration of bicycle policy and planning in overall mobility and urban planning policies and initiatives has varied greatly over the years. While the Dennis Agreement of the 1990s and the congestion charging scheme of the 2000s had an impact on how important actors assessed the possibility to cater for cycling by means of a distinct infrastructure in the city centre, cycling policy was hardly an important or integral part of the packages. The traffic zoning programs of

83 The influence of smaller parties is partly a result of a 1994 reform concerning majority rule in the City of Stockholm. Instead of representation in committees being based on the number of seats held in the Municipal Council, the majority now gets to appoint the chairs of all committees, which in elections with small margins gives ample room for smaller parties to negotiate whether to grant their support to one minority coalition or the other.

the 1970s did integrate bicycle planning, but they failed to take duly notice of cyclists' actual preferences of using the main streets. Especially during the last five-years, bicycle planning and policy have become increasingly integrated in and empowered by overall mobility-related policies (encouraging the efficient use of space) and urban planning documents (favouring the compact city). It is still largely bypassed, however, when larger and often regionally conceptualized infrastructural schemes are discussed and implemented. Here, cyclists are still invisible. Although this chapter has focused on Stockholm, this limitation is most probably shared with other European cities.

CHAPTER 5

# Making the Bicycle Dutch: The Development of the Bicycle Industry in the Netherlands, 1860–1940

*Sue-Yen Tjong Tjin Tai and Frank Veraart*

## 1 Introduction

Urban policymakers around the world flock to the Netherlands to learn from the Dutch cycling culture.[1] In their home cities, they are battling with traffic jams and air pollution caused by the ever-increasing number of motorized vehicles. They see cycling as an ideal and sustainable solution, because it is clean, cheap, fast and healthy. These policymakers are inevitably referred to the Netherlands, a country that continued everyday cycling despite the rise of the car in the twentieth century. In 2013, the Dutch nation of 16.8 million people owned 22.3 million bicycles, and made 26 per cent of their journeys by bicycle.[2] Dutch schools, offices, shopping centres and railway stations have facilities for cyclists. Dutch television commercials feature cyclists to appeal to their viewers.[3] Moreover, cycling is so commonplace in the Netherlands, that until recently, there were not enough cycling experts to advise foreign policy makers.[4]

Cycling may be truly Dutch, but their bicycles are not. "The Dutch bicycle is a lie", claims Alfred Thun, a German crankshaft manufacturer.[5] In a 2014 interview, he pointed out that during the last decade, about 50 European manufacturers of bicycles and parts had either closed or offshored their production to

1 Margreet Vermeulen, "Fietsologie", *Volkskrant, Sir Edmund*, April 18, 2015, 46–48.
2 This is the total number of bicycles in 2013, including e-bikes, as estimated by BOVAG/RAI and GFK Panel Services, see: "Mobiliteit in cijfers, Tweewielers, 2014/2015, RAI/BOVAG". Note that Statistics Netherlands stopped producing bicycle ownership statistics after 2007, identified by Ruth Oldenziel as an accounting trick which influences mobility policies (see her keynote speech for T2M, Philadelphia, 2014); Transfer statistics are from "Mobiliteit in Nederland; vervoerwijzen en motieven, regio's", September 17, 2014, https://opendata.cbs.nl, accessed February 21, 2015. The 2007 statistics show that 84 per cent of the Dutch population owned one or more bicycles, see: "Mobiliteit; voertuigenbezit naar achtergrondkenmerken, 1985–2007".
3 See the blogposts at: http://cyclingacademics.blogspot.nl/search/label/commercials.
4 Margreet Vermeulen, "Fietsologie", *Volkskrant, Sir Edmund*, April 18, 2015, 46–48.
5 Ariana Kleijwegt, "De Hollandse fiets is een leugen", *NRC*, July 18, 2014.

© KONINKLIJKE BRILL NV, LEIDEN, 2019 | DOI:10.1163/9789004289970_006

South East Asia. Production statistics from 2000 show that the top four bicycle-producing countries are indeed Asian: China with 52.2 million bicycles, India 11 million, Taiwan 7.5 million and Japan 4.7 million.[6] The entire European Union produced 14.5 million bicycles, of which 3.2 million in Germany, 3.2 million in Italy, 1.9 million in France and 1.2 million in the UK.[7] Nevertheless, Dutch statistics confirm Thun's statement: in 2012, domestic production was 900,000 bicycles, which means 67 per cent of the new bicycles for sale in the Netherlands were imported.[8] Thun's statement is no surprise, as offshoring production has been happening for several decades.[9] Moreover, due to its geography and size, the Netherlands was always more a nation of merchants rather than manufacturers.[10]

Around 1939, the Dutch bicycle was not a lie. The bicycle industry in the Netherlands was supplying 99 per cent of national demand by producing 389,000 bicycles a year.[11] In those days, the Dutch population of 8.7 million owned 3.3 million bicycles.[12] This domestic dominance was not unique. Several other European countries also imported hardly any bicycles: the UK, Germany, Italy, France, Sweden, Belgium and Czechoslovakia.[13] Furthermore, in the late 1930s, the popularity of the bicycle was not just a distinct Dutch characteristic. Germany estimated it had about 20 million bicycles, around one for every three people.[14] Belgium counted around 3 million bicycle ownership tax payers in 1939.[15]

Notwithstanding its modest international status, the Dutch industry proudly presented its products. Bicycle manufacturer Gazelle's advertisements conveyed the message that Gazelle bicycles were real Dutch products

---

6 See: http://www.ibike.org/library/statistics-data.htm which refers to the following sources: Bicycle Retailer & Industry News Directory, Cycle Press, European Bicycle Manufacturers Association, Japan Bicycle Promotion Institute, Bike Europe, and Bicycle Retailer & Industry News.

7 Country statistics: see previous note. EU statistics, see: https://issuu.com/coliped/docs/european_bicycle_market__industry_profile_2010

8 This statistic is based on total 2012 sales, consisting of domestic production (900,000) and imports (1,836,000). See: "Mobiliteit in cijfers, Tweewielers, 2014/2015, RAI/BOVAG."

9 Sluyterman, *Dutch Enterprise*, 241–242.

10 Gerwen and Goey, *Ondernemers in Nederland*, 25–27; Sluyterman, *Dutch Enterprise*, 4–8.

11 Tjong Tjin Tai, Veraart and Davids, "How the Netherlands," 19–20.

12 Centraal Bureau voor de Statistiek, *Tweehonderd jaar statistiek*; Jong, *Geschiedenis eener Nederlandsche vereeniging*, 116.

13 Andrew Millward kindly provided the original datasheets from his PhD research. See: Millward, *Factors*.

14 There are no statistics, Ebert quotes two sources (1936 and the late 1930s), see: Ebert, *Radelnde Nationen*, 286.

15 Estimate based on figure 12 in: Albert de la Bruhèze and Veraart, *Fietsgebruik*, 118.

by emphasising that all its bicycle parts were made in-house. The manufacturer thus counteracted accusations that Gazelle's products were rebranded imports.[16] Thus we see that in the past, the Dutch certainly did produce most of their own bicycles. However, the bicycle is not a Dutch invention: it was introduced in France in 1867.[17] This leads us to the related questions of how the bicycle became a Dutch product, and how the Dutch industry succeeded in dominating its home market.

According to economic historian Andrew Millward, after 1870 the UK became the leading bicycle-producing country. European follower countries copied bicycle production from the French and then from the British. Some countries like the United States, Germany and Belgium, had manufacturers in armaments and sewing machines, which gave them an advantage in this process. Other countries like Russia and Austria benefited from UK companies setting up factories there to avoid tariff barriers. These new industries had to compete with bicycles imported from the UK that had more status thanks to their superior quality. Consequently, they sold their bicycles at low prices.[18] Increasing competition and the economic crisis led to protectionist policies and by the 1930s, the Netherlands, Poland and Switzerland protected their markets with import quotas.[19] Historian Anne-Katrin Ebert concludes that people of all classes in the Netherlands perceived the bicycle as Dutch; in the early 1900s, the *Algemeene Nederlandsche Wielrijders-Bond* (ANWB, Dutch Cyclists Union) had transformed cycling into a collective nation-building activity to support political stability.[20] A study on the history of technology and business history by Sue-Yen Tjong Tjin Tai, Frank Veraart and Mila Davids shows how interactions between users, firms and intermediaries created the Dutch cycling nation. From 1920, its cartelized bicycle industry produced Dutch-style bicycles that everyone wanted and used—not only because these utilitarian vehicles were considered an integral part of Dutch culture, but also the Netherlands' infrastructure enabled cycling for utility as well as recreational purposes.[21]

From the start, the bicycle was a global good appropriated by local industries around the world. Colonial historian David Arnold demonstrates how

16 Gazelle published these advertisements around 1920. See: Hanenbergh and Röben, *Ons stalen ros*, 193.

17 Herlihy, *Bicycle*, 75–76.

18 Austria, France, Germany and Italy are examples of countries that built up their assembly industries by importing parts from the UK, thereby gaining experience to start up in the cycle industry. See: Millward, *Factors*, 239–240.

19 Ibid., 412.

20 Ebert, *Radelnde Nationen*.

21 Tjong Tjin Tai, Veraart and Davids, "How the Netherlands."

bicycles were assimilated and appropriated by India's industry and society. Both producer and user knowledge were shaped in this process of diffusion.[22] Historian Margrit Schulte Beerbühl illustrates how appropriation happened in a different context via German migrants. They introduced sausages from their native country to the UK, supported by relationships from their country of origin, which provided the relevant knowledge and supplies. In this process, the British had to adapt their tastes, while immigrant butchers had to adapt their recipes to local taste preferences.[23] Business historian Johzen Takeuchi concludes that small Japanese firms localized imported equipment by replacing brass and steel parts with locally available materials. This was the start of a learning-by-doing process which enabled these small firms to catch up with foreign industries.[24]

Innovation researchers characterize the diffusion of innovation as a learning process involving knowledge circulation. According to the economist Bronwyn Hall, diffusing innovation means learning, imitation and feedback that enhance the original innovation. She views diffusion as a critical process for technologically laggard countries and firms wanting to catch up. Her suggestion is that governments consider temporary regulations to protect infant industries.[25] In a historical study about Dutch firms and innovation, Mila Davids, Harry Lintsen and Arjan van Rooij conclude that around 1900, Dutch firms mainly innovated by copying products and processes from abroad or from other firms. According to the authors, this was part of an artisan innovation pattern, in which knowledge was tacit and experiential, transferred from master to apprentice during training in the workshop. Entrepreneurs acquired knowledge through colleagues, suppliers and customers by using local networks or learning from the knowledge embedded in artefacts, such as machines and products. To gain know-how and implement innovations from abroad, they made study tours or invited foreign experts for visits.[26]

This chapter builds on previous research by posing the question, how did the bicycle become a Dutch product between 1860 and 1940?, and the subquestion, how did Dutch bicycle firms develop their knowledge base and their skills? Whereas a previous paper by Tjong Tjin Tai et al. studied how the Dutch bicycle nation developed as a result of interactions between users, firms and intermediaries, this chapter will focus on firms, knowledge and materials.[27]

22 Arnold, *Everyday Technology*.
23 Schulte Behrbühl, "Migration, Transfer and Appropriation."
24 Takeuchi, "Historical Features," 207–208.
25 Hall, "Innovation and Diffusion," 459–460.
26 Davids, Lintsen and van Rooij, *Innovatie en kennisinfrastructuur*, 63–64.
27 Tjong Tjin Tai, Veraart and Davids, "How the Netherlands."

It will trace international and national flows of knowledge and materials to reconstruct how Dutch firms advanced their knowledge base and skills. Moreover, studying the diffusion of the bicycle in Dutch industry enables a comparison with cases of diffusion in other countries. To date, researchers have studied the invention and development of the bicycle, national bicycle histories and the rise of well-known bicycle manufacturers like Raleigh.[28] Studies of the Dutch bicycle industry are limited to commemorative books, publications of the Dutch veteran bicycle club *De Oude Fiets* (The Old Bicycle), and the above-mentioned paper by Tjong Tjin Tai, et al.[29] This chapter is based on archival research in the Dutch bicycle museum *Velorama*, the ANWB archives, the Dutch business association of main bicycle manufacturers and traders (RAI) and the government agency for small and medium sized enterprises (*Rijksnijverheidsdienst*, RND). Furthermore, we studied the literature from a wide range of sources including academic works, commemorative books and veteran bicycle clubs' publications.

The chapter consists of four parts. The first reviews the origins of the Dutch bicycle industry between 1860 and 1900. The second covers the period between 1900 and 1920, including the impact of the First World War and the major opportunities this created for the Dutch bicycle industry. The third part investigates the period between 1920 and 1940, when Dutch suppliers dominated the domestic market. The final part is a reflection and conclusion on the knowledge and material flows which developed the Dutch bicycle industry and helped make the bicycle Dutch.

## 2 Starting Up the Dutch Bicycle Industry, 1860–1900

When Pierre Michaux introduced the velocipede in Paris in 1867, Dutch wagonmakers and blacksmiths were already experimenting with human-powered vehicles.[30] Attracted by the possibilities of fast and animal-independent transport, wagonmakers and blacksmiths in Europe copied models from each other

---

28 Other studies include: Herlihy, *Bicycle*; Lloyd Jones and Lewis, *Raleigh*; Millward, *Factors*; Rosen, *Framing Production*; Steele, *Betting on the Wheel: The Bicycle and Japan's Postwar Recovery, 1945–1958*.

29 For a review of the historiography of Dutch bicycle studies, see Stoffers and Oosterhuis, "Ons populairste vervoermiddel." There are two academic bicycle studies in addition to the paper by Tjong Tjin Tai, et al.: an MSc thesis and a monograph. See: Veraart, *Geschiedenis van de fiets*; Vinne, *Eysink*. A recent, well-illustrated overview of Dutch bicycles and the Dutch cycling history for a wide public is by Hanenbergh and Röben, *Ons stalen ros*.

30 Herlihy, *Bicycle*, 75–76.

and made local improvements.[31] Around 1850, a Dutch manufacturer of instruments and binoculars built tricycles.[32] About fifteen years later, blacksmith Durk Driebergen built tricycles inspired by a picture smuggled from London by a local fisherman. These machines were manually powered by a construction of rods. Users tested the vehicles in races organized by local pubs.[33]

Immediately after Michaux introduced his pedal-driven bicycle, users and traders brought it to the Netherlands, where local blacksmiths and wagonmakers copied the new vehicle. Michaux's velocipede had an iron frame and wooden wheels with iron tyres.[34] It was not difficult for experienced blacksmiths and wagonmakers to replicate this. Moreover, except in the USA, the velocipede was not patented, which further stimulated diffusion to other countries.[35] Besides, there was no patent law in the Netherlands.[36]

It is not possible to give a quantitative overview of the Dutch bicycle industry in these years due to a lack of statistics. Instead, some estimates can give us an indication. In the first decades after 1860, Dutch users preferred to buy foreign bicycles because of their reputation and superior quality. At first, Dutch traders imported bicycles from France, and then mainly from the UK. Dutch production must have been very limited, in the order of several hundred bicycles per year.[37] Imports of bicycles were probably in the same range. Cycling started as a pastime for young, adventurous and well-to-do males. In the early 1890s, British manufacturers introduced the safety model and a number of other innovations, like pneumatic tyres and improved brakes and gears. This attracted a wider group of users to cycling, assuming that they could afford the still relatively expensive vehicles. Consequently, cycle sales and production boomed all over the world.[38] Several major Dutch manufacturers then increased their production or built new facilities. So, from the mid 1890s, the estimated Dutch production capacity was around 20,000 bicycles per year.[39] To

31 Ibid., 15–71.

32 His name was Hommema. See: Jongsma and Pijper, *Wielrennen in Friesland*, 10.

33 Ibid.

34 Herlihy, *Bicycle*, 75–96.

35 Ibid., 102–126. US manufacturer Pope succeeded in holding all main patents and a monopoly of the American market. See also: Epperson, "Failed Colossus"; Herlihy, *Bicycle*, 189–204, 208–209, 229; Norcliffe, "Popeism and Fordism."

36 From 1867 till 1912, Dutch law did not recognize patents. See: Agentschap NL, *Een terugblik*; Gerzon, *Nederland*.

37 Based on early production statistics from Burgers, a Dutch manufacturer (122 bicycles in 1873). See: Berlage, *De Fiets*, 21.

38 Herlihy, *Bicycle*, 225–282.

39 This estimate is an extrapolation of the production capacity overview in Table 4 in: Tjong Tjin Tai, Veraart and Davids, "How the Netherlands," 9.

put this in perspective: the UK produced 750,000 bicycles in 1896, Germany's production was 350,000 in 1897.[40]

In the late 1890s, traders also imported cycles from the United States and Germany.[41] The only export statistics for this period are from the UK. In 1893, the UK exported 11,106 cycles to the Netherlands, which increased to 25,820 in 1896 and dropped to 6,445 in 1899.[42] These were considerable numbers, taking into account that the total number of bicycles in the Netherlands was 113,000 in 1900.[43] In 1897, the end of the bicycle boom led to an overcapacity in American cycle factories. The international bicycle industry then had to compete with a flood of low-priced American bicycles, resulting in new challenges for bicycle manufacturers.[44]

An example of a Dutch blacksmith and wagonmaker who started bicycle production was Henricus Burgers (1843–1903). Around 1869, he read about the velocipede in the newspapers and decided he might start making these wooden vehicles as well. Burgers was further stimulated by a well-known local who learnt cycling and then ordered a velocipede from him. After that, more orders followed.[45] Several of these orders came from sons of the local elite in Deventer who also founded one of the first Dutch cycling clubs in 1871.[46] Finally, the increasing demand made Burgers decide to start a bicycle factory, which produced 122 bicycles a year in 1873 already.[47] His first bicycles had an iron frame, wooden wheels without bearings and a saddle connected to a blade spring.[48]

Dutch bicycle entrepreneurs not only consisted of blacksmiths, wagonmakers and traders. Several new users became so engrossed and enthusiastic about cycling that they decided to turn their favourite pastime into a business. These entrepreneurs were of course very young, as were most users. Some of them were also involved in the national cycling union ANWB, which was

40 Millward, *Factors*, 159–160, 265.

41 Ibid., 238–243.

42 Department of Trade and Industry (UK), *Annual Statement (1892–1937)*.

43 Centraal Bureau voor de Statistiek, "De Nederlandsche rijwielmarkt," 21.

44 Millward, *Factors*, 238.

45 There are several anecdotes of how Burgers started, see http://www.rijwiel.net/fietsfabrikanten_n.htm. This is one of the more neutral versions, from: Eigen Haard no. 24 (1898): 266.

46 "Een en ander over de Burgers-fabriek." *De Kampioen* (1901): 727.

47 Berlage, *De Fiets*, 21.

48 For details of the early Burgers bicycle: "Leeuwarder Jeugd Herinneringen Hendrik Burger (1864–1957)", an unpublished manuscript on the childhood of doctor and professor Hendrik Burger, https://historischcentrumleeuwarden.nl/component/content/article/11-import/356-hendrik-burger.

established in 1883.[49] H.W. Bayer, at only 19 years of age, started up a firm importing high wheelers from the UK in 1880.[50] The Englishman D. Webster, a founding member of the ANWB, became an agent for Singer, Rudge and Bayliss Thomas bicycles in 1884. G.C. Lelsz, the ANWB secretary, was an agent for Rudge and Enfield cycles. Charles Bingham, the first ANWB chairman, founded the Simplex bicycle factory in 1887, at the age of 26.[51]

Traders and producers also promoted cycling in order to sustain and expand their business. Around 1870, retailer H.H. Timmers began renting bicycles to make these expensive vehicles more accessible to new users. He also founded a bicycle riding school where new users could practise before hitting the streets of Amsterdam.[52] The combination of bicycle sales and cycling school was common until the early twentieth century. Manufacturer Burgers also rented bicycles to customers to stimulate sales.[53]

Some entrepreneurs started importing and trading bicycles and then extended their business with bicycle production. From 1868, trader J.T. Scholte (1827–1891) imported velocipedes from Paris to Amsterdam. In the same year he started producing these in his metal works.[54] B.A. Jansen, a sportsman who began trading bicycles at the age of 20 in 1881, was an excellent networker and salesman: he advertised, participated in exhibitions, sponsored bicycle races and sportsmen, and built up a large retail network. Like some other traders, he started producing his own brand of bicycles, Hecla, which were assembled in Birmingham. He traded in bicycle parts from various countries, such as Dunlop tyres and Westwood rims from the UK, parts from Neckarsulmer Fahrradwerke AG (later named NSU Germany), and his own BAJ brand of fittings and hubs. Furthermore, Jansen exported to France, Germany, India and South Africa.[55]

One of the main challenges facing the Dutch cycle industry was its lack of manufacturing capabilities. The UK cycle industry could rely on experienced engineers and skilled craftsmen who had earned their credits in the sewing machine and armament industry. They applied their skills to improve the bicycle's roadworthiness.[56] In the Netherlands there was no equivalent for this

49 Ebert, "Cycling towards the Nation"; Linders-Rooijendijk, *Gebaande wegen*, 350.

50 Hogenkamp, *Een halve eeuw wielersport*, 541.

51 http://www.rijwiel.net/simplexn.htm; Linders-Rooijendijk, *Gebaande wegen*, 397–399.

52 *40 Jaar—uit het archief der Kampioen*, 1; Hogenkamp, *De geschiedenis van Burgers*, 39.

53 H. Kuner, Fietsfabrikanten in Nederland in de 19e eeuw (2014), http://www.rijwiel.net/fietsfabrikanten_n.htm

54 Ibid.

55 Steendijk-Kuypers, Koos. "Bernardus Antonius Jansen (1861–1932), een pionier in de fietsbranche." *De Oude Fiets* no. 1 (2008): 4–11; https://nl.wikipedia.org/wiki/BAJ_Rijwielen.

56 Herlihy, *Bicycle*, 161.

type of machine industry nor did it have sufficiently up-to-date technical education for industrial production in the late 1800s.[57] ANWB magazine *De Kampioen* ('The Champion') described Dutch bicycle production in these early decades as "blacksmiths in obscure workshops (who) ordered parts and assembled, enamelled and nickel-plated bicycles without understanding how to deliver quality".[58] Nevertheless, the ANWB supported bicycles produced in the Netherlands, provided their quality was good and they competed with foreign bicycles on their own merits, without quotas or import tariffs.[59]

Dutch bicycles could not compete with the quality of UK imports and were sold at lower prices.[60] Traders used this quality difference in an opportunistic way, as illustrated by manufacturer Burgers' complaint about his former agent J.W. Koopman in 1885. According to Burgers, Koopman used Burgers' bicycles as scapegoats to sell more expensive English bicycles by referring to the vehicles' different origins. Interestingly, Burgers admitted that the quality of English bicycles was higher, however, he found it "not chivalrous to start up with my (bicycles) and later denigrate them".[61]

Between 1860 and 1890, bicycle design and production underwent many changes: innovations by UK producers transformed the cycle from velocipede, to high-wheeler then safety cycle, with many variations along the way. These changes also had an impact on the Dutch producer base. Some early producers did not continue when all-metal bicycles and high wheelers succeeded the original velocipede in the 1870s. The abovementioned J.T. Scholte only produced velocipedes for about two years.[62] Another example is blacksmith Hendricus Johannes Simons Jr., who presented himself as a velocipede manufacturer in Arnhem for three years, from 1879 to 1882. After that he reverted to producing carriages and decorative ironwork.[63] One explanation for these

57 Dutch industrialization was late and mainly took place in textile production, the food sector and heavy machinery (shipbuilding). See: Sluyterman, *Dutch Enterprise*, 23, 26–31. Informal training was more important for entrepreneurs than formal education, see: Gerwen and Goey, *Ondernemers in Nederland*. The Dutch government did not see formal technical education as a state responsibility, see: Lente, *Techniek en ideologie*, 81–122. Dutch professional engineers entered the industry around 1900, see: Lintsen, *Ingenieurs in Nederland*, 349–354.

58 *De Kampioen*, February 1887, April 1887, November 16, 1894.

59 *De Kampioen*, March 3, 1893, January 19, 1894, March 2, 1894.

60 *De Kampioen*, April 1887.

61 ANWB Archives (hereinafter: ANWBA), Box BE, Correspondence book Edo Bergsma, Letter from H. Burgers, January 16, 1885.

62 H. Kuner, Fietsfabrikanten in Nederland in de 19e eeuw (2014), http://www.rijwiel.net/fietsfabrikanten_n.htm.

63 Ibid.

brief ventures is that commercial prospects were limited, as the bicycle was still a very expensive product without utilitarian uses. Another reason could be that new models and technical innovations were beyond the capabilities of early Dutch manufacturers.[64]

The Dutch extended their knowledge and skills through training and internships in UK factories, visits to exhibitions, and hiring skilled craftsmen from that country. J.W. Koopman worked for four years in UK bicycle factories before establishing a repair workshop in the Netherlands in 1886. From 1889, he manufactured his own Sint Bavo bicycles with American and British machinery. In advertisements, he proudly referred to his work experience in the UK.[65] Twenty-year old sportsman B.A. Jansen opened his bicycle trading and retailing firm in 1881, after he became fascinated by bicycles while visiting an exhibition in London.[66] It was probably during another UK exhibition that he became acquainted with the newly established Sparkbrook Cycle Company. He then started importing Sparkbrook's vehicles from 1883. Another manufacturer, H.A. Samuels, hired UK craftsmen to produce bicycles. He began importing bicycles from the UK in 1886, and founded a bicycle factory in Amsterdam in 1887. To prevent being associated with low quality Dutch products, he advertised that he had UK craftsmen in his factory.[67] Samuels was also the first Dutchman to exhibit his bicycles at the Stanley bicycle club show in London, the world's leading bicycle exhibition. His products were admired for their unexpectedly high quality.[68]

Another way to catch up in the bicycle manufacturing process was by learning from using imported parts. This was also an opportunity to improve the original design as needed. In this way, Dutch manufacturers used their existing skills and knowledge to learn bicycle production. Bicycle manufacturer Fongers began production in 1884, using mainly parts from the UK.[69] Manufacturer Burgers did the same. However, in the 1880s he had difficulty procuring these parts, so he improvised and used gas piping to make the backbone for high

---

64 Herlihy, *Bicycle*, 144, 161–163.

65 *De Kampioen*, November 1886, May 1889, July 1, 1889, March 1, 1895.

66 Stadsarchief Den Bosch, Aart Vos, "Geschiedenis van B.A. Jansen"; Steendijk-Kuypers, Koos. "Bernardus Antonius Jansen (1861–1932), een pionier in de fietsbranche." *De Oude Fiets* no. 1 (2008): 4–11; https://nl.wikipedia.org/wiki/BAJ_Rijwielen. Note that these sources have inconsistencies, as Sparkbrook only started in 1883 and according to https://www.gracesguide.co.uk/1880_Stanley_Cycle_Show, the 1880 show took place in Holborn Town Hall, not Crystal Palace.

67 *De Kampioen*, November 1886, 188, February 1887, April 1887, May 1887, September 1888.

68 "The Stanley exhibition of cycles, 1890." *The Engineer*, February 14, 1890, 138.

69 Brusse, *De Groninger rijwielenfabriek*, 23.

wheelers. He even tried to make ball bearings.[70] Later, when Burgers copied the Facile bicycle, an 1881 British patented model, he added Aeolus ball bearings which were absent in the original design.[71] Burgers and Fongers were able to expand their facilities. Many other firms continued producing bicycles by assembling purchased parts.

In the last decade of the nineteenth century, the Dutch bicycle industry had several mature and skilled producers. Top sales years in 1894 and 1895 attracted investors to this industry, so bicycle manufacturers were able to expand their facilities and build new factories. In 1896, Burgers increased production by 5,000 to 6,000 bicycles a year, Simplex added 5,000 bicycles to its production capacity and Hinde built a new factory. In 1897, Eysink and Gruno built new factories, while Fongers increased its production by 2,000 to 3,000 bicycles a year.[72] These manufacturers installed machinery from the US and the UK, also copying these countries' production methods and organization.[73] A professional engineer managed the production of Simplex bicycles.[74] Burgers had about 200 employees in his factory, producing its own tyres, saddles and wheels in several workshops.[75] Fongers also had a tool workshop.[76] These facilities enabled mass production by mechanization, standardization and task division. The ANWB magazine concluded that "Bicycle production had become an established and mature industry (…) The Dutch did not have to go abroad anymore to buy good quality bicycles".[77]

A spin-off of the bicycle boom in the 1890s was the establishment of 'The Bicycle Industry' business association *Vereeniging 'De Rijwiel-Industrie'* (RI).[78] This was set up because the boom had led to an increasing number of bicycle clubs, which all organized exhibitions where manufacturers and traders were expected to show up with a booth. As this became too expensive, six bicycle

70 Zellekens, Willy A. "De Nederlandsche rijwiel-industrie." *Eigen Haard*, no. 24 (1898): 266; "Een en ander over de Burgers-fabriek." *De Kampioen* (1901): 727.

71 ANWBA, Box BE, Correspondence book Edo Bergsma, Letter from H. Burgers, January 16, 1885. Herlihy, *Bicycle*, 218.

72 http://www.rijwiel.net; *De Kampioen*, July 3, 1896; Brusse, *De Groninger rijwielenfabriek*; Hogenkamp, *De geschiedenis van Burgers*; Vinne, *Eysink*.

73 Bingham, Burgers, Bayer, Simplex, Fongers and Koopman had UK and US machinery, see: *De Kampioen*, March 1, 1895.

74 *De Kampioen*, November 16, 1894.

75 Zellekens, Willy A. "De Nederlandsche rijwiel-industrie." *Eigen Haard*, no. 24 (1898): 245–249, 263–266.

76 http://fongers.net/bedrijf/.

77 *De Kampioen*, November 16, 1894.

78 Jong, *Geschiedenis eener Nederlandsche vereeniging*.

manufacturers and twelve traders decided to establish the RI in 1893.[79] The RI had an exclusive advertising arrangement with the ANWB and its members gave customers who were also ANWB members a six-month guarantee-period. It organized its first sales exhibition in 1895, which became an annual event from 1899.[80]

To summarize, in 1900, about thirty years after the introduction of the velocipede, the Netherlands had a diverse bicycle industry that included a number of mature manufacturers and a business association: RI. Dutch producers followed the UK industry. They developed their capabilities by importing bicycles and parts, first from France and then from the UK, using these in assembly and production or by copying them. They also learnt by travelling to the UK for exhibitions and training purposes, and by inviting skilled UK craftsmen to the Netherlands. Dutch producers and traders marketed their products by renting bicycles, operating bicycle riding schools and sponsoring races. In the twentieth century, new challenges and opportunities arrived with the advent of the automobile and downward pressure on bicycle prices.

## 3 Increasing the Dutch Production Base, 1900–1920

Between 1900 and 1920, cycling became an activity for all Dutch classes. Bicycle ownership rose from 2 per cent of the population in 1900 to 12 per cent in 1919, increasing the number of bicycles from 113,000 to 861,000.[81] However, bicycles were still relatively expensive. The ANWB continued building a touring infrastructure and educating cyclists about appropriate behaviour in traffic.[82] The First World War created an opportunity for the Dutch bicycle industry to expand and gain market share from the UK and Germany.

The scarcely available statistics show that in these decades, both domestic production and imports increased to fulfil Dutch demand. Whereas the Netherlands had an estimated annual production of 20,000 bicycles around 1900, this increased to about 100,000 bicycles in 1920.[83] The number of import-

79 The six manufacturers were: Burgers, Fongers, Simplex, Eysink, Bayer and Hinde.

80 Stadsarchief Amsterdam, 1302 (RAI Archives), inventory number (inv.no.) 115, Minutes of meeting exhibition committee, December 7, 1898; Inv. no. 116, Minutes of meeting RI members, October 26, 1899; *De Kampioen*, May 3, 1895, 339, March 2, 1900, 179.

81 Centraal Bureau voor de Statistiek, "De Nederlandsche rijwielmarkt," 21.

82 Linders-Rooijendijk, *Gebaande wegen*, 161, 236.

83 The Netherlands produced 117,000 bicycles in 1922, see: Centraal Bureau voor de Statistiek, "De Nederlandsche rijwielmarkt," 23. Ebert quotes 49,300 bicycles in 1919, and around 60,000 bicycles in 1920, see: Ebert, *Radelnde Nationen*, 293, 434.

ed cycles increased even more: from an estimated 10,000 cycles around 1900 to about 160,000 bicycles in 1920, mainly from Germany.[84] In the period before 1914, Germany took over the UK's role as main supplier of foreign bicycles, but not as supplier of parts. Between 1907 and 1913, imports of German bicycles increased from 26,723 to 49,056 per year, whereas imports from the UK remained more or less constant, between 4,203 and 6,416 bicycles per year, with a peak of 10,521 in 1910. The UK partially compensated this by exporting more parts. The value of parts imported to the Netherlands from the UK doubled between 1907 and 1913, from 111,171 to 235,275 pound sterling. Imports of German parts showed less growth, and were on average 4,287,000 Marks between 1907 and 1913.[85]

During this period, two changes had a major impact on the global bicycle industry. First, the introduction of a new mobility technology, the automobile, which made cycling a more mundane activity, as elite and well-to-do cyclists changed their interests to motorized vehicles. However, in sharp contrast with many other countries, in the Netherlands, cycling remained a suitable mode of transport for all classes, primarily thanks to the activities of the ANWB.[86] At the same time, bicycle entrepreneurs extended and transferred their activities to the new automobile industry.[87] Second, the bicycle bust around 1897 and the ensuing dumping of heavily discounted American bicycles were soon followed by equally low priced UK and German bicycles into the open Dutch market. Both changes forced Dutch bicycle producers to lower their prices.

The increase in sales stimulated more entrepreneurial activities in the Netherlands, which also made the bicycle more Dutch. New Dutch entrepreneurs started producing bicycles, bicycle parts and accessories, because they saw a growing domestic market. The changes also affected bicycle design and production due to price pressures. Originally, the bicycle was an expensive product made with attention to detail, whether produced by small or large firms. Lower prices forced manufacturers to reconsider bicycle design and include simpler, less costly details.[88] Consequently, there was no place any

84 The number of imported German bicycles is extrapolated. In 1922, the Netherlands imported 166,000 bicycles, of which 147,000 were German. See: Centraal Bureau voor de Statistiek, "De Nederlandsche rijwielmarkt," 25–26.

85 Department of Trade and Industry (UK), *Annual Statement* (*1892–1937*); Statistisches Reichsamt, *Der auswärtige Handel* (*1897–1913*).

86 Ebert, *Radelnde Nationen*.

87 Two examples of Dutch bicycle entrepreneurs who extended their activities to the automobile industry: Eysink and B.A. Janssen. This was a global trend, as demonstrated by other manufacturers: Pope (USA), Rover (UK), Sparkbrook (UK), Opel (Germany).

88 Kuner, Herbert. "De rijwielhandel in de jaren '20." *De Oude Fiets*, no. 4 (2005): 6–11; Tjong Tjin Tai, Veraart and Davids, "How the Netherlands," 14.

more for artisan bicycle manufacturers.[89] The availability of more domestic suppliers of bicycles, parts and fittings and the growing Dutch user base who preferred the bicycle for touring and utilitarian purposes, also gave it a more Dutch character.[90] This eventually evolved into the Dutch style bicycle, a local variation of an original British model.[91]

The First World War effectively halted bicycle imports, first from Germany and later from the UK, thereby stimulating Dutch entrepreneurs to fill this gap by producing bicycles, parts and fittings.[92] Imports of bicycles from the UK dropped from 6,730 in 1914 to 668 in 1917 while the number of German bicycles dropped from 49,056 in 1913 to zero during the war.[93] The First World War also had another positive effect on sales of Dutch cycles. Buying Dutch products became a patriotic act. In order to stimulate this behaviour, manufacturers could request a VNF hallmark to prove that their products were manufactured in the Netherlands.[94]

The increasing popularity of cycling also created business opportunities for repairmen. Early cyclists touring in the countryside who suffered a breakdown originally resorted to the local blacksmith for repairs or sent their bicycle to the manufacturer. From 1895, cyclists could use ANWB aid boxes. These boxes, which were installed at pubs and restaurants, contained tools and materials for cyclists to fix their vehicle and take care of their personal injuries.[95] Meanwhile, entrepreneurs started local repair workshops, so that by the early 1900s, the aid box was no longer necessary, as "there was a bicycle repairman in every village".[96] In the early days, repairmen needed several types of wooden and steel rims, tyres, spokes, handlebars, brakes, wheels and chains, which made repairs

89 For example bicycle manufacturer Schwager made high quality bicycles including self-constructed frames. See: http://www.rijwiel.net/schwager_n.htm.

90 Three examples: Durabo, established in 1898, produced carriers. Hopmi, established in 1917, produced a patented bicycle lock and other parts and accessories. Blacksmith Boddendijk and tanner Ohmann successfully manufactured saddles from 1915, but struggled after WW I. See: Van der Horst, Jos and Herbert Kuner. "Durabo." *De Oude Fiets*, no. 3 (2005): 3–9; Kuner, Herbert. "De Hollandse Patent Metaal Industrie." *De Oude Fiets*, no. 4 (2002): 4–8; http://www.encyclopediedrenthe.nl/Ohmann%20&%20Boddendijk.

91 Ebert, *Radelnde Nationen*; Tjong Tjin Tai, Veraart and Davids, "How the Netherlands," 15, 17.

92 See the note above for two examples: Hopmi and Ohman & Boddendijk.

93 Department of Trade and Industry (UK), *Annual Statement (1892–1937)*; Statistisches Reichsamt, *Der auswärtige Handel (1897–1913)*.

94 Tjong Tjin Tai, Veraart and Davids, "How the Netherlands," 17.

95 Linders-Rooijendijk, *Gebaande wegen*, 145–147, 150–151.

96 ANWBA, Box 1097, Map 12–131, 1.03.31, File Rijwielhulpkist, Letter to the ANWB Executive Committee, J.C. Redelé, April 11, 1914.

expensive, difficult and cumbersome.[97] Bicycle innovation made their work easier: the safety model became the standard and bicycle parts were standardized. However, competition was fierce. Repairmen, who often combined their workshop with bicycle sales, had to deal with price-cutting, bicycle auctions and the unreliable workmanship of other repairmen. All of this tarnished their reputation. That is why in 1903, repairmen and retailers established the *Bond van Rijwielherstellers en—Handelaren in Nederland* (BRHN, Dutch Association for Bicycle Repairmen and Retailers) to regulate the competition and improve their reputation by stimulating craftsmanship.[98] The BRHN conducted its first activities in an aggressive style. In order to prevent supplying bicycles to non-BRHN members, the BRHN threatened to blacklist uncooperative members of the RAI (formerly RI).[99] Relations with the RAI were thus suspended and it took until 1910 before they agreed to cooperate.[100] The BRHN had similar difficulties in building relations with the ANWB—its members only received ANWB recognition for repairmen in 1913.[101]

The First World War improved the cooperation between the bicycle industry's business associations. Until the war, the Dutch government followed a policy of minimum economic intervention. During the war, the Netherlands remained neutral, so in order to ensure political and economic stability, it had to instate regulations for distribution and pricing. It also established an organization to prevent Dutch firms from selling imported raw materials and products to Germany.[102] It therefore became more difficult for Dutch bicycle firms to do business, due to material shortages and new regulations. For example, in 1916, imports of tyres stopped and a government committee set maximum prices to prevent speculation.[103] As a result, bicycle firms used their business association to negotiate with governmental bodies, committees and other business associations to overcome these difficulties.[104]

---

97 Bond van Rijwiel- en Motorhandelaren in Nederland, *Een kwart eeuw organisatie.*, 13–15.

98 Later renamed *Bond van Rijwiel- en Motorhandelaren in Nederland* (BRHN, Association of Bicycle and Motorcycle Retailers in the Netherlands). For example, the BHRN regulated the competition by agreeing standard price lists. See: *Metallicus: Orgaan van den Bond van Smedenpatroons in Hollands-Noorderkwartier* 1, no. 50 (1904); ibid.

99 In 1900, the RI renamed itself *Nederlandsche Vereeniging 'De Rijwiel- en Automobiel-Industrie'* (RAI, Dutch Association 'Bicycle and Automobile Industry') to reflect the changes in mobility technology.

100 Bond van Rijwiel- en Motorhandelaren in Nederland, *Een kwart eeuw organisatie*, 47; Jong, *Geschiedenis eener Nederlandsche vereeniging*, 40–41, 45.

101 Bond van Rijwiel- en Motorhandelaren in Nederland, *Een kwart eeuw organisatie*, 55.

102 This was the *Nederlandsche Overzee Trust Maatschappij* (Dutch Overseas Trust Company).

103 Linders-Rooijendijk, *Gebaande wegen*, 162.

104 Jong, *Geschiedenis eener Nederlandsche vereeniging*, 46–48.

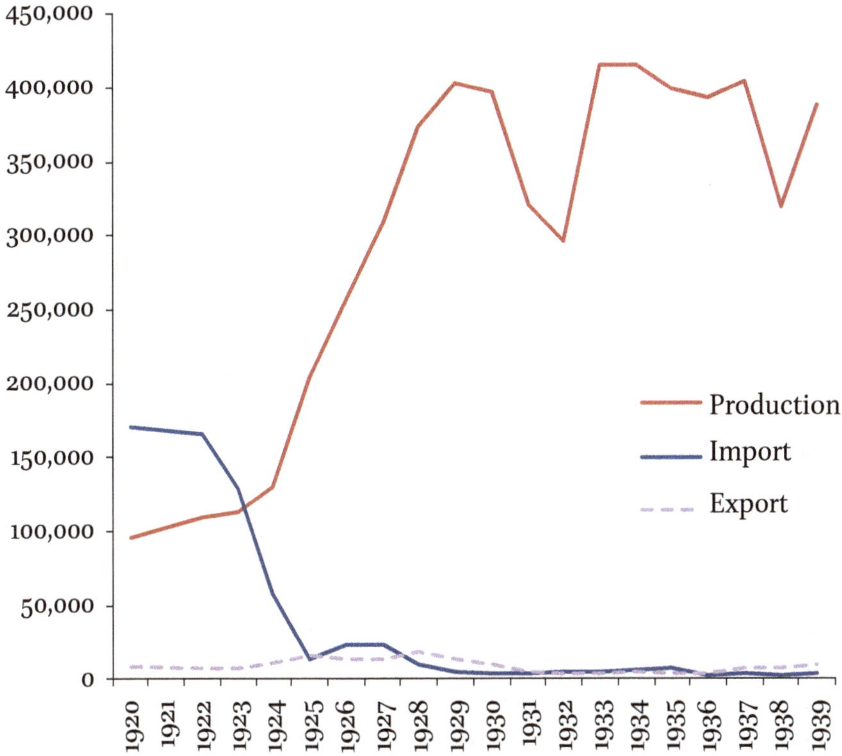


GRAPH 5.1 The Netherlands annual bicycle production, in number of bicycles, 1920–1939
SOURCES: CENTRAAL BUREAU VOOR DE STATISTIEK, "DE NEDERLANDSCHE RIJWIELMARKT."; DALMULDER, "DE VRAAG NAAR RIJWIELEN."

Dutch bicycle firms continued learning as in the previous century. The main differences were not only the circulation of Dutch frames or parts and experienced local staff and craftsmen, but also they could rely more on education in vocational schools. According to manufacturer Simplex, their workers did not need vocational training, but it enabled them to learn their mass-production work faster. Only production managers and supervisors needed technical education.[105] New entrepreneurs still depended on their experience as a blacksmith, carriage maker, or salesman, or their cycling experiences and contacts.

If necessary, firms could hire experienced craftsmen and production managers from other Dutch bicycle factories to fill the skills gaps. When bicycle firm Union started production in 1911, it used Gazelle frames. The first Union bicycles were a failure due to lack of production knowledge and skills. Union then hired production managers from Burgers, Gazelle and Gruno to develop

105 Waerden, *Geschooldheid en techniek*, 91–94.

its production capabilities.[106] Others were self-taught. In 1896, carpenter Johan L.A. Seitzinger Sr. (1864–1943) was fascinated by the new bicycles, and learnt more about the technology through self-study. In 1901, he started his own bicycle firm, which first of all provided bicycles for hire, followed by a repair workshop. Later, he extended this to a construction and manufacturing workshop.[107]

The growing popularity of the bicycle, together with lower prices and the import shortage created by the First World War, resulted in a larger and more diverse Dutch bicycle industry. It consisted of major manufacturers like Burgers, Fongers and Gazelle, who mass-manufactured bicycles and produced their parts mainly in-house. On the other side of the spectrum were bicycle manufacturers and assemblers who depended entirely on bicycle parts from UK and German suppliers. In addition, they could purchase saddles, bicycle locks, carriers and fittings from new Dutch manufacturers as well as from foreign suppliers.

The Dutch bicycle industry did not receive government support. Although its agencies used bicycles for utilitarian purposes, the government did not show a preference for Dutch bicycles, because it wanted the right quality. Around 1900, the army purchased bicycles from Simplex (between 1897 and 1904) and Fongers (between 1904 and 1908). In 1915, a governmental committee investigation concluded that standard consumer bicycles were not fit for military purposes. The army needed more robust models, so started producing its bicycles in-house.[108] Another reason the government did not buy Dutch was cost. Initially the Dutch Postal Service had purchased Burgers bicycles. However, in order to save maintenance costs, in 1901 this service started renting bicycles from German manufacturer Brennabor, who won the tender.[109]

After the First World War, Dutch bicycle manufacturers continued their wartime cooperation in a cartel. During the war, the Dutch bicycle industry had flourished in the absence of foreign imports. Consequently, when German bicycles reappeared in the Dutch market at extremely low prices, Dutch entrepreneurs decided to protect their market position. Accordingly, the RAI, BRHN and the association of wholesale traders of bicycles and parts (NEVGRO)

106 Timmerman, *Rijwielfabriek Union*, 7–11.

107 Herbert Kuner, "Haagse fietsenmakers Käuderer en Seitzinger: twee fietsenmakers met passie (1/3)." *De Oude Fiets* no. 2 (2010): 18–27.

108 "De Simplex." *Nieuws van de Dag*, October 16, 1903; Rietveld, Jos. "Fongers 1884–1922." *De Oude Fiets* (2004); Murk, Rutger. "Naar de wielen! Militair gebruik van rijwielen in Nederland." *Het Rijwiel*, no. 1 (2012): 4–19.

109 *De Kampioen*, March 1, 1901, 182–183.

established a bicycle cartel in 1919.[110] It started operating in 1920 and controlled bicycle production and sales with quota and prices, also managing a list of recognized suppliers and retailers.[111]

To summarize, between 1900 and 1920, the Dutch bicycle industry became more diverse, mature and organized, thanks to the increasing popularity of the bicycle and the consequences of the First World War. The bicycle evolved into a utilitarian vehicle, with a more local character on account of the expanded Dutch supplier base providing parts. Nevertheless, diffusion of innovations and imports of bicycles, parts and fittings from the UK and Germany continued. Dutch business associations cooperated well during the First World War. They established a cartel in 1919, because after the war, market protection seemed to be the only way to make sure the Dutch bicycle industry survived.

## 4 A Cartelized and Dutch Bicycle Market, 1920–1940

Between 1920 and 1940, the bicycle became a mass consumer product. In spite of the economic crisis, bicycle ownership in the Netherlands increased from 861,000 in 1919 to 3,300,000 in 1939 for a population of 8.7 million.[112] Dutch bicycle ownership therefore increased from 12 to 38 per cent of the total population. Furthermore, from 1925 onward, bicycle imports were successfully blocked by the Dutch cartel, so that virtually all bicycles sold in the Netherlands originated from Dutch suppliers. To meet the increasing demand, Dutch bicycle production almost trebled between 1924 and 1928: from 139,000 to 374,000 bicycles per year (Graph 5.1). This was quite an achievement, as imports were about 170,000 a year between 1920 and 1923, at a time when domestic production was only about 110,000 bicycles. After 1928, production remained in the range of 300,000 to 400,000 bicycles a year.[113] This development led a Dutch engineer to conclude that the competition with imported cycles was over, as "the public now agrees that the Dutch cycle is a first-class product".[114]

The ANWB initially opposed the cartel, but the government did not intervene. ANWB chairman Edo Bergsma, who was also a member of the Senate of

110 *Nederlandsche Vereeniging van Grossiers in Rijwielen en Onderdelen* (NEVGRO, Dutch Association of Wholesalers in Bicycles and Parts).

111 *De Rijwiel- en Automobiel-Industrie*, June 1920, 7 and August 1910, 5–6.

112 Centraal Bureau voor de Statistiek, "De Nederlandsche rijwielmarkt," 21; Jong, *Geschiedenis eener Nederlandsche vereeniging*, 116.

113 Centraal Bureau voor de Statistiek, "De Nederlandsche rijwielmarkt," 25.

114 Erven Dorens, W.K. van, "De rijwielindustrie in Nederland." *Algemeen Handelsblad*, June 27, 1933.

ILLUSTRATION 5.1 Workshop, H.T. Specker bicycle manufacturer, Musselkanaal, around 1910
SOURCE: COLLECTION HANS SPECKER

the Dutch Parliament, spoke in the Senate in March 1921, defending the rights of cyclists to buy bicycles on an unregulated and open market.[115] The Minister replied that he did not believe the government could stop cartels in or outside the Netherlands. More importantly, he did not see retailers making high profits. He also expected the cartel to be advantageous for consumers, as it would ensure reliable retailers.[116]

Ultimately, the Dutch cartel created a profitable situation for both producers and consumers. It created business for Dutch bicycle firms, while not halting a drop in prices, so more people could afford a bicycle. Thus, the Dutch bicycle industry remained under pressure to keep production costs low. In 1922, the average price of a major manufacturer's bicycle was 89 guilders, compared to an agricultural worker's monthly income of 120 guilders. In 1935, a similar bicycle could be bought for 28 guilders, while incomes had hardly changed due to the crisis.[117]

115 *Handelingen Eerste Kamer*, 21ste vergadering, March 2, 1921.

116 *Handelingen Eerste Kamer*, 25ste vergadering, March 9, 1921.

117 Centraal Bureau voor de Statistiek, "De Nederlandsche rijwielmarkt," 23; Hoepen, ed., *Van veertig zegenrijke melkjaren*, 198–205.

The Dutch industry was able to rapidly upscale its output with bicycle assembly. The available import statistics show that, between 1924 and 1940, imports from Germany and the UK shifted from bicycles to parts. Cycle imports from the UK dropped from 7,629 in 1924 to 1,258 in 1938. Imports from Germany started at 44,289 cycles in 1924, and stopped completely after 1934 (Table 5.1).[118] Graph 5.2 shows that the Netherlands continued importing bicycle parts, in particular from Germany. The graph also shows that the number of parts imported from Germany follows the trend in Dutch bicycle production (Graph 5.2).[119] The data illustrates that Dutch bicycles consisted of a fair percentage of German parts and a much lower percentage of UK parts. According to Ebert, the Dutch imported about two thirds of their bicycle parts from Germany.[120] Thus, many "Dutch" bicycles were actually made up of foreign parts, assembled in the Netherlands.

TABLE 5.1 Dutch bicycle imports, in number of bicycles per year, 1924–1938

| Year | UK | Germany | Belgium | France | Japan |
|---|---|---|---|---|---|
| 1924 | 7,629 | 44,289 | 3,164 | | |
| 1925 | 5,210 | 9,499 | 2,127 | | |
| 1926 | 3,901 | 12,846 | 5,361 | | |
| 1927 | 3,492 | 19,306 | 2,045 | 216 | |
| 1928 | 3,498 | 6,763 | 1,692 | 53 | |
| 1929 | 2,296 | | 1,266 | 457 | |
| 1930 | 1,647 | 1,340 | 830 | 197 | |
| 1931 | 1,715 | 1,368 | 635 | 193 | 38 |
| 1932 | 3,725 | 427 | 455 | | |
| 1933 | 2,580 | 818 | 785 | 199 | |
| 1934 | 3,670 | | 686 | 220 | 1,370 |
| 1935 | 5,074 | | 822 | 12 | 587 |
| 1936 | 1,404 | | 660 | 15 | 285 |
| 1937 | 1,799 | | 479 | 20 | 360 |
| 1938 | 1,258 | | 760 | 55 | |

SOURCES: DEPARTMENT OF TRADE AND INDUSTRY (UK), *ANNUAL STATEMENT* (*1892–1937*); STATISTISCHES REICHSAMT, *DER AUSWÄRTIGE HANDEL* (*1922–1927*); DATASHEETS FROM MILLWARD, *FACTORS*.

118 Datasheets from Millward, *Factors*.

119 The difference in value between 1924 and 1925 has not been explained but may be due to fluctuating exchange rates.

120 Ebert, *Radelnde Nationen*, 299–300.

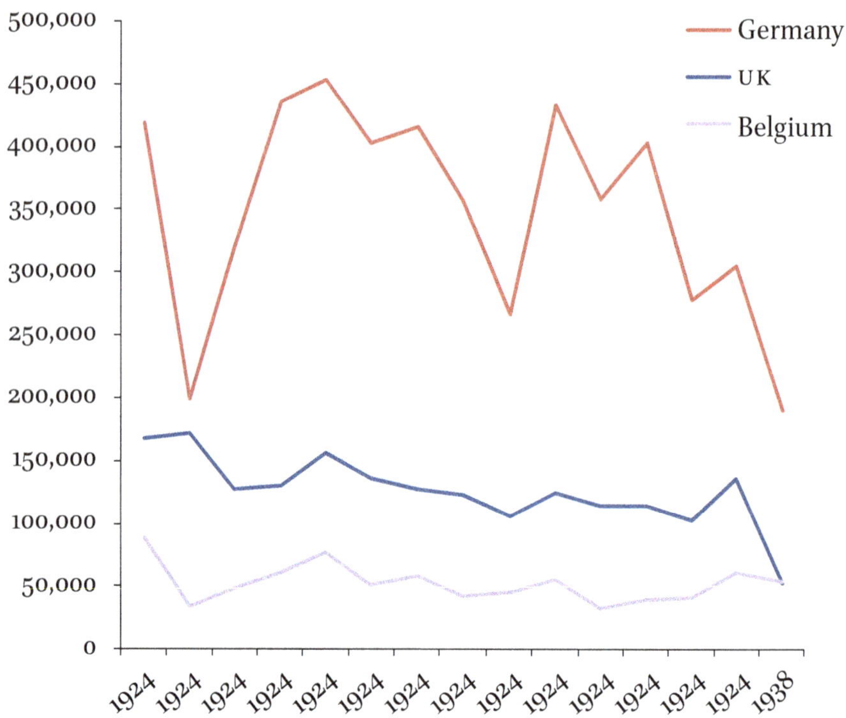


GRAPH 5.2 The value of bicycle parts in the Netherlands imported from Germany, the UK and Belgium, in pounds sterling, 1924–1938
SOURCES: DEPARTMENT OF TRADE AND INDUSTRY (UK), *ANNUAL STATEMENT (1892–1937)*; MILLWARD, *FACTORS*; STATISTISCHES REICHSAMT, *DER AUSWÄRTIGE HANDEL (1922–1927)*

Between 1925 and 1939, the Dutch bicycle industry assembled about 50 per cent of its output using mainly purchased parts.[121] These parts were imported or produced by Dutch manufacturers. Assembly was a competitive way of producing bicycles, as parts were standardized and cheap. Moreover, these low-priced vehicles attracted consumers who otherwise could not afford a bicycle. The downside was that these bicycles were less reliable due to low coating quality and less production control.[122] Furthermore, many assembly firms were small-scale, which concerned one Dutch engineer who felt this relatively inefficient and expensive process would prevent the Dutch bicycle

121 Dalmulder, "De vraag naar rijwielen."
122 Erven Dorens, W.K. van, "De rijwielindustrie in Nederland." *Algemeen Handelsblad*, June 27, 1933.

ILLUSTRATION 5.2 Workers in the grinding shop, Fongers bicycle factory, Groningen, Netherlands, around 1929
SOURCE: COLLECTION RHC GRONINGER ARCHIEVEN

industry from expanding into the export market.[123] There were 35 manufacturers (with an average of 52.8 employees) and 58 assemblers producing more than 500 bicycles a year by 1929. The larger manufacturers accounted for 51 per cent of the production (209,700 bicycles) and 22 per cent came from the larger assemblers (89,400 bicycles), so that the average large producer supplied 3,216 bicycles.

At least 200 smaller firms produced the remaining 105,900 bicycles.[124] Many of these small firms were actually repair shops that presented themselves as bicycle manufacturers, even though they assembled bicycles from purchased parts or packages. The fuzzy distinction between bicycle manufacturers, assemblers and repair shops is reflected in the number of Dutch bicycle brands. An inventory in Utrecht, a major Dutch city, between 1890 and 1940 revealed 41 brands, of which 21 were held by wholesalers and manufacturing firms,

123 Hoogstraten, "De rijwielindustrie in 1929," 681.

124 Ibid. In 1930, the Dutch Census counted 84 bicycle manufacturers (of which 42 per cent had between 1 and 10 employees) and 6,157 bicycle repair shops and retailers (100 per cent between 1 and 10 employees). See: Centraal Bureau voor de Statistiek, *Bedrijfstelling 1930*.

and 20 by repair shops. As the city had no major bicycle manufacturer, all the wholesalers and manufacturing firms depended on purchased frames and parts. For example, the oldest wholesaler in Utrecht, Willem Gerth, produced Fama bicycles from 1911 onward. He purchased frames from Burgers and a manufacturer in Amsterdam.[125]

The cartel was not supported by government regulation. Furthermore, not all bicycle firms upheld import tariffs or quotas because they depended to different degrees on imported bicycle parts. Major Dutch bicycle manufacturers lobbied for many years for higher tariffs and import quotas, but were not successful until 1935, in the midst of the economic crisis.[126] In 1935, import tariffs on bicycles and all parts were 10 per cent. Furthermore, import quotas dictated that the number of imported bicycles per quarter was not to exceed the number imported the previous year. For frames and parts, the import quotas were set at 50 per cent.[127]

As mentioned above, the cartel did not result in artificially high prices. The main reason was that, from the start, Dutch bicycles were already more expensive than imported ones. In 1923, the average price of an imported bicycle was 34 guilders as opposed to 75 guilders for a Dutch bicycle. Only in 1929 was this difference bridged: an imported bicycle was then 54 guilders against 53 guilders for a Dutch bicycle.[128] The cartel protected the Dutch industry so that it had time to learn how to reduce production costs. If the cartel had kept its prices high, it would have made fewer sales and been bypassed more.

The Dutch bicycle cartel became powerful thanks to its coverage and the rule that only cartel members could get supplies from other members. The three business associations in the cartel represented the entire Dutch bicycle industry, consisting of manufacturers and traders, wholesalers, retailers and repairmen. Repairmen who were usually also retailers, only sold Dutch bicycles, because this ensured that they could purchase spare parts and fittings from wholesalers and traders. As a result, owners of foreign bicycles found it difficult to get them fixed by Dutch repairmen. Similar restrictions applied to traders, who could not find retailers willing to sell foreign bicycles. Despite this strong coverage, Dutch manufacturers did not always keep to their production quotas and ran the risk of fines.

---

125 "Themanummer De Utrechtse rijwielhandel" *De Oude Fiets* (2008): 6–7, 12.

126 "De Tariefwet." *Algemeen Handelsblad*, August 24, 1924;

127 Tweede Kamer. Kamerstuk 153 ondernummer 3. 1932–1933; Tweede Kamer. Kamerstuk 355 ondernummer 1 (23 april 1935). 1934–1935; Eerste Kamer. Minutes of meeting, 57ste vergadering. July 18, 1935.

128 Centraal Bureau voor de Statistiek, "De Nederlandsche rijwielmarkt," 23.

ILLUSTRATION 5.3 Fongers bicycle model BDG for women, 1929
SOURCE: COLLECTION RHC GRONINGER ARCHIEVEN

The Netherlands was one of the few countries which successfully cartelized its bicycle market. In the 1920s and 1930s, confronted with international competition, price-cutting and economic crises, bicycle industries abroad also tried to establish cartels. This attempt failed in the UK because some manufacturers were not willing to stop supplying parts to non-members and major bicycle manufacturer Hercules refused to join the cartel. In Germany, manufacturers with a diverse product range in bicycles, typewriters, sewing machines and automobiles did not see the need to join. Another reason the cartel failed was the lack of support from retailers and repairmen.[129]

Cartelization and increasing mass production and standardization affected the Dutch bicycle model. It became a cheap, mass-produced vehicle based on a standard design, the so-called 'Dutch-style' bicycle. It had a standard British cycle frame of the early 1900s: a diamond frame for men and a loop frame for women. Details and accessories were not only influenced by national user preferences, but also by government regulations. For example, bicycle tyres were required to be 40 mm wide to enable comfortable cycling on bumpy roads. Black moleskin chain covers made the bicycle

129 Millward, *Factors*, 267, 379–404. Lloyd-Jones and Lewis, *Raleigh*, 225.

look attractive and protected people's clothes while they used the bicycle to commute, to visit friends and relatives and to tour. Government regulations specified the design of rear mudguards and the use of front and rear lights.[130]

TABLE 5.2 Dutch bicycle exports, in number of bicycles per year, 1924–1938

| Year | UK | Germany | Belgium | France | Dutch EI[a] | Other |
|---|---|---|---|---|---|---|
| 1924 | | 765 | | | 8,499 | 943 |
| 1925 | 191 | 956 | 61 | 44 | 13,114 | 684 |
| 1926 | 63 | 155 | 49 | 25 | 11,649 | 1,109 |
| 1927 | 183 | 158 | 30 | 6 | 11,196 | 869 |
| 1928 | 245 | 165 | 46 | 43 | 17,152 | 623 |
| 1929 | 60 | 183 | 51 | 18 | 11,383 | 975 |
| 1930 | 22 | 185 | 47 | 8 | 8,327 | 543 |
| 1931 | 17 | 115 | 38 | 15 | 4,484 | 563 |
| 1932 | 10 | 179 | 72 | | 1,973 | 443 |
| 1933 | 9 | 170 | 50 | 11 | 2,667 | 452 |
| 1934 | 14 | 128 | 72 | 14 | 2,671 | 682 |
| 1935 | 16 | 119 | 86 | | 2,355 | 614 |
| 1936 | | 99 | | | 2,392 | 935 |
| 1937 | 143 | 96 | 75 | | 6,102 | 1,004 |
| 1938 | 16 | 101 | 48 | | 4,948 | 1,503 |

a Dutch East Indies

SOURCE: DATASHEETS FROM MILWARD, *FACTORS*

The Dutch-style bicycle had one disadvantage for domestic industry: it limited its export volume. Export statistics show that the number of Dutch bicycles sold in other countries was in stark contrast with domestic production figures (see Table 5.2). The Netherlands mainly exported bicycles to one of its colonies, the Dutch East Indies. Exports ranged from a maximum of 17,152 bicycles per year in 1928 and a minimum of 1,973 bicycles in 1932. However, in 1928, the UK exported 17,537 bicycles to this colony and Germany exported 12,727 bicycles, as the Dutch could still not compete on pricing. This lack of competitiveness is also apparent in other statistics: the Dutch exported between 96 and 956 bicycles per year to Germany, between 9 and 245 bicycles to the

130 Kuner, "Fahrräder."; Millward, *Factors*, 330–340; Tjong Tjin Tai, Veraart and Davids, "How the Netherlands," 23–24.

UK, and between 30 and 86 bicycles to Belgium. Dutch exports to other countries were much lower.[131] Table 5.3 gives an overview of the 1938 export data. It illustrates that the UK, Japan and Germany had major international bicycle industries, with 576,458, 216,327 and 193,999 bicycles exported, respectively. They were followed by Italy, France and Sweden. In comparison, exports from Belgium, the Netherlands (6,616 bicycles), Czechoslovakia and Switzerland were negligible.

TABLE 5.3 Several countries' bicycle exports in 1938

| Country | Number of bicycles |
|---|---|
| UK | 576,458 |
| Japan | 216,327 |
| Germany | 193,999 |
| Italy | 52,015 |
| France | 27,347 |
| Sweden | 16,381 |
| Belgium | 9,710 |
| Netherlands | 6,616 |
| Czechoslovakia | 2,754 |
| Switzerland | 56 |

SOURCE: DATASHEETS FROM MILWARD, *FACTORS*

In addition to organising the cartel, adopting assembly production and producing a standard Dutch-style bicycle, Dutch firms continued their previous modes of learning. Manufacturers still looked to the UK bicycle industry for inspiration. For example, Eip Meerbeek, owner of the Durabo factory, introduced sports cycles with narrow 26 inch wheels, which he had noticed during his visit to the UK in 1933. Although the model did not sell very well, as it could not compete with the Dutch-style bicycle's 28 inch wheels which were the standard, the Dutch manufacturer Simplex followed Durabo in 1937.[132] Dutch manufacturers also copied other foreign producers. For example, in the 1930s, Dutch firms started producing electric lighting after this was developed in the

131 Millward, *Factors*.

132 Horst, Jos van der, and Herbert Kuner. "Durabo." *De Oude Fiets*, no. 3 (2005): 3–9.

UK, Germany and Switzerland.[133] Dutch manufacturers copied not only each other's bicycle models, but also construction details.[134] Sales and marketing were inspired by UK precedents. Following the UK industry, Fongers introduced new and cheaper models every year.[135]

A similar trend in copying and implementing existing knowledge took place in production. Firms implemented product and production innovations from elsewhere, illustrated by their requests for advice from government agency *Rijksnijverheidsdienst* (RND, Government Technical Information Agency). This agency employed consultants with a degree in mechanical engineering who advised firms in technical and business matters. Their reports show that bicycle firms asked for advice about electrical installation, sandblasting equipment, material selection and usage for shafts and cones, testing metals and chemicals, testing bicycle parts for tensile strength, information about a coating process to prevent rusting and wear, procurement of a soldering furnace, etc.[136]

RND also helped the BRHN to organize formal training for repairmen. As both RND and BRHN were concerned about the repairmen's level of technical skills and expertise, they organized training to enhance the repairmen's reputation in the early 1920s.[137] In addition, the BRHN circulated knowledge via its journal. It was not until 1937 that Dutch law required repairmen to hold a certification in trade and business skills before they could apply for a business permit. This was also to limit entry into the sector. Until then, repairmen had depended on vocational schools and learning-by-doing.

To summarize, by 1939, the Netherlands had a mass-production bicycle industry which mainly delivered low cost, standard, Dutch-style bicycles for a cartelized domestic market. The Dutch bicycle production base consisted of major manufacturers and assembly or production firms of various sizes. Assembly firms used many bicycle parts from Germany, so the Dutch bicycle was to a large extent foreign-made and assembled in the Netherlands. Innovations continued to be based on copying from the UK, from other manufacturers and imported parts. RND and BRHN activities formalized the technical advice and training.

---

133 Rietveld, Jos. "Ontwikkeling van de elektrische rijwielverlichting (1889–1936)." *De Oude Fiets*, no. 1 (1999): 3–7.

134 Tjong Tjin Tai, Veraart and Davids, "How the Netherlands," 21.

135 Veraart, *Geschiedenis van de fiets*.

136 Brabants Historisch Informatie Centrum (Noord-Brabant Provincial Archive, hereinafter: BHIC), 169, inv.no. 93, Maandverslagen Rijksnijverheidsdienst Deventer.

137 BHIC, 169, inv. no. 55, Verslagen Nijverheidslaboratorium Delft; Kuner, Herbert. "Käuderer en Seitzinger: Twee fietsenmakers met passie (deel 1)." *Het Rijwiel*, no. 2 (2010): 22.

## 5 Conclusion

From the time of its first invention, the bicycle was a global good. It was introduced in France and further developed in the UK. This chapter has sought to demonstrate how a process of diffusion of innovation and localization resulted in the production of a type of bicycle that was distinctively Dutch but which was also the product of a Dutch bicycle industry that relied heavily on copying and integrating parts from abroad. This process consisted of international and national flows of knowledge and materials. It was influenced by several external changes: in particular the shifting user base after the introduction of the automobile and downward price pressures due to the bicycle bust and the First World War.

The diffusion of the innovation process was three-fold: learning, imitation and feedback, as defined by Hall.[138] In the mid-nineteenth century, the Dutch industry's lack of technical capabilities and education facilities meant that Dutch firms had to learn by acquiring and importing knowledge, and by copying products and processes. Thus Dutch entrepreneurs visited exhibitions, trained in UK factories and invited UK craftsmen to their factories. Many entrepreneurs had existing technical capabilities and some formal education which they could apply while learning bicycle production in their own workshops. A similar on-the-job learning process took place for bicycle repairmen. Between 1860 and 1900, regular design and technical product changes in the UK industry forced Dutch manufacturers to update their skills and product range. Once the Dutch had several mature producers, knowledge also circulated between Dutch firms, for example through production managers hired by the new bicycle factories. Dutch manufacturers imitated French and British bicycles, either by copying bicycles from foreign producers, or by assembling purchased parts into bicycles. In their own way, Dutch manufacturers were contributing with their feedback to the overall bicycle innovation process because they needed locally made parts for original models, to compensate for the lack of parts supply. The Dutch bicycle sector also contributed to bicycle innovation by developing the Dutch-style bicycle, under the influence of Dutch user preferences and traffic regulations.

This pattern of innovation, whereby entrepreneurs rely mainly on personal networks, learn in the workshop, and copy products and processes from elsewhere, is a typical example of the artisan innovation pattern defined by Davids, et al.[139] Only around 1920 did more formal activities appear, especially

138 Hall, "Innovation and Diffusion," 459–460.

139 Davids, Lintsen and van Rooij, *Innovatie en kennisinfrastructuur*, 63–64.

from external parties: the business association BRHN organized some training activities and the government agency RND advised bicycle entrepreneurs on technical and business matters and supported the BRHN in organising training.

The changes in knowledge and material flows documented in this chapter illustrate both the diffusion of innovation and localization of the bicycle. The Dutch always depended on foreign producers, first from France, then the UK, and later Germany. After 1920, when the Dutch market was cartelized, local influence and innovations had the greatest impact. In the first decades, knowledge and materials mainly originated from France and the UK. Around 1900, when the Dutch bicycle industry had a number of mature manufacturers, more national circulation of knowledge and materials took place, through production managers and craftsmen. Although the cartel did not stop imports, it transformed bicycle imports from Germany and the UK into imported German parts. Dutch firms used these in the Dutch-style bicycle, together with Dutch parts, fittings and accessories.

The appropriation of the bicycle by the Dutch was not unique. Other countries took the same steps of copying bicycles and using imported parts. The uniqueness of the Dutch case may be the combination of continued bicycle use by all classes, and a bicycle industry that did not start from an existing industrial base, unlike countries such as the United States, Germany and Belgium. The latter factor enabled cartelization of the industry and a protected market, even though the government did not actively support this with tariffs or quotas. Whether this was a unique situation remains to be investigated through further studies. The value of an international perspective is already shown in this paper by comparing Dutch production data with international bicycle production. This led to the conclusion that although the Dutch dominated their market in the late 1930s, its scale and efficiency were low compared to the UK, Germany and Japan.

The development of the Dutch bicycle industry is not, therefore, a story of manufacturers expanding production because their product became more and more popular. It is a story about the diffusion of innovation, product transformation and cartelization. First of all the Dutch had to learn from foreign industries and survive in a global market that was continually putting pressure on prices. Second, the bicycle was transformed during the first decades of the twentieth century from an elite status product to a low-cost utilitarian vehicle for everyday use, stimulated by Dutch user organizations. Third, the First World War was a timely opportunity for the Dutch industry to capture its domestic market from foreign rivals, after which it cartelized. Through this chain of events and various international and national flows of knowledge and materials, firms in the Netherlands were able to make the bicycle a Dutch product.

## 6 Acknowledgements

This research was financially supported by the Dutch Ministry of Economic Affairs and Syntens Innovations Centre.

# PART 3

## *Bicycle in the Practices*

CHAPTER 6

# Betting on the Wheel: The Bicycle and Japan's Post-War Recovery

*M. William Steele*

## 1 The Bet

Bicycles are an integral part of everyday life and mobility in all corners of the world, but as this volume points out, their presence and their past often go unheralded. At one time, bicycle riding stood out: it was extraordinary. But, as the number of users grew in the early twentieth century, the bicycle became ordinary: a simple machine for everyday use that made it virtually invisible, even to those interested in mobility studies and urban planning. In Japan, one of the great bicycle nations of the world, studies of the history and social and economic importance of bicycle riding pale in comparison with work done on the automobiles and trains.[1] And the study of bicycle racing, a gambling sport in Japan, is so pale as to render it invisible.

Bicycles are nonetheless important to the history of modern Japan. In 1927, with more than five million bicycles on the road and a bicycle export industry generating over ¥200 million in revenue, the government declared 11 November to be "Bicycle Commemoration Day", describing Japan as "the foremost bicycle riding country in the world."[2] Bicycle usage continued to expand throughout the 1930s, despite the coming of the automobile: when Japan entered the Pacific War in 1941, there were some 8.6 million bicycles used by a population of roughly 70 million. Japan's pre-war bicycle industry also flourished: in 1937, for example, production exceeded one million units, nearly half intended for

---

1 For a general history of the bicycle in Japan, especially during the period 1868–1945, see M. William Steele, "The Speedy Feet of the Nation: Bicycles and Everyday Mobility in Modern Japan," *Journal of Transport History* 32, no. 1 (2010). In Japanese, the major work is Jitensha sangyō shinkō kyōkai, *Jitensha no isseki: Nihon jitensha sangyōshi* [The Bicycle Century: The History of Japan's Bicycle Industry] (Tokyo: Jitensha sangyō shinkō kyōkai, 1973). For the post-war period, see M. William Steele, "The Making of a Bicycle Nation: Japan", *Transfers* 2, no. 2 (2012), 70–94; Tatsuzō Ueda, *The Development of the Bicycle Industry in Japan after World War II* (Tokyo: United Nations University, 1981).

2 "Jūichi gatsu ichi nichi wa jitensha kinenbi—waga kuni te kotoshi ga saisho", *Doyō yūkan*, 10 October 1927.

© KONINKLIJKE BRILL NV, LEIDEN, 2019 | DOI:10.1163/9789004289970_007

export (primarily to China), making the bicycle Japan's number one machinery export commodity.[3] The war years, however, exacted a heavy toll. After defeat in 1945, Japan's major cities lay in ruins and some three million bicycles had been destroyed while the remaining vehicles were in poor shape.[4] As in other areas of Japan's economy and society, the post-war bicycle recovery was slow. Miyata, Dai Nippon Jitensha, Okamoto Jitensha and other pre-war bicycle-manufacturing giants were joined by an explosion of small bicycle workshops and factories eager to produce bicycles for peacetime Japan, but it was not until the end of the Allied Occupation of Japan in 1952 that production began to alleviate Japan's post-war "bicycle famine".[5] Twenty years later, in 1975, Japan was briefly the world's largest producer and exporter of bicycles, to be overtaken by China in the 1980s.[6] Japanese bicycle production peaked in 1990, with nearly eight million units, and has since declined, but the bicycle remains today an indispensable component of Japanese everyday mobility.

This chapter examines the rebirth of the Japanese bicycle industry in the immediate post-war period. At the same time, it seeks to reveal the hidden history of the role bicycle racing played in Japan's post-war reconstruction. Taking advantage of correspondence between bicycle race planners and officials of the American-led occupation of Japan, it focuses on the origin of promotional subsidies derived from legalized betting at bicycle races (*keirin*), a track cycling event that originated in Japan in 1948. Between 1949 and 1978, the bicycle

---

3 *Jitensha no isseki*, 329–31.

4 According to statistics of bicycle ownership in Japan in 1913–1971 (Jitensha sangyō shinkō kyōkai, *Jitensha no isseki*, 518), the number of bicycles reached 8.6 million in 1942, falling to 5.7 million in 1945. Earlier on, the Bicycle Association of Japan presented a different set of numbers: "before the war the number of the bicycles in use was approximately 10,000,000, or one bicycle for every eight persons, but during the war, due to the limitation enforced upon its manufacture and concurrently to a large quantity of the loss caused by the natural wear and tear as well as war calamity, the number of the vehicles now available for use remarkably decreased and it is authentically assumed that the number stands at about 4,000,000, or 40% of the number we had before the war" (Bicycle Association of Japan to the General Headquarters of the Supreme Command for the Allied Powers, "Petition Re: Rehabilitation of the Bicycle Manufacturing Industry in Japan with an Illustration of the Importance of the Industry as One of the Leading People's Industries of Japan", October 1946, in *Supreme Command for the Allied Powers Declassified Document 775019*, National Diet Library of Japan, Tokyo, 3).

5 Bicycle Association of Japan, "Petition", 3. For more on the post-war re-birth of the Japanese bicycle industry, see Ueda, *The Development*; Steele, "The Making of a Bicycle Nation".

6 See statistics in Jitensha sangyō shinkō kyōkai, *Jitensha tōkei yōran* 10 (Tokyo: Jitensha sangyō shinkō kyōkai, 1976), especially the chart on world bicycle production on p. 230. In 1975, Japan produced six million bicycles, the United States 5.4 million, and China five million. The spectacular growth of the Chinese bicycle industry quickly ended Japanese prominence: by 1985, China was producing 32 million bicycles, more than four times the Japanese output.

industry received subsidies amounting to a total of ¥38.8 billion (approximately $108 million),[7] primarily for research and development to improve production techniques and the promotion of bicycle exports.[8] In addition, during the same period, bicycle racing generated some ¥8.9 trillion (approximately $24.8 billion) for use in urban housing, education, public works, and social welfare projects.[9] These subsidies proved indispensable to the rebuilding of Japan and specifically the revival of the bicycle industry. Nonetheless, there have been few academic studies, either in Japanese or in English, on the history of *keirin* and the role played by "betting on the wheel".

## 2 The Race

In Japan, all instances of private gambling are strictly prohibited. After 1945, however, gambling was allowed in four so-called "public sports": horse racing, bicycle racing (*keirin*), powerboat racing, and speedway motorcycle racing.[10] Special laws were enacted between 1947 and 1950, legalizing such gambling, but at the same time imposing strict controls to guard against corruption and ensure that profits served the public good.[11] In the case of *keirin*, a certain percentage of the takings were to be devoted to the reconstruction of cities that had been bombed during the war, as well as for bicycle and small machinery research and development, and export promotion.

---

7 The conversions here are approximate. The Japanese yen was fixed at 360 yen to 1 US dollar in 1949, and was only allowed to float again in 1970. In 1975, the rate was roughly ¥296 to 1 US dollar, while in 2015 the average rate was ¥121 to 1 US dollar; at the time of this writing (August 2016), it stood at ¥102 to 1 US dollar.

8 Ueda, *The Development*, 25.

9 See chart on annual distribution race proceeds in Nihon Jitensha Shinkō-kai, *Keirin sanju-nen shi* (Tokyo: Nihon Jitensha Shinkō-kai, 1979), 58–59, back matter.

10 Takuya Yamazaki and Yuki Mabuchi, "Sports Betting and the Law in Japan", in *Sports Betting: Law and Policy*, edited by Paul M. Andersen, 512–15. Den Haag: T.M.C. Asser Press, 2012.

11 The gambling world of powerboat racing (*kyōtei*) is studied in Tom Gill, "Those Restless Little Boats: On the Uneasiness of Japanese Power-Boat Gambling", *The Asia-Pacific Journal* 11, issue 42, no. 2 (2013), http://www.japanfocus.org/-Tom-Gill/4012. As Gill notes, gambling, despite its illegal status, is big business in today's Japan: "Kyōtei is just one branch of the mighty Japanese empire of gambling. It is the second most popular form of race gambling after horse-racing with sales of ¥2,625 billion in 2011. Another ¥620 billion was spent on bicycle racing, and ¥84 billion on motorbike racing. The total spent on race gambling in 2011 was ¥4.25 trillion."

*Keirin* is essentially a form of sprint racing lasting for eight laps of a 250-metre circuit.[12] During the early laps, the cyclists (normally six in number) remain behind a pacer bike and steadily accelerate before the final 625 meters, during which the pacer leaves the track and the cyclists sprint for victory, reaching speeds of 70 or 75 km/h. This exciting racing event, born in post-war Japan, has spread overseas. Beginning in 1980, the Union Cycliste Internationale (UCI) sponsored annual world championship *keirin* track cycle tournaments, and the sport became one of the most popular track cycle events in the Olympics, beginning with the 2000 Sydney Games. As a betting sport, however, *keirin* has only spread to South Korea.

The first *keirin* race was held in Kokura, a city in northern Kyushu, in 1948, and rapidly attracted a devoted fandom. Currently it is the most popular gambling sport in Japan after horseracing, with 43 velodromes (as of 2016) located throughout the country. Local governments, largely cities, serve as organizers of *keirin* events, while national organizations, the Japan Keirin Association (JKA) Foundation and the Japan Cycling Federation oversee the operation of the event to guarantee fairness and safety.[13] The JKA Foundation (before 2007 the Japan Bicycle Promotion) also manages the distribution of subsidies to

12 There is little literature available on *keirin* in English. The Wikipedia article on the phenomenon (http://en.wikipedia.org/wiki/Keirin), however, is useful. Somewhat more detailed information may be found in William Fotheringham, *Cyclopedia: It's All about the Bike* (Chicago: Chicago Review Press, 2011), 213–16. For general information, see Hiroyuki Kouno, "Bicycle Racing in Japan and the History of Keirin", in *Cycle History 11: Proceedings, 11th International Cycling History Conference*, ed. Andrew Ritchie and Rob van der Plas (San Francisco: Van der Plas, 2000). For the role *keirin* played in Japan's post-war reconstruction, see James Longhurst, *Bike Battles: A History of Sharing the American Road* (Seattle: University of Washington Press, 2015), 178–84. In Japanese, aside from a plethora of books, magazines, and manga on *keirin* in the country's popular culture, the best source is the series of books on the *keirin* history authored and published by the country's bicycle promotion association Nihon Jitensha Shinkō-kai in Tokyo every ten years: *Keirin ju-nen shi* [A Ten-Year History of *Keirin*] (1959); *Keirin niju-nen shi* [A Twenty-Year History of *Keirin*] (1969); *Keirin sanju-nen shi* [A Thirty-Year History of *Keirin*] (1979); *Keirin yonju-nen shi* [A Forty-Year History of *Keirin*] (1989); *Keirin goju-nen shi* [A Fifty-Year History of *Keirin*], 1999; and *Keirin rokuju-nen shi* [A Sixty-Year History of *Keirin*] (2009). For this chapter, I rely primarily on the fifty-year overview, especially its Chapter 1 on the origins of *keirin* and Chapter 2 on the construction of velodromes.

13 For details on the operation of publicly managed gaming in Japan, including *keirin*, see Yoshinori Ishikawa, "Japanese Publicly Managed Gaming (Sports Gambling) and Local Government" (Papers on the Local Governance System and its Implementation in Selected Fields in Japan No. 16, Council of Local Authorities for International Relations, Institute for Comparative Studies in Local Governance of the National Graduate Institute for Policy Studies, Tokyo, 2010). http://www.clair.or.jp/j/forum/honyaku/hikaku/pdf/BunyabetsuNo16en.pdf.

ILLUSTRATION 6.1 *Keirin* racers at the Keiōkaku Velodrome, Chofu, Tokyo
PHOTOGRAPH BY AUTHOR

small machine industries and various public works; monies that formerly went primarily to support research and development for the post-war bicycle industry and to help with the construction of Japan's post-war urban infrastructure.

## 3 At the Gate

Japanese cyclists were active in the world of bicycle racing before the Second World War: long distance road racing was popular from the 1920s. *Keirin*, however, was a new bicycle sport born in Japan in the immediate post-war period. Its origins can be traced to the "dream" of three enterprising men—Yanuma Isao, Kurashige Sadasuke, and Ebizawa Seibun—who in 1946 established the International Sporting Company with the goal of promoting sports and setting up an international resort town to be called "Leisure Land."[14] It was a daring plan, especially considering the timing: less than one year after Japan's total defeat in the Pacific War and when most cities in Japan were in ruins caused by napalm and atomic bombs. It was enough just to survive, and hardly a propitious time to set up a "Leisure Land" resort. Moreover, the costs involved were considerable, and seemingly impossible at a time when there was little capital available for investment.

Ebizawa's plan to gain money for their project was to hold "bicycle races for financial rewards": in other words, to secure money from the proceeds of betting on bicycle racing, using the model of pre-war horse racing. In September 1947, he sought permission from the governor of Kanagawa Prefecture to allow people to bet on bicycle races, naming the new form of bicycle racing *keirin*

14 Nihon Jitensha Shinkō-kai, *Keirin goju-nen shi*, 9.

(a word that literally means "racing wheels"). The governor was immediately interested.

Ebizawa drew up a prospectus for the project, explaining that profits from gambling could be used to support the bicycle manufacturing industry and thus help meet domestic demand as well as serve to expand bicycle exports. Moreover, racing would expand popular interest in cycling and provide a source of income for public works, especially for the reconstruction of cities bombed during the war. He argued that bicycles had been a major export industry and that the post-war recovery of Japan—indeed, the birth of a new Japan—depended on the reconstruction of the bicycle industry.[15] Finally, the prospectus proclaimed that bicycle racing would not only contribute to Japan's economic health, but to its physical health as well.[16]

The International Sports Company proposed to manage revenue derived from admission fees, while the prefectural government would manage the gambling proceeds. Ebizawa's plan was that their company would sponsor two events each year, one in the spring and another in the autumn. He predicted that each event would realize some ¥1.5 million, of which ¥800,000 would be earmarked for public works, ¥350,000 would cover winnings and operating expenses, and ¥360,000 would result in pure profit.[17]

It was well beyond his imagination that three years later, in 1950, that proceeds from *keirin* events would produce some ¥30 billion for public works projects and that six years later, in 1953, it would double to ¥60 billion (roughly $160 million), and that eventually more than 60 velodromes would dot the landscape throughout Japan. By the end of the Allied occupation of Japan in 1952, *keirin* had developed into an industry worth more than one trillion yen, thereby providing an extraordinary (but largely invisible) stimulus to the economy.[18]

15 An interesting comparison may be made with arguments to resume the Tour de France in 1947. On the one hand, the race would symbolize the re-birth of France. At the same time, however, by promoting the French cycle industry, the race would contribute to France's economic revival. Moreover, local communities throughout France vied to host stages of the race, seeking media attention and economic benefits that would help them recover from horrific damage sustained during the war years. See Christopher S. Thompson, *Tour de France: A Cultural History* (Berkeley: University of California Press, 2008), 178–85.

16 Nihon Jitensha Shinkō-kai, *Keirin goju-nen shi*, 9–10.

17 Nihon Jitensha Shinkō-kai, 10.

18 Nihon Jitensha Shinkō-kai, 10.

## 4 The Law

The Governor of Kanagawa found the proposal intriguing, but told Ebizawa that such a project would require national legislation. Unfazed, Ebizawa and his colleagues immediately set about lobbying for a new bicycle racing law. Through their connections, they managed to gain the support of Prime Minister Katayama Tetsu and the Minister of Commerce and Industry, Mizutani Chosaburō. On 13 October 1947, one month after the initial proposal was submitted, a national parliamentary "League to Enact a Bicycle Racing Law" had begun to lobby support for the new law. Hayashi Daisaku, a member of the Social Democratic Party elected to the lower house, played a leading role. He argued on behalf of the new law: "The war experience was so devastating, that it is important to give people some new source of enjoyment and enable them to feel refreshed."[19] He was won over by petitions of support by pre-war bicycle champions, including Murakami Jisaku and Yokota Takao, cyclists who achieved international fame in the 1930s. Hayashi reasoned that bicycle racing would help Japan re-enter the international sporting world. Finally, he argued that bicycle racing would stimulate the bicycle industry and contribute to local economic development.

The International Sports Company circulated a statement designed to gain the support of Japanese politicians and bureaucrats:

> The bicycle is a central part of our country's export industry. Moreover, as an instrument of transportation, it plays a vital role in the daily lives of Japanese people. Moreover, the contribution of the bicycle to the world of sports should not be overlooked. However, compared with pre-war bicycle ownership at more than 8 million bicycles, after the war, their numbers had been reduced to a mere 2 million. And, given the rate at which people in Asian countries imported manufactured goods from Japan in the pre-war years, we can expect a close-to infinite demand for bicycles manufactured in Japan [in the future].
>
> Although we need to overcome all obstacles and make plans to increase the production of bicycles, as a result of the war, our current production of bicycles can only produce one-tenth of pre-war levels.
>
> That is why we seek the enactment of a bicycle racing law similar to the horse racing law, and why we are so concerned to allow the purchase of bicycle betting coupons similar to horse race betting coupons.

19 Nihon Jitensha Shinkō-kai, 12.

> The reasons why we support the enactment of a bicycle racing law is that it will stimulate the increased production of bicycles, and directly serve to improve the quality of bicycles. Moreover, it will clearly differentiate between professional and amateur cyclists and ensure the healthy development of bicycle racing as a sport.
>
> We feel certain that the speedy enactment of this law will help to overcome difficulties standing in the way of the emergence of a new Japan.[20]

Effective lobbying produced results. By December, the new bicycle law had the support of ranking members of all parties in the Diet with the exception of the Communist Party. The International Sports Company next approached the bicycle industry seeking its cooperation and support. Contrary to expectations, not all bicycle manufacturers were eager to support their movement. These were years of favourable economic activity for the bicycle industry. Many industrialists who were making money in the black market selling bicycles and bicycle parts were reluctant to place themselves under government scrutiny.[21] Nonetheless, when the bicycle racing law sponsors managed to gain the understanding of the head of the Yamaguchi Bicycle Company, other industrial leaders came on board, including Miyata, Kayano, Okamoto, Nichbei Shōkai, and the League of Bicycle Manufacturers.

Emperor Hirohito was even reported to be enthusiastic: at the 1947 National Athletic Meet (*kokutai*) held in Ishikawa prefecture, the emperor was asked about what he thought of bicycle racing. He replied that he was greatly in favour of *keirin*, much to the delight of the promoters of the new law.[22] Finally, by the end of December, the proposed bicycle racing bill won the approval of the Ministry of Commerce and Industry. By the end of January 1948, the bill was ready to submit to the Diet, with the full support of Japan's 'holy trinity': big business, party politicians, and the bureaucracy.[23]

---

20 Nihon Jitensha Shinkō-kai, 12.

21 For details on the post-war black market for bicycle and bicycle parts, see Jitensha sangyō shinkō kyōkai, *Jitensha no isseki*, 373–77. In order to control inflation in the early post-war years, the government attempted to impose a system of rationing and price controls. Bicycle and bicycle parts were in great demand and sometimes more easily purchased outside legal channels.

22 Nihon Jitensha Shinkō-kai, *Keirin goju-nen shi*, 15.

23 The chronology of events leading to the enactment of the Bicycle Law is included in Ministry of Commerce and Industry to the Supreme Command of the Allied Powers, "Memorandum of Current Situation of the Bicycle Race Project", 7 May 1949, in *Supreme Command for the Allied Powers Declassified Document 775019*, 1–5.

Article 1 explained the purpose of the proposed new law: "By virtue of this law, the Metropolis, Hokkaido, and all prefectures, in addition to the cities designated by the competent Minister, taking their population and finance into consideration … may hold bicycle races for the purpose of improving bicycles, including their production, increasing their exportation and meeting domestic demands, besides increasing the revenues of the local finances."[24] Despite the International Sporting Company's success in gaining support for the bill, the new law was not in fact promulgated until 1 August 1948. Before any voting could take place, the Bicycle Racing Law required the approval of the Supreme Command of the Allied Powers (SCAP). Japan was occupied from 1945 until 1952, during which time all legislation had first to be submitted for consideration by occupation authorities before a vote could take place.

## 5 The Chief Official

The bill was formally presented to the Lower House on 3 March 1948, jointly sponsored by four parties (Social Democratic, Democratic Liberal, Democratic, People's Co-operative). Its chief spokesman was Hayashi Daisaku who was joined by 48 other members of the Diet. However, before any action could take place, the proposed legislation had to be translated into English and submitted to the Government Section of SCAP. Within SCAP, the bill was referred to the Economic and Scientific Section (ESS) for comment and approval. The ESS responded stating that:

1. Under no circumstances should the central government be placed in charge of the proposed bicycle racing scheme; rather all authority should be placed in the hands of local bodies.
2. Accordingly, no recognition shall be given to a national Bicycle Promotion Association as a legal body with special authority. SCAP is opposed to national bodies and will not permit the formation of any association with nation-wide authority.
3. SCAP reserves the right to reconsider the rate of proceeds to be derived from sales.[25]

---

24 The English translation of the Bicycle Race Law is included in *Supreme Command for the Allied Powers Declassified Document 775019* as "Law No. 209: Bicycle Race Law", promulgated by Emperor Hirohito on 1 August 1949. An earlier typescript of the proposed legislation, dated 28 April 1949, also exists.

25 Nihon Jitensha Shinkō-kai, *Keirin goju-nen shi*, 16.

At that time, in 1948, occupation policy was still opposed to Japanese economic revival beyond subsistence—it thought that the former *zaibatsu* contributed to the rise of Japanese militarism, and sought instead a more equitable sharing of wealth and power. The policy was to destroy the concentrations of wealth held by industrial conglomerates and landlords, and to make Japan into a country of small merchants and small farmers. Decentralization was the cure to excessive centralization, with powers in the hands of one person, the emperor. In order to guard against the concentration of power in government organizations, an anti-monopoly law was passed in November 1947 governing the formation of trade associations.[26] This law was very much in the mind of the occupation authorities in the Economic and Scientific Section as they examined the draft of the law on bicycle racing.

Hayashi Daisaku met with Occupation representatives to explain the draft legislation. Other politicians and leaders in the bicycle industry used what connections they had to secure SCAP approval. On 13 April 1948, SCAP sent approval to the House of Representatives that the bicycle law could go forward with the proviso that local regulatory bodies be established. The bicycle racing law was immediately referred to the Industrial Committee of the House of Representatives. SCAP retracted its approval on the next day; the legislation had been approved by the Economic and Scientific Section, but other sections remained opposed. Finally, on June 8, the bill was cleared by SCAP. From there it went to the House of Representative and after debate was approved on 29 June: the bill was finalized by vote in the House of Councillors on 3 July. The Bicycle Racing Law was promulgated by Emperor Hirohito on 1 August, but in a much-amended form, largely due to points raised by SCAP.

## 6 The First Event

In April, while negotiations over details of the bicycle law were underway, the mayor of Kokura, Hamada Yoshio, travelled to Tokyo to offer his city as the site of the first *keirin* race.[27] He proposed to use the velodrome that was under construction for the third National Athletic Meet in Fukuoka; nearby Kokura had been chosen as the site for baseball and cycling events at the end of October. He was especially attracted by the possibility that a percentage of the takings

26 For background information on the Occupation of Japan, including the operation of black markets, see John W. Dower, *Embracing Defeat: Japan in the Wake of World War II* (New York: W.W. Norton, 2000).

27 Nihon Jitensha Shinkō-kai, *Keirin goju-nen shi*, 22.

could be used to help with post-war reconstruction. North Kyushu was an important mining and industrial area and target of successive air raids. All cities in the area, including Kokura, not only lay in ruins, but lacked capital for reconstruction. Hamada saw in the *keirin* project a chance to gain much needed revenue to help re-build his city. He mobilized local contractors to build a first-class concrete bicycle-racing track for the National Athletic Meet. This in itself was a sort of gamble; the city had to borrow money. Hamada's hopes were that the velodrome would be used by professional racers in the future. In effect, he had set about to construct Japan's first *keirin* velodrome.

Hamada's mission to Tokyo was successful. He announced that the first *keirin* event would be held at Kokura for three days beginning on November 20, 1948. Some 130 cyclists applied as contestants. Some were pre-war racers such as Yokota Takeo and Mori Takeo, while others were well-known amateurs. Some, however, were completely new to the world of competitive racing, but saw in *keirin* an opportunity for gainful employment. A women's event was also planned, but since only two professional women racers (Ohashi Toshiko and Akimoto Toshiko) applied, the organizers were forced to recruit several female secretaries to help fill out the ranks.[28]

Lacking funds, the city borrowed an additional three million yen from the North Kyushu Branch of the Fukuoka Bank in order to prepare for the race in November. The fate of the city was at stake. The organizers hoped to realize ¥15 million over the four-day event, but were uncertain if this goal could be reached. They sought advice from the local horse track.[29] The *keirin* organizers were advised that a take of ¥1.4 to ¥1.5 million per day in admissions and sales of betting coupons would be considered successful; they knew, however, that they needed around 3.8 million a day to break even. As the opening day grew closer, the organizers became increasingly worried. It was early winter, chilly and raining. The track was in good condition and the National Athletic meet had gone well, but it was uncertain whether people would be willing to pay money to watch (and bet on) a cycling race.

The opening ceremony was scheduled for 10:30 in the morning of 20 November. It had rained earlier in the week, but the day was clear. The Chairman of the Board of Trustees of the Fukuoka Bicycle Association, Endo Sasuke, proclaimed: "Today we hold Japan's first *keirin* meet!" On that day, some 7,000 people gathered at the Mihagino *Keirin* Stadium in the outskirts of

---

28 Nihon Jitensha Shinkō-kai, 25.

29 Nihon Jitensha Shinkō-kai, 27. In the pre-war era, horse racing was under central government supervision. It was halted during the war years, but resumed in 1948 under a new horse racing law. As such, horse racing served as a model for *keirin* operations and betting.

Kokura. The second day, Sunday, attracted 20,000 people, the third day 8,000 people, and another 20,000 on the final day for a total of 55,000 spectators.[30] Many were miners from the nearby coal mines, eager to spend what little money they had in the hopes of a better life, or at least, a little extra spending money.

On the first day, ten races were held and on the second day, twelve races. To the great relief of the organizers, sales for the four days totalled ¥19,730,000, well beyond the break-even goal of ¥15,500,000.[31] News of the success at Kokura focused on the money that was passed over to local governments for recovery purposes. By that time, Osaka, Kanagawa, Hyogo, Miyazaki, Kyoto, and Chiba had all formed local Bicycle Promotion Associations and begun plans to construct *keirin* facilities. Osaka, the centre of Japan's bicycle industry, was especially enthusiastic. The Osaka Government vied with Kokura to hold the first event, seeking financial stimulus for the local bicycle industry and help with wartime recovery costs.

Osaka's first *keirin* event was held for six days beginning on 11 December 1948. The weather was good. In Kokura, the cost of one ticket was ¥100, whereas the Osaka organizers charged only ¥50 per ticket, thereby making the race attractive to people even with few means. The organizers also knew how to advertise, placing *keirin* posters in train stations (and in the trains themselves), department stores and cafes. Large banners were hung from the windows of office buildings and strings of hanging lanterns in the city's central covered shopping mall advertised the event. Advertisements were placed in newspapers and magazines.[32]

The result was a major success. The six-day event attracted some 67,000 spectators and the total intake was ¥36,820,000.[33] Osaka quickly held a second six-day event just after New Year's Day (from 4 January); this time the total proceeds of the betting coupons reached ¥57,016,000.[34] After the success at Kokura and Osaka, 18 more cities applied to hold *keirin* events. Some 19 tracks were completed in 1949 and another 35 in 1950. The first 11 *keirin* meets held between November 1948 and April 1949 netted a total of ¥569,460,950 in proceeds.[35] But more was to come. In 1949, some 19 tracks were completed,

30 Nihon Jitensha Shinkō-kai, 28–29.
31 Nihon Jitensha Shinkō-kai, 29.
32 Nihon Jitensha Shinkō-kai, 30.
33 "Appendix I: Incomes Accrued from Execution of Bicycles Races", November 1948–May 1949, in *Supreme Command of the Allied Powers Declassified Document 775019*, 6, gives the attendance at the first Osaka bicycle race as 52,290.
34 "Appendix I: Incomes Accrued", 6, gives the attendance at the second Osaka bicycle race as 39,683.
35 "Appendix I: Incomes Accrued", 8.

including five cantered or in nearby Tokyo, in Omiya to the north, Kawasaki to the south, Chiba to the east, Keiōkaku (Chōfu) in the Western suburbs, and Kōrakuen, in the very centre of the city that remained largely in ruins. Another 35 tracks were constructed in 1950 in regional cities throughout Japan. By 1953, five years after the first track was opened in Kokura, some 63 *keirin* velodromes were in operation and bicycle racing had become one of Japan's most popular sporting events.

The two Tokyo velodromes were particularly successful. Keiōkaku was an extraordinary construction project, considering the existing restrictions on steel and cement. It remains, with its 23,291 seats and plenty of standing room, one of the largest bicycle racing tracks in Japan. Constructed on the site of a pre-war amusement park on the banks of the Tama River, from its opening on 24 September, 1949 it quickly became a weekend family destination. On 3 January 1972 Keiōkaku attracted some 73,000 fans, setting a *keirin* attendance record. Its chief rival was the Kōrakuen track that held its first race on 2 November 1949. Built next to the Kōrakuen Baseball Stadium (home of the Yomiuri Giants since 1937), the synergy between the two sports, baseball and *keirin*, helped to create a fresh post-war leisure culture.[36] Kōrakuen velodrome, filled to capacity throughout the 1950s and 1960s, was also the most profitable venue, earning the name of the "Mecca of Bicycle Racing". Even though it was closed in 1967, it continues to hold the record for sales of betting coupons, year by year taking in about 10 per cent of the *keirin* revenues nationwide.[37]

## 7 The Payoff

Of course, the *keirin* story is much more complex, including episodes of corruption, graft, and periods of public disfavour. Nonetheless, it makes visible the role bicycles played in Japan's post-war economic recovery. According to the provisions of the Bicycle Race Law, 75 per cent of proceeds from betting coupons were to be used as prize money. The remaining 25 per cent was counted as gross revenue, from which operating expenses were derived and stipulated contributions made to public bodies (urban reconstruction and welfare) and

36 The Kōrakuen velodrome featured in several motion pictures, including Ozu's 1952 *Ochazuke no Aji* (The Flavour of Green Tea over Rice).

37 Precisely because of its profitability, in 1967, Minobe Ryōkichi, the new progressive mayor of Tokyo, determined to end revenues derived from gambling. The Kōrakuen velodrome was initially turned into a golf driving range and later, in 1988, dismantled along with the Kōrakuen Baseball Stadium as the site for Tokyo Big Dome, Tokyo's premier sporting venue. For details, see Nihon Jitensha Shinkō-kai, *Keirin yonjū nenshi*, 538–53.

ILLUSTRATION 6.2 Racers line up for the first event at the Kokura Velodrome, 20 November, 1948
PHOTOGRAPH COURTESY OF KOKURA VELODROME

ILLUSTRATION 6.3 Poster for the August Races at the Nara Velodrome, 1953
PHOTOGRAPH COURTESY OF AUTHOR

to bicycle and other small and medium industries for sales promotion and research and development. In March 1949, only five months after the first *keirin* meet at Kokura, the Ministry of International Trade and Industry (MITI) sought SCAP approval for the construction of additional bicycle racing tracks, claiming the following:

- The popularity of *keirin* had led to 40 requests nationwide to construct bicycle racing tracks.
- Bicycle racing would stimulate the domestic bicycle industry, promote exports, and provide needed capital for urban reconstruction projects.
- To stimulate the bicycle industry was to directly and indirectly stimulate overall economic growth.
- In 1937 and 1938, bicycles and bicycle parts were a major export item to countries in Asia; the bicycle industry could once again spearhead Japan's exports to Asia.
- The bicycle industry qualified as a post-war peace industry; the ruins of defunct wartime factories could be put to peacetime use by converting them into bicycle race tracks.
- Bicycles were indispensable to the Japanese way of life: the association of "Japanese and bicycles" was much the same as that with "Americans and automobiles"—so deeply ingrained were bicycles into the web of Japanese society. Increasing ownership of bicycles would add further stimulus to the Japanese economy.[38]

The arguments advanced were effective, quickly winning the approval of the occupation authorities. Americans, pleased with the superior status seemingly afforded them as an automobile people, were willing to advance Japanese desires to rebuild the bicycle industry, convinced that it would lead to peace and prosperity. By the mid-1950s, nearly every city devastated by wartime bombing boasted a velodrome. By the 1960s, there were more than 60 bicycle racing tracks nationwide. Currently (as of 2016) there are 43.

Popular demand for cheap entertainment and the chance to win at the betting booth was one reason for *keirin*'s explosive growth. More instrumental were campaigns by local officials, especially mayors, to bring *keirin* to their city as a way to help pay the costs of post-war reconstruction. In September 1949, the Ministry of International Trade and Industry sought SCAP approval for the use of revenue from bicycle racing to construct low-cost housing. In summary, the "Bicycle Race of Love Movement" (*Ai no keirin undō*) "has as its main purpose the popularization of the new bicycle race track industry in the public eye through furnishing low-cost housing from profits. This will also aid

38 Nihon Jitensha Shinkō-kai, *Keirin goju-nen shi*, 34–36.

the concerned officials in securing permissions and support to construct more tracks in the future."[39] In arguing its case, the MITI document emphasized the expected financial success of the bicycle racing law:

> Actual figures up to 25 July 1949:
> Sales proceeds of coupons: ¥1,779,127,650
> Revenue to local governments: ¥197,000,000 (approx.)
> Revenue to national treasury: ¥98,000,000 (approx.)
>
> Predicted figures for 1949:
> Sales proceeds of coupons: ¥7,700,000,000
> Revenue to local governments: ¥900,000,000
> Revenue to national treasury: ¥450,000,000
>
> Predicted figures for 1950:
> Sales proceeds of coupons: ¥19,500,000,000
> Revenue to local governments: ¥2,100,000,000
> Revenue to national treasury: ¥1,100,000,000

A 27 September 1949 memorandum from MITI followed these rosy predictions with even more "utterly fantastic" figures relating how profits from bicycle races in 1950 would be used.[40] The list below is nonetheless valuable to envision how monies were expected to be used to promote Japan's bicycle industry:

- Construction expense of Bicycle Foreign Trade Hall: ¥100,000,000.
- Expense of sending inspectors of bicycle businesses overseas: ¥107,711,000.

39 "Movement for Construction of Houses for Everybody by Proceeds from Bicycle Races: On 'Bicycle Race of Love' Movement'", 19 September 1949, in *Supreme Command of the Allied Powers Declassified Document 775019*. The request was approved, but not all Occupation officials were in favour of the project. A handwritten note on the document describes the purpose of the proposal: "The primary objective of this proposed legislation is obviously for the promotion of securing approval and support of the pertinent agencies for the contemplated expansion of bicycle racing with a view towards future exploitation of the Japanese public by the gambling element and related interests."

40 Vehicles Section, International Trade and Industry Ministry, memorandum, "On Functions of Bicycle Business Development Associations and Use of Incomes from Bicycle Races", 27 September 1949, in *Supreme Command of the Allied Powers Declassified Document 775019*, 1–8, especially pages 7–8 that list the various ways in which bicycle race profits would be used. At the top of the list is a handwritten comment "utterly fantastic!!" presumably by a SCAP official. Drawing attention to the total figure of ¥637,829,000 on page 8 there is also the handwritten comment "silly". Yet another note, also by hand, states that "I am of the opinion that the attached would in no way interest me. I feel that my age would be a bar to participation in this 'sport'".

- Research expenses on overseas affairs: ¥28,080,000.
- Expense of bicycles to be shipped abroad for publicity purposes: ¥42,465,000.
- Subsidies for the production of publicity newsreels and movies to be sent abroad: ¥10,000,000.
- Subsidy for the construction of schools for bicycle racers: ¥7,500,000.
- Subsidy for inspection centres of bicycles and inspection expenses: ¥4,170,000.
- Subsidy for local government officers for checking numbers of bicycles in possession of residents: ¥32,740,000.
- Subsidy for local government offices in their allocation business of bicycles and rear-cars: ¥3,552,000.
- Expenses required in carrying out the Bicycle Race Law: ¥2,312,000.
- Expenses in diagnosing and guiding medium and small bicycle businesses: ¥16,256,000.
- Expenses for the construction of a composite national Research Institute of Bicycles: ¥204,640,000.
- Subsidy for industrial research on bicycles: ¥50,000,000.
- Expense for standardization of the bicycle industry: ¥9,946,000.
- Subsidy for experimentation and invention of bicycle improvement: ¥5,691,000.
- Expense for the establishment of a bicycle inspection system: ¥7,750,000.
- Expenses required by district bureaus of international trade and industry in carrying out the Bicycle Race Law: ¥3,419,000.

  Total: ¥637,829,000

As Tatsuzō Ueda notes in his study of the post-war Japanese bicycle industry, the Bicycle Racing law created a legal basis to channel state funds into the bicycle industry.[41] He describes the promotional subsidies derived from *keirin* as "a major event that cannot be overlooked in tracing the reconstruction and development of the [bicycle] industry after World War II". These subsidies sought 1) the improvement of bicycle production techniques and quality control; 2) the establishment of a bicycle testing system; 3) the promotion of bicycle exports; 4) guidance for medium and small bicycle industries; and 5) loans to the bicycle industry.[42]

Between 1949 and 1978, the bicycle industry received subsidies amounting to a total of ¥38.8 billion (over $100 million), primarily for research and

---

41 Ueda, *The Development*, 24. Ueda's source for how the bicycle race profits would be used as promotional subsidies for the bicycle industry is Jitensha sangyō shinkō kyōkai, *Jitensha no isseki*, 40–43, 409–23.

42 Ueda, 24–25.

development to improve production techniques and the promotion of bicycle exports.[43] The amount is not as large as MITI projected in 1949, but it did contribute substantially to the revival of the Japanese bicycle industry.

Far more money (¥8.9 trillion, or approximately $24.8 billion) was generated from *keirin* revenues during the same period for post-war urban reconstruction projects throughout Japan.[44] Table 6.1 below breaks down the uses to which these funds were put.

TABLE 6.1 Distribution of revenues from sales of betting coupons in Japan, 1948–1976

| Spending item | JPY | Percentage |
|---|---|---|
| Operating expenses | 1,108,117,690,000 | 12.34 |
| Public housing | 86,761,591,000 | 7.83 |
| Education | 285,922,588,000 | 25.8 |
| Public works | 348,991,501,000 | 31.49 |
| Social welfare | 133,294,304,000 | 12.03 |
| Small/medium industries | 25,790,727,000 | 2.33 |
| Unemployment benefits | 21,405,901,000 | 1.93 |
| Other monies | 176,984,980,000 | 15.97 |
| Total | 8,981,464,434,000 | |

SOURCE: NIHON JITENSHA SHINKŌ-KAI, *KEIRIN GOJŪ-NEN SHI*, BACK MATTER, 58–59

The percentages of revenue distribution are an average of the 28 years covered. In fact, between 1948 and 1950, more than 40 per cent of *keirin* revenues were devoted to infrastructure and public works projects (including roads and

43 For further details on the origin and role of subsidies for the promotion of the post-war bicycle industry, see Jitensha sangyō shinkō kyōkai, *Jitensha no isseki*, 409–23. Using these subsidies, an Open Research Center for Bicycle Production Techniques was established in 1954, reorganized in 1958 as the Bicycle Research Technology Foundations. In 1964, it was reorganized again as the Japan Bicycle Promotion Institute. Even today, it continues to rely upon the Japan Keirin Association (JKA) for funding. For JKA's on-going activities and data resources, visit http://www.keirin-autorace.or.jp/index.php. See also the Japan Bicycle Promotion Institute homepage at http://www.jbpi.or.jp/.

44 For a breakdown of *keirin* revenues and disbursements in the 55 years between 1948 and 2004, see the report on *keirin* finances submitted to the Takamatsu City government: http://www.city.takamatsu.kagawa.jp/file/2962_L11_keirinshikumi1.pdf. According to this report, as of 2004 some ¥883 billion had been made available for educational facilities, ¥120 billion for public housing, ¥913 billion for social welfare projects, and ¥965 billion for public works.

bridges), and in the early 1950s, nearly 30 per cent of revenues were earmarked for the construction of public housing. The Occupation had not encouraged economic recovery or projects of basic infrastructure reconstruction. Air raids during the last year of the war targeted Japan's urban centres; over four million homes, were destroyed. And yet, reconstruction was slow. In Tokyo, for example, by 1948 only 6.8 per cent of those areas burnt by repeated napalm bombing raids had been rebuilt. Millions of people were in living in slum-like conditions. The coming of the Cold War and the very hot outbreak of the Korean War in 1950 fundamentally changed Occupation policy, sparking a reconstruction boom in public housing (often taking the form of blocks of multi-story apartment buildings), office buildings, roads, bridges, sewers, river embankments, and dams.

It may be no accident that the birth of *keirin* overlapped with the "reverse course" in Occupation policy; in any case, its payoff was put into immediate use in a multitude of physical reconstruction projects. The post-war baby boom meant that for a period in the 1950s some cities, the industrial city of Kawasaki, for example, had to build a new school every month; revenue derived from bicycle racing was more than welcome.[45] Another case in point is Hakodate, a port city in Japan's northern island of Hakodate. It was bombed during the war primarily to disrupt ferry traffic between Hakodate and Aomori on the main island of Honshu. The first race at the Hakodate velodrome was held on 30 June 1950. An elaborate advertisement campaign, involving newspaper ads, banners, neon signs, decorated archways, and loud-speaker announcements, helped to create a curious and in the end committed crowd of *keirin* fans. Off-site betting began in August as the city began to realize the payoff from bicycle racing, including a variety of infrastructure rebuilding projects. Between 1950 and 1956, some ¥190 million (roughly $530,000) went directly into Hakodate City accounts for public works: ¥20 million for port facilities, ¥53 million for housing, ¥64 million for schools, and ¥53 million for the North Sea Industrial Exhibition held in Hakodate in 1954.[46] Other cities, large and small, throughout Japan have similar stories to tell. In 1967, amid much controversy, Governor Minobe closed down the nation's most popular and profitable velodrome, Kōrakuen in downtown Tokyo, declaring that the city could do without money derived from public betting. By that time Tokyo, especially after the 1964 Olympics, had achieved a reputation as one the great cities of the world. This reputation, however, was built upon the more than ¥15.6 trillion

45 *Sankō shiryō* (Tokyo: Zenkoku Keirin Shikōsha Kyōgikai, 1969), 3.

46 "Hakodate-shi shi", Hakodate City (website), accessed 6 August 2016, http://archives.c.fun.ac.jp/hakodateshishi/tsuusetsu_04/shishi_07-02/shishi_07-02-28.htm.

(roughly $435 million) that entered city coffers between 1955 and 1964 from Kōrakuen proceeds alone.[47] These payoffs from legalized gambling helped to finance housing, schools, and public works projects at a time when they were most needed.

## 8 Futures

In 1955, a government white paper declared an end, economically at least, to the post-war period. One year later, in 1956, bicycle production at 1.3 million finally topped the 1940 figure. And increased production allowed more and more people to realize the dream of bicycle ownership. In 1950, the bicycle census stood at 10 million, jumping to 14 million in 1955, nearly double the pre-war peak of eight million. Advances in technology and design allowed for the appearance of light city bikes and mini-bikes that targeted women and the urban middle class. Bicycle use continued to rise despite the rapid spread of motorized transport and the decline of Japan's bicycle industry in the 1990s.[48] Bicycle ownership grew from 71 million in 1991 to an estimated 90 million in 2015. In 2016, the bicycle census stood at just under 90 million, confirming Japan, at 70 bicycles per 100 people, to be one of the great bicycle nations of the world.

This story of Japan's rise to bicycle fame (or infamy as many pedestrians would have it) is well known, as is the complementary story of the rise of Japan as one of the leading economies of the world. This role played by *keirin* bicycle racing in these success stories is less well known. This invisibility is perhaps understandable. *Keirin* today is a shadow of its former self, its fans a slowly diminishing crowd of old men in fishing vests, studying their betting sheets in velodromes that have obviously seen better times. The golden age of bicycle racing has long past. Total attendance at *keirin* events peaked in 1974 at 46.2 million; in 2014 the number had declined to 3.6 million. The decline in revenue from betting coupons, thanks to television and off site (and online) betting, has been less dramatic: from around ¥1.1 trillion in the early 1970s to a peak of ¥1.9 trillion in 1991, before falling to ¥616 billion in 2015.[49] Year by year, moreover, velodromes have gone out of business. Nonetheless, subsidies from

---

47 *Tōkyō zaisei-shi, gekan*, vol. 3 (Tokyo: Tokyo Metropolis, 1969), 703.

48 This is the subject of Steele, "The Making of a Bicycle Nation".

49 For statistics between 1948 and 1998, see Nihon Jitensha Shinkō-kai, *Keirin gojunen-shi*, back matter, 12; for the years up to 2015, see "'Gakeppuchi' kōei keirin ni kaifuku no kizashi", *Cyclist*, 20 December 2015, http://cyclist.sanspo.com/221491. The latter article notes that online betting boosted *keirin* revenues in 2015 to ¥616 billion, the first time revenues had gone up in 23 years.

*keirin* and other public sports, in diminished amounts, continue to be paid to local governments and to small and medium industries. The Japan Keirin Association annual report for 2015 indicates that subsidies totalling ¥2.06 billion (roughly $17 million) were awarded for the promotion of bicycle and small machine industries, while ¥2.63 billion (roughly USD 22 million) were given to local municipalities for various sports, health, welfare, and educational projects.[50]

Given *keirin* diminished and somewhat unsavoury present, one worries about its future. As one of only four legal channels for betting in Japan today, however, there is little chance that it will entirely disappear. The futures for gambling will always be bright. This chapter, however, is more concerned with *keirin*'s past, with reclaiming it from the margins of history and documenting the role it has played in the reconstruction of post-war Japan. The study of bicycle racing in Japan and elsewhere is to be encouraged for the unique perspective it offers on social and economic development. In Japan's case, betting on the wheel has left a very visible mark in its modern history.

---

50 See JKA Annual Report for 2015, available online: http://www.keirin-autorace.or.jp/work/report/h26_report.pdf (accessed 23 June 2016).

CHAPTER 7

# Modernizing the Bicycle: The International Human-Powered Vehicle Movement and the "Bicycle Renaissance" since the 1970s

*Manuel Stoffers*

## 1 The "Bicycle Renaissance" and the Image of the Bicycle*

In 1970, the Dutch writer Bob den Uyl (1930–1993), much appreciated for his tragicomic short stories and travelogues, published a witty little essay on cycling in which he contrasted his own idiosyncratic enthusiasm for bicycle touring with the more common depreciation of the bicycle by his fellow Dutchmen:

> On the Dutch roads, even when the weather is fair, *one hardly sees a bicyclist*.... The time that astonished foreigners threw up their hands in wonder, observing the hordes of cyclists rushing home in a frenzied pace, is gone forever.... The few cyclists that remain, mount their vehicles reluctantly and with a sense of shame. They are the biggest losers of our time, they have as yet not been able to buy a car.[1]

"A new era has come", Den Uyl concluded, "nowadays the masses sit in cars and the single individual rides a bicycle".[2]

Den Uyl's eye-witness account of the decline of cycling in the Netherlands may come as a surprise to those present-day visitors who are impressed by the

* This chapter originated as a paper presented at the 7th International Conference on the History of Transport, Traffic and Mobility, Luzern, November 5–8, 2009. Since then, two related articles on the topic have been published by me: Manuel Stoffers, "The Human Powered Vehicle Movement and the Changing Image of the Bicycle at the End of the Twentieth Century", in *Cycle History 22. Proceedings of the International Cycling History Conference, Paris, May 2011* (Birmingham: Cycling History (Publishing), 2012), 211–219; Manuel Stoffers, "The Politics of Bicycle Innovation: Comparing the American and Dutch Human-Powered Vehicle Movements, 1970s—present", In *Cycling and Recycling. Histories of Sustainable Practices*, ed. Ruth Oldenziel and Helmuth Trischler (New York, Oxford: Berghahn, 2016), 33–57.

1 Bob den Uyl, *Wat fietst daar?* (Den Haag/Rotterdam: Nijgh & Van Ditmar,1970), 11–12 (italics added, MS).

2 den Uyl, *Wat fietst daar?* 92.

© KONINKLIJKE BRILL NV, LEIDEN, 2019 | DOI:10.1163/9789004289970_008

Dutch "bicycle mania".[3] But in fact, just as in other European countries bicycle use in the Netherlands had decreased strongly (although not as strongly as elsewhere) in the decades after the Second World War, and the early 1970s marked here as elsewhere a turning-point in the appreciation of the bicycle: abandoned as a poor man's vehicle of necessity, it was starting to become a vehicle of choice again, connected with issues of lifestyle and ethics. Ten years after Den Uyl's essay, bicycle recreation was booming and new social movements were advancing cycling into a political issue—not only in the Netherlands, but in many other western countries. Most remarkable of all, the downward trend in bicycle use was checked and in some cases reversed, leading in the following decades to substantial increases of bicycle use in some cities.[4] For instance, in München bicycle use more than doubled between 1976 and 1992, from 6 per cent to 15 per cent of the modal share.[5] Berlin also experienced a strong rise of bicycle use since its local low in 1973, reaching as much as half of its top level of the early 1950s in 2001.[6] And this development continued further: in 2007 the bicycle share in the modal split had doubled since 1990 to 10 per cent.[7] Similarly, Pooley and Turnbull, analysing 1,834 individual life histories covering a century of British mobility history, have found that whereas nationally the level of commuting by bicycle increased only marginally after 1980, in their London sample there was both a deeper low of the bicycle share between 1960–1979 and a steeper rise again between 1980–1998, increasing

3 *Bicycle Mania* is the title of a book by the American photojournalist Shirley Agudo on Dutch cycling culture (Den Haag: XPat Media, 2009).

4 Data on the revival of cycling can be found in: Adri A. Albert de la Bruhèze and Frank C.A. Veraart, *Fietsverkeer in praktijk en beleid in de twintigste eeuw: overeenkomsten en verschillen in fietsgebruik in Amsterdam, Eindhoven, Enschede, Zuidoost-Limburg, Antwerpen, Manchester, Kopenhagen, Hannover en Basel* (Den Haag: Ministerie van Verkeer en Waterstaat / Stichting Historie der Techniek, 1999), now translated, updated and expanded as *Cycling Cities: The European Experience. Hundred Years of Policy and Practice*, ed. Ruth Oldenziel, Martin Emanuel, Adri Albert de la Bruhèze and Frank Veraart (Eindhoven: Foundation for the History of Technology, 2016); John Pucher, "Bicycling Boom in Germany: A Revival Engineered by Public Policy", *Transportation quarterly: an independent journal for better transportation* 51 (1997): 31–46; John Pucher, Charles Komanoff, and Paul Schimek, "Bicycling Renaissance in North America? Recent Trends and Alternative Policies to Promote Bicycling", *Transportation Research Part A* 33, no. 7/8 (1999): 625–54; John Pucher and Ralph Buehler, "Making Cycling Irresistible: Lessons from the Netherlands, Denmark and Germany", *Transport Reviews* 28, no. 4 (2008): 495–528; John Pucher, Jennifer Dill and Susan Hardy, "Infrastructure, programs and policies to increase bicycling: An international review", *Preventive Medicine* 50 (2010), S106–S125.

5 Pucher, "Bicycling Boom in Germany", 35.

6 Edward L. Fischer et al., *Pedestrian and bicyclist safety and mobility in Europe* (Washington: US Department of Transportation, 2010), 43.

7 Pucher, Dill, and Hardy, "Infrastructure, programs and policies to increase bicycling", 118.

from 1.5 per cent to 5.85 per cent (which was even above the local 1950 level in their sample).[8] Again, this trend has continued in the twenty-first century: between 2000 and 2008 the number of bicycle trips in London doubled.[9] Data on Danish bicycle levels between 1990 and 2005 indicate an average decrease of bicycle use on a national level (after a substantial increase from the 1970s onwards), but a strong increase for Copenhagen.[10]

Also bicycle policy—or rather bicycle policy plans—have witnessed a revival, perhaps more so than actual bicycle use. Since the 1970s, but more pronounced since the 1990s, many if not all western governments have at least paid lip service to the idea that bicycle traffic should be promoted as a serious mode of (urban) transport. One of the first countries to do so was the Netherlands, where in 1976 bicycle traffic returned as a subject of national government policies. After several subsequent policy documents and measures supportive of cycling were introduced, an ambitious national *Bicycle Master Plan* was launched in 1990.[11] Many other western countries followed in the 1990s. In Denmark bicycle policy was incorporated in a national traffic plan from 1993.[12] In 1990, the United States Department of Transportation for the first time called attention to "the forgotten modes" of bicycling and walking in a national policy document and consequently started to monitor and facilitate cycling (and walking).[13] In Germany, the *Bundestag* required the federal government in 1994 to report on bicycle traffic every five year. After its first report in 1999, the German government started to work on a national bicycle plan that was presented in 2002.[14] In the United Kingdom a trend breaking *National Cycling Strategy* was launched in 1996, the same happened in Australia in 1999, and again several years later in Norway, following a 2001 request from

8 Colin G. Pooley, Jean Turnbull and Mags Adams, *A mobile century? Changes in everyday mobility in Britain in the twentieth century* (Aldershot: Ashgate 2005), 116 & 118.

9 Pucher, Dill, and Hardy, "Infrastructure, programs and policies to increase bicycling", S118.

10 *Flere cykler på sikker vej i staten. Transport- og Energiministeriets cykelstrategi* (København: Transport- og Energiministeriet 2007), 20.

11 Ton Welleman, *The Dutch Bicycle Master Plan: Description and Evaluation in an Historical Context* (The Hague: Ministry of Transport Public Works and Water Management Directorate-General for Passenger Transport, 1999), 42.

12 De la Bruhèze and Veraart, *Fietsverkeer*, 151.

13 See Department of Transportation, the United States, *The National Bicycling and Walking Study: A Ten-Year Status Report.* (Washington, DC: Department of Transportation 2004). n.p.

14 *Erster Bericht der Bundesregierung über die Situation des Fahrradverkehrs in die Bundesrepublik Deutschland* (Berlin, Bonn: Bundesministerium für Verkehr, Bau- und Wohnungswesen, 1999), 1; *Nationaler Radverkehrsplan 2002–2012, FahrRad! Maßnahmen zur Förderung des Radverkehrs in Deutschland* (Berlin, Bonn: Bundesministerium für Verkehr, Bau- und Wohnungswesen, 2002).

the Norwegian parliament.[15] In 1999 the European Commission produced the handbook *Cycling: the way ahead for towns and cities* and sent it in an unusual move directly to local decision makers to convince them of the benefits of cycling and do away with existing prejudices against cycling—which was considered especially necessary in southern and eastern EU-member states.[16]

The values attached to the bicycle in all these policy documents are more or less the same. The bicycle is viewed as an environmentally friendly, healthy, economical, and comparatively quick mode of urban transport, that provides a partial solution to urban congestion and can improve both accessibility and livability of modern cities. This "official" evaluation of the bicycle seems to be a largely western phenomenon. Most reports on the status of cycling in Africa and Asia stress that cycling is considered either by governments or substantial parts of the population, or both, as backward, low-status, and, sometimes, undignified (especially for women). Demonstrating perhaps the international power of the car as a superior status symbol, even in Africa, where bicycle ownership rate is reported to be the lowest of all continents, the bicycle suffers from the image of being a poor man's vehicle, apart from other disqualifications.[17]

In Asia, where cycling seems far more widespread than in Africa, authorities have at least until the end of the twentieth century acted against pedal driven vehicles, denouncing them as unsafe or inhumane for the drivers and holding them responsible for a persistent air of backwardness.[18] In the fastest growing Asian economies, human-powered vehicles seem increasingly incompatible with dominant conceptions of progress and modernity. Even in "bicycle kingdom" China, where the communist regime since the 1950s consistently stimulated bicycle use as a way to modernise the country, the tables have turned since the 1990s and negative views of cycling are gaining ground, resulting

15 *Australia Cycling 1999–2004. The National Strategy* (Sydney: Austroads, 1999); *National Cycling Strategy 2006–2015* (Oslo: Norwegian Public Roads Administration, 2003).

16 J. Dekoster and U. Schollaert, *Cycling. The Way Ahead for Towns and Cities* (Luxembourg: European Commission, 1999). See also the plenary presentation on this EC-handbook at the Velo-City conference in Graz of 1999 by European Commission representative Claude Bochu: http://kamen.uni-mb.si/velo-city99/docs-velo-city99/Bochu.pdf.

17 David Simon, *Transport and development in the Third World* (New York: Taylor & Francis, 1996), 24–26; Luuk Eickmans and Imelda Nasei, *Sustainable Mobility for African Cities* (Nairobi: UN-HABITAT, 2011), 3; Margaret Grieco, Jeff Turner, and E.A. Kwakye, "A tale of two cultures: ethnicity and cycling behaviour in urban Ghana", *Transportation Research Record* 1441 (1994): 101–107; compare however Hans-Peter Hahn, "Use and Cycling in West Africa", in *Cycling and Recycling*, ed. Ruth Oldenziel and Helmuth Trischler, 15–32.

18 Marcia D. Lowe, *The Bicycle: Vehicle for a Small Planet* (Washington, DC: Worldwatch Institute, 1989), 27–28; Michael A. Replogle, *Non-Motorized vehicles in Asia: Lessons for Sustainable Transport Planning and Policy*. (Washington, DC: World Bank, 1991), 12–14.

in policies that restrict cycling and declining use of the bicycle in some cities (although there are also some indications of counter-developments).[19] As one high Shanghai official declared in the mid-1990s, cycling to work might represent the height of enlightenment in north west Europe: in Shanghai the bicycle was to be eliminated as a nuisance and a reminder of past poverty.[20] The cycle rickshaw represents best perhaps the divergence of recent western and non-western views of cycling. Whereas in Asia the rickshaw predominantly is associated with poverty and backwardness, as forcefully expressed by the award winning Vietnamese movie *Cyclo* (*Xich lo*, 1995, directed by Anh Hung Tran), in many western cities, including New York, London, Berlin and Amsterdam, modern versions of the Asian rickshaw known as pedicabs or bike taxis, were introduced in the 1990s and promoted as an "innovative", "sustainable" and "environmentally friendly" mode of (tourist) transport.[21]

The negative images of cycling in non-western countries resemble very much earlier European disqualifications of the bicycle as a mode of transport from the decades after the Second World War. Indeed, the symbolism of the rickshaw in *Cyclo*, representing the vicious circle of poverty and crime, recalls the very similar role of the bicycle in Italian director Vittorio De Sica's famous movie *Bicycle Thieves* from 1948. After the Second World War, if not earlier, European traffic engineers and urban planners, partly influenced by American examples, were convinced that cities had to prepare for the future dominance of the car and considered the bicycle both as an obsolete mode of transport and as an obstacle to a modern and car friendly city. Whereas in present day policy documents the bicycle is presented as a potential substitute for the car (at least for the 50 per cent or so of car trips that are within cycling range), in the 1950s and 1960s it was just the other way around: the car was viewed as the much appreciated successor of the bicycle as a vehicle for individualized mass

---

19 Amir Moghaddas Esfehani, "The Bicycle's Long Way to China—the Appropriation of Cycling as a Foreign Cultural Technique", *Cycle History: Proceedings of the International Cycling History Conferences* 13 (2002): 94–102; Rongfang Liu and Jiahua Song, "Adversary bicycle policies and their impact on urban transportation in China", *TRB 85th Annual Meeting Compendium of Papers CD-ROM* (Washington, DC: Transportation Research Board, 2006): 6–11; Edward J.M. Rhoads, "Cycles of Cathay: A History of the Bicycle in China", *Transfers* 2 (2012): 95–120; Pan Haixiao, "Evolution of urban bicycle transport policy in China", in *Cycling and sustainability*, ed. John Parkin (Bingley: Emerald, 2012), 161–180.

20 Quoted by Deyan Sudjic, "Megalopolis Now: Hong Kong, Shanghai, Jakarta", *City* 1, no 1–2 (1996): 30–37, here 35.

21 See e.g. Sylvia Westall, "Rickshaws pick up speed in Europe's top cities", *www.reuters.com*, last modified 3 October 2007, http://www.reuters.com/article/us-europe-rickshaws-idUSL0343085520071003.

mobility. Investments in bicycle facilities were reduced or stopped altogether and bicycle facilities themselves were removed to accommodate the increasing car traffic. In public awareness cycling increasingly came to be seen as an outmoded and low status mode of transport.[22] With a growing number of cars on streets with little bicycle facilities left, it also increasingly and with reason came to be seen as an unsafe mode of transport.[23] In the United States, where cycling never had become a mode of mass transport in the first place, it was considered to be a youthful pastime and, as such, at its best a preparation for car driving as the "grown up" mode of transport.[24] In 1968, of the fewer than 7 million bicycles sold in the United States, 5.5 million were children's bicycles.[25] Typically, some of the most popular of these American children's bicycles were designed to look like motor-cycles.

The change in the "official" (and popular) appreciation of the bicycle from an outmoded to a modern vehicle in western countries at the end of the twentieth century is remarkable for several reasons. It is not uncommon, for historians of mobility and traffic engineering experts alike, to believe that increasing speed is a fundamental factor in the past and future development of transport and mobility. This view seems to be particularly strong because of its connection with dominant ideas of modernity. Most contemporary commentators—from Paul Virilio (*Essay on Dromology,* 1977) and Zygmunt Baumann (*Liquid Modernity,* 2000) to Hartmut Rosa (*High-Speed Society,* 2008)—consider the fast (and increasing) pace of life, both physically and socially, a fundamental characteristic of modernity, next to and in combination with the other major processes of social differentiation, rationalization, individualization, and the domestication of nature.[26] Being modern seems to imply being mobile, and

---

22 See on the bicycle in the 1950s and 1960s: Bruhèze and Veraart, *Fietsverkeer*, 136, 174, 184, 190–191; Thomas Fläschner, "Out of Date, out of Mind—Public Awareness of the Bicycle During the 1950s in Germany's Saarland State", *Cycle History* 13 (2002): 46–50; the German version of this article gives a more detailed analysis: Thomas Fläschner, "Stahlroß auf dem Aussterbe-Etat. Zur Geschichte des Fahrrades und seiner Verdrängung in den 50er Jahren", *Eckstein. Journal für Geschichte* 9 (2000): 4–22.

23 Pucher and Buehler, "Making Cycling Irresistible", 508.

24 Robert A. Smith, *A Social History of the Bicycle: Its Early Life and Times in America* (New York etc.: American Heritage Press, 1972), 246–248; David V. Herlihy, *Bicycle: The History* (New Haven, CT etc.: Yale University Press, 2004), 319ff; see e.g. the educational movie *Drive your bike* (Glendale, Cal.: The Sullivan Company, 1954), available online: www.youtube.com/watch?feature=player_detailpage&v=fDm6jiQz5BY.

25 Frank Berto et al., *The Dancing Chain: History and Development of the Derailleur Bicycle*. 3rd rev. & exp. ed. (San Francisco: Cycle Publishing / Van der Plas Publications, 2009), 230

26 See Hartmut Rosa and William E. Scheuerman, *High-Speed Society. Social Acceleration, Power, and Modernity* (University Park, PA: Penn State University Press, 2008), 77–78. On Virilio and Baumann, among others, see Dorthe Gert Simonsen, "Accelerating modernity.

the greater someone's "aptitude for movement" (called "motility"), the more "modern" he or she is considered to be.[27]

How does the bicycle fit into this picture? Indeed, some historians have emphasized and demonstrated that the introduction of the bicycle contributed to modern motility by hugely and efficiently increasing the range and speed of individual human-powered transportation.[28] The fascination with the possibilities of the bicycle to increase human speed undoubtedly contributed to the excitement and interest cycling aroused in the nineteenth century.[29] But from a purely technological and speed-oriented perspective, the bicycle could not match the train and was soon superseded by the motor car.[30] Not surprisingly, therefore, the bicycle quickly disappeared from the dominant discourse on the development of modern mobility and transport technology and came to be seen by many traffic engineers as no more than a niche mode of transportation.[31]

The western bicycle revival of the last decades seems to qualify the standard account of mobility history dominated by speed, technological progress and innovation. For one thing, the continuing uses of the bicycle for modern urban mobility illustrate the general point made by David Edgerton and others about

---

Time-space compression in the wake of the aeroplane", *Journal of Transport History* 26 (2005): 98–117, here 98–104 and Peter F. Peters, *Time, innovation and mobilities: travel in technological cultures* (London, New York: Routledge, 2006), 51, 59–63.

27 Vincent Kaufmann, "Mobility: trajectory of a concept in the social sciences", in *Mobility in history. The State of the Art in the History of Transport, Traffic and Mobility,* ed. Gijs Mom, Gordon Pirie and Laurent Tissot, 41–60, here 51, 58 (Neuchâtel: Éditions Alphil–Presses universitaires suisses, 2009).

28 E.g. Jim Fitzpatrick, *The Bicycle and the Bush. Man and Machine in Rural Australia* (Oxford [etc.]: Oxford University Press, 1980); Glen Norcliffe, *The Ride to Modernity: The Bicycle in Canada, 1869–1900* (Toronto [etc.]: University of Toronto Press, 2001).

29 Rüdiger Rabenstein, *Radsport und Gesellschaft: Ihre Sozialgeschichtlichen Zusammenhänge in Der Zeit Von 1867 Bis 1914*. 2nd, corrected ed. (Hildesheim, München, Zürich: Weidmann, 1996), 26–47.

30 Rabenstein, *Radsport und Gesellschaft,* 43.

31 E.g. Ruud Filarski, *The rise and decline of transport systems: changes in a historical context* (Rotterdam: 2004), 55; John Forester, "The place of bicycle transportation in modern industrialized societies [paper] presented to: Preserving the American Dream Conference, 2005", www.johnforester.com/Articles/Social/place_of_bicycle_transportation.htm. Compare Joachim Radkau, "Das Fahrrad in den Technikvisionen der Jahrhundertwende, oder: das Erlebnis in der Technikgeschichte", in *Wege zur Fahrradgeschichte,* ed. Volker Briese, Wilhelm Matthies, and Gerd Renda (Bielefeld: BVA Bielefelder Verlag, 1995), 9–32, here 12–14; Hans-Jürgen Teuteberg, "Entwicklung, Methoden und Aufgaben der Verkehrsgeschichte", *Jahrbuch für Wirtschaftsgeschichte* (1994), 173–194, here 190.

the persisting relevance of "old" technologies in modern society.[32] The fact that it is perfectly possible for modern western cities to have more than one third of all inner city trips made by bicycle, as the Dutch case demonstrates, raises questions about the presupposed inevitability and omnipresence of modern high-tech and high-speed mobility, at least in urban settings. Furthermore, the bicycle not only continues to be used, but it is even increasingly used in at least some western cities, and has (re)gained a place in western transport policies and transport visions. This indicates that cycling is perceived today by a growing number of people as an appropriate and *modern* way of mobility. This raises the question how the comparatively slow and low-tech bicycle, that only a few decades ago was considered "obsolete" and "outdated", has become "modern" again in the West.

It is not my ambition to cover all aspects of this question in this chapter, which is more about the changing western *images* of cycling than about changing practices, urban structures, facilities, or policies—although the latter are, of course, equally important. For instance, the revival of cycling is unquestionably related to the utter success of automobility: the congestion to which the proliferation of cars has led, has reduced the average door-to-door speed of car trips within cities to such an extent that the bicycle has again become a serious competitor to the car even when speed, for many the hallmark of modern life, is the sole criterion. According to an EU handbook for local decision makers, cycling is generally faster than car driving on trips of fewer than 5 km.[33] Even *Top Gear*, the popular BBC series made by and for passionate car-lovers, had to stand first place to an untrained cyclist when it staged a race through London between car, bicycle, public transport and boat.[34] The revival of commercial bicycle messenger services in many American cities dominated by cars, is another indication of the competitive speed of cycling in congested urban environments.

Instead of addressing such "practical" changes, this chapter will focus on the relationship between the conceptions of the bicycle and modernity and deal with the question how the image of cycling has been brought in accordance with (evolving) ideas of modernity since the 1970s. Whilst being concerned with images rather than practices, this question itself is not without practical

---

32 David Edgerton, *The Shock of the Old. Technology and Global History since 1900* (London: Profile Books, 2008), esp. xii, 45–58.

33 See Dekoster and Schollaert, *Cycling. The Way Ahead*, 11.

34 Episode 5 of the 10th series of *Top Gear*, first broadcasted on 11 November 2007, see: http://www.bbc.co.uk/programmes/b008czf4. The Top Gear team was clearly relieved when a similar race through St. Petersburg ended in a "defeat" of the cyclist (22nd series, episode 1, first broadcasted 25 January 2015, see http://www.bbc.co.uk/programmes/b0516rld).

relevance. Available research suggests that next to policies, and geographical and urban characteristics that create barriers to car driving or facilitate cycling, higher levels of bicycle use can be attributed to positive collective ideas and public images of cycling in society and among policymakers.[35] Cycling facilities will only be used by many if many feel cycling is a preferable way of transportation—and, conversely, policy makers will only invest in cycling facilities if they feel cycling is important and considered important by the electorate. Of course, public images also depend on practices. For instance, the idea that urban cycling is an unduly dangerous activity, quite common in countries with low levels of bicycle use, may find confirmation in accident statistics and newspaper reports. But at the same time, as a public image this idea itself will have a limiting effect on the practice of bicycle use, and may act as a justification for policymakers to intervene in existing practices—favouring or further hindering cyclists.[36] Similarly, American cycling transport engineer John Forester has suggested that in the United States, widespread perceptions of utility cycling as "an activity of odd people", forbids massive use of cycling facilities, even where these are available.[37] Recent research on Dutch bicycle history suggests that the high overall levels of cycling in the Netherlands can partly be explained by the way cycling already before the Second World War was perceived and promoted as part of Dutch national identity.[38] In cycling as in other areas of human activity collective ideas and public images play a distinct role of their own.

## 2 The Bicycle's New Frame: Green, Hip, and Innovative

When it comes to investigating the ways collective ideas of cycling and modernity were reconciled since the 1970s, some historical phenomena immediately spring to mind, while others are less evident. I have identified three partly interlocking developments: the "greening" of the bicycle, the bicycle becoming a life-style product signifying a healthy, hip and active life-style, and finally the

35 De la Bruhèze and Veraart, *Fietsverkeer*, 197–198; Anne-Katrin Ebert, *Radelnde Nationen. Die Geschichte des Fahrrads in Deutschland und den Niederlanden bis 1940* (Frankfurt a.M.: Campus Verlag, 2010); John Parkin, Tim Ryley and Tim Jones, "Barriers to cycling: an exploration of quantitative analyses", in *Cycling and Society*, ed. Dave Horton, Paul Rosen, and Peter Cox (Aldershot: Ashgate, 2007), 67–82, here 71, 80.

36 Dave Horton, "Fear of cycling", in *Cycling and Society*, ed. Horton, Rosen, and Cox, 133–152.

37 Forester, "The place of bicycle transportation".

38 Ebert, *Radelnde Nationen*.

re-establishment of the bicycle as a product worthy and capable of technological innovation.

Clearly, the renewed popularity of cycling for transportation is connected to the rise of environmentalism and other post-materialist movements criticizing the "project of modernity" since the end of the 1960s.[39] Since then, many "counter-culture" movements have propagated the bicycle as an ideal vehicle to oppose and counteract certain characteristics of modernity, not only environmental pollution and urban congestion, but also social and political issues such as (global) inequity, the dominance of industrial technologies and the lack of "conviviality". This new public image of the bicycle was arguably most profoundly and spiritedly promoted by bestselling 1970s publicist Ivan Illich (1926–2002), especially in his *Tools for Conviviality* (1973) and *Energy and Equity* (1974). Linking transport issues with a more general plea for a "low-energy, convivial modernity", one of Illich's characteristic suggestions was that the American defeat in Vietnam could be seen as the well-deserved victory of bicycles over motorized transport. According to Illich:

> High speed is the critical factor which makes transportation socially destructive ... Participatory democracy demands low-energy technology, and free people must travel the road to productive social relations at the speed of a bicycle.... The bicycle lifted man's auto-mobility into a new order, beyond which progress is theoretically not possible. [40]

Of all the new social movements cherishing this new image of the bicycle, environmentalism probably was the most successful in reaching a general audience in the course of the following decades. Environmentalism can also be considered as a major stimulus for the establishment or revival of bicycle advocacy organizations in many western countries from the 1970s onwards.[41] These new cyclists' organizations stood up for the interests of cyclists, but at least

---

39 See Z.M. Furness, "Put the Fun between Your Legs! The Politics and Counterculture of the Bicycle" (Ph.D. thesis, University of Pittsburgh, 2005); Benoît Lambert, *Cyclopolis, ville nouvelle. Contribution à l'histoire de l'écologie politique* (Genève: Editions Médecine & Hygiène, 2004).

40 Ivan Illich, *Energy and Equity* (London: Calder & Boyars, 1974), 24, 73 and 75; see also Illich's *Tools for Conviviality* (London: Calder & Boyars, 1973). Illich's remark about the role of the bicycle in the Vietnam war was repeated by Edgerton, *Shock of the old*, 152.

41 The Dutch Fietsersbond, founded in 1975, reached the number of 30,000 members in the early nineties and has stayed above this level ever since. The German ADFC, founded in 1979, welcomed its 100,000st member in 1999. The League of American Wheelmen (since 1994: League of American Bicyclists) was revived in 1965 and represents at this moment over six hundred affiliated organizations that together have more than 300,000 members.

initially their aim and inspiration was broader than just interest representation: they wanted to promote those "means of transport that are safe and environmentally friendly", as the 1975 founding statutes of the Dutch Fietsersbond read.[42] Other examples of this "cycle-ecologist movement" were the French-Canadian Le Monde à Bicyclette (1975), the German ADFC (1979), the British Cycle Campaign Network (1980), and the Swiss IG/Pro Velo (1985).[43] Typical for the "greening of the bicycle" was that the environmentalist organization "Friends of the Earth" became in 1983 one of the twelve founding partners of the European Cyclists' Federation (ECF), by now representing sixty member organizations in 37 countries. As John Forester has observed, many of these organizations were "advocates of the bicycle"—propagating the use of a certain type of vehicle—rather than "cycling advocates", defending the rights of existing cyclists.[44]

How, then, are modernity and the bicycle reconciled in the environmentalist discourse? Ethnographic research by David Horton among (British) environmentalists in the early 2000s suggests that they, except for viewing the bicycle as a "zero-emission", "vehicle of sustainability", also very much see it as a "vehicle of opposition"—against the dominant car-culture that epitomizes modernity.[45] For many environmentalists the bicycle represents an *antidote* to modernity. Characteristically, the bicycle is often valued among environmentalists as a deliberately *slow* means of transportation and typically, as Horton has found, they do not prefer excessively "modern", trendy or innovative bicycles.[46] On the other hand, one could argue that environmentalists espouse an alternative ideal of modern life—one that is "post-materialist" and compatible with cycling for short range transportation, in Illich's words "low-energy, convivial modernity". Environmentalism reconciles the conceptions of modernity and the bicycle by changing both: modernity has become a battlefield for a greener future, to which the "green" bicycle can contribute. Organized in a huge variety of pressure groups both environmentalism and bicycle advocacy were able to influence policymakers and the general public, at least in changing the politically correct discourse in many western countries. To illustrate the effect that this change in public discourse has had on the image of cycling,

42 Bram Duizer, *'In Het Nut Van Actie Moet Je Geloven': Dertig Jaar Actievoeren Door De Fietsersbond* (Utrecht: Wetenschapswinkel Letteren Universiteit Letteren, 2005), 10. Only at the end of the 1990s did the Fietsersbond change its statutes and expressly restricted itself to interest representation (Duizer, *'In Het Nut Van Actie Moet Je Geloven,'* 28).

43 The term "cycle-ecologist movement" is taken from Lambert, *Cyclopolis, ville nouvelle.*

44 John Forester, *Bicycle Transportation* (Cambridge, MA, London: MIT Press, 1983), 19.

45 Dave Horton, "Environmentalism and the bicycle", *Environmental Politics* 15 (2006), 41–58.

46 Horton, "Environmentalism and the bicycle", 45, 51.

one can point to the fact that Volkswagen in 2009, before the car manufacturer was exposed as cheating the emissions tests for their cars, tried to sell cars by presenting them *as bicycles* in a series of Dutch commercials, thus (cynically) hoping to convince viewers of their "green" qualities.[47]

Secondly, the marketing of the bicycle as an adult lifestyle product and status symbol, especially through its incorporation in the increasingly important outdoor recreation market, has made the bicycle a trendy vehicle that has become strongly associated with the modern values of youthfulness, enduring health, and ageless fitness—though not necessarily with the idea of cycling as transport.[48] Already at the end of the nineteenth century the bicycle was marketed as a life-style product providing status through the public display of leisurely activities, but after cycling had become a common mode of everyday transport, this image of cycling gradually lost its appeal. The bicycle as status symbol returned as such in the 1960s, arguably with the launch of the small wheeled unisex folding bicycle in the United Kingdom. The folding bicycle was advertised as a lifestyle product, fitting for the modern man or woman, "a mini-bike to go with mini-skirts and mini-cars".[49] Indicating its success, by 1975, in Germany, one third of all bicycles sold were folding bicycles.[50]

At about the same time in the United States, the number of adult bikes sold—mostly touring bicycles—increased by a factor of twenty between 1965 and 1974, establishing a permanent change in the segmentation of the US bicycle market.[51] The introduction of the "mountain bike" at the very end of the 1970s—a sturdy rather than lightweight bicycle with fat tires, specifically designed for off-road use—confirmed and strengthened this trend.[52] Rapidly spreading to the rest of the western world, the mountain bike became an international sales-hit. In general, the sale of bicycles increased sharply at the end

47 See the 2009 commercials for "Volkswagen Sustainable BlueMotion Technologies", developed by the Amsterdam advertising company DDB: https://www.youtube.com/watch?v=u7yJKT-yEpw.

48 Cf. Trine Agervig Carstensen and Anne-Katrin Ebert, "Cycling cultures in northern Europe: from 'Golden Age' to 'Renaissance'", in *Cycling and sustainability*, ed. Parkin, 23–58.

49 Tony Hadland about the Moulton mini-bike, quoted in Jim McGurn, *On your bicycle: the illustrated story of cycling* (York: Open Road Publishers, 1999), 166. See also the 1966 UK advertisement in De la Bruhèze and Veraart, *Fietsverkeer*, 135 and Paul Rosen, *Framing Production: Technology, Culture, and Change in the British Bicycle Industry* (Cambridge: MIT Press, 2002), 102.

50 Gunnar Fehlau, *Das Modul-Bike. Faltbare Fahrräder* (Kiel: Moby Dick Verlag, 1997), 26–31.

51 Berto, *The Dancing Chain*, 228–230, 225.

52 Frank J. Berto, *The Birth of Dirt: Origins of Mountain Biking*, 2nd ed. (San Francisco: Cycle Publishing/ Van der Plas Publishing 2009).

of the 1960s and has stayed at a higher level since.[53] As a result, bicycle ownership among adults in the UK in 2000 was twice as high as in 1976.[54] In Germany likewise, bicycle ownership rose by 50 per cent between 1969 and 1993, from 50 per cent to 75 per cent of all households.[55] Even in the Netherlands, bicycle ownership increased further, from (less than) 56.3 per cent in 1970 to 84.0 per cent of all inhabitants in 1997.[56]

The successful product innovations just mentioned, and others that followed in the wake of these, were facilitated by changes in production methods and organization, making bicycle production more flexible to adapt to a segmented market.[57] But arguably more important was the changed demand that made these bicycle designs into a commercial success. The small wheeled folding bicycle of the 1960s, for instance, was developed to fit into the trunk of a car and meant for adults in possession of a car, who wanted to combine car and bicycle trips. Thus, riding a folding bike suggested car ownership, rather than lack of money to afford a car. Later, it became popular in combination with aeroplanes and boats for touring or with trains for commuting—both adding to, rather than diminishing someone's social standing. The mountain bike too developed out of specific uses and is by now recognized in the economics research literature as a classic example of demand-side innovation.[58] The first mountain bikers, using modified old bicycles, sought excitement and thrill in downhill racing and an outdoor physical workout in cross country rides.[59] Exploring the natural environment (not just enjoying the landscape as in

---

53 For the UK see: De la Bruhèze & Veraart. *Fietsverkeer*, 229; Rosen, *Framing Production*, 120. In the US, bicycle sales increased gradually from the 1940s onwards, but steeply only in the first half of the 1970s, followed after a short fall back, by another steep rise in the early 90s, see Frank J. Berto, "The Great American Bicycle Boom", *Cycle History: Proceedings of the International Cycling History Conferences* 10 (1999): 133–41, here 133.

54 Rosen, *Framing Production*, 122.

55 *Erster Bericht der Bundesregierung*, p. 3.

56 Numbers based on: *Zicht op Nederland Fietsland* (Amersfoort: Stichting Landelijk Fietsplatform, 2009), 6; CBS Statline (statline.cbs.nl), and De la Bruhèze & Veraart, *Fietsverkeer*, 51. The first percentage is based on the proportion of bicycles and inhabitants, not accounting for those who have more bicycles than one; the second percentage gives the proportion of bicycle owners within the total population.

57 Rosen, *Framing production*, 152–154.

58 See e.g. Guido Buenstorf, "Designing Clunkers. Demand-Side Innovation and the Early History of the Mountain Bike", in *Change, Transformation and Development*, ed. J. Stanley Metcalfe and Uwe Cantner (Heidelberg: Physica, 2003), 53–70. This was also the case with the "high riser bike" of the sixties and the not yet mentioned BMX (Bicycle Motor Cross) of the late seventies: both motor cycle imitations were developed by users before commercial producers took over. See Berto, *The Dancing chain*, 228, 241.

59 Buenstorf, "Designing Clunkers", 65.

touring) became another new motive for cycling, made possible by the mountain bike. Even before the introduction of the mountain bike, recreational cycling increased in popularity, reflecting a growing interest in physical fitness and outdoor recreation.[60] In the US, for instance, the percentage of people who participated in cycling for recreation more than tripled from 1960 to 1982/3 (to 37 per cent of the American population older than 12 years), establishing touring bicycles as an important segment of the adult bicycle market at the end of this period.[61]

Made possible by a further rise of affluence (allowing for an increase in bicycle ownership) and available leisure time, the renewed interest in recreative cycling was not so much associated with ideals of changing the course of modernity, as cycling was in the environmentalist discourse.[62] Rather it was experienced as an integral part of being modern: an individual consumer's way to relax, be in control and compensate for the lack of physical exercise that characterizes modern daily life. The main focus was on staying young, fit and healthy—and it had little to do with the idea of cycling as transport. Only in the first decade of the twenty-first century, it seems, *urban* cycling, again helped by product styling and differentiation, has acquired the same fashionable image among a young and hip audience in many western cities.[63] Not only for environmentalists, but for a broader group of consumers, the bicycle has increasingly become a "vehicle of distinction" (Horton), used to publicly demonstrate a preferred identity and purposefully distinguish oneself from others. The status issues involved in this development explain why in American neighbourhoods subject to gentrification "low-income and minority communities see cycling culture as accompanying processes of rising living costs, displacement, and the undermining of established local cultures".[64]

---

60 A full history of recreative cycling since the 1960s has not yet been written; see for a start Herlihy, *Bicycle*, 363–375.

61 H. Ken Cordell et al., "Outdoor recreation participation trends", in *Outdoor recreation in American life: a national assessment of demand and supply trends,* ed. H. Ken Cordell and Susan M. McKinney (Champaign, IL: Sagamore Publishing, 1999), 219–321, here 235; Berto, *The Dancing Chain*, 245–246.

62 I feel, Paul Rosen is overstressing the "de-modernizing impulse" and "nostalgia" involved in mountain biking as well as "mountain bikers' rejection of the modernization of the city", see Paul Rosen, "The Social Construction of Mountain Bikes", *Social Studies of Science* 23 (1993): 479–513, here 499, 501 and Rosen, *Framing Production*, 135.

63 See the proliferation of the websites devoted to "cycle chic" since 2007, now including websites from Hollywood, New York, London, Barcelona, Sidney, Malmö, Copenhagen, Helsinki, Vancouver, Lisbon and Vienna, see: www.copenhagencyclechic.com.

64 Elizabeth Flanagan, Ugo Lachapelle and Ahmed El-Geneidy, "Riding tandem: Does cycling infrastructure investment mirror gentrification and privilege in Portland, OR and

A third and last aspect of the changing image of cycling, next to the greening of the bicycle and the increased importance of the bicycle as signifying a healthy, hip and active life style, is perhaps less evident: the re-establishment of the bicycle as a technological artefact with an open future, capable of being innovated and worthy of scientific research and development. This view of the bicycle as a potentially innovative product stands in contrast to the idea that the bicycle essentially had reached its final technological development at the end of the nineteenth century and could not substantially be improved any further. The commercial motor behind this new development was undoubtedly the growth and diversification of cycling as a global sport since the 1970s, which has led to an intensified competition in vehicle development. The introduction and development of mountain-biking especially has brought a wave of technological innovations that were subsequently transferred to other types of bicycles. There is, however, also another, lesser known aspect to the re-establishment of the bicycle as potentially innovative product, that I will discuss in the rest of this chapter: the rise of the Human-Powered Vehicle movement since the 1970s.

## 3 The Human-Powered Vehicle Movement

To start with, imagine the freeway traffic on a busy Californian morning. The double lanes of the freeway are filled with motorcars and trucks, most of them driving at a speed well above the legal minimum of 45 miles per hour (70 km/h). One vehicle is different, however, and attracts the attention of drivers passing by. Its slim and smooth shape makes it look miniscule among the voluminous and hard-edged contours of most other vehicles on the road. With less than half the width of other cars, its roof does not even reach one meter above the tarmac that passes swiftly below its three skinny and barely visible wheels. The vehicle moves silently with an average speed of just above 80 km/h (50.5 m/h), hitting 95 km/h (59 m/h) on downslopes. Its passengers are shielded from the outside world, not by steel and glass, but by a synthetic hull. It is a shiny, 3.85 m long bullet-like vehicle, unlike any other car on the freeway. The most remarkable difference with the other cars, however, is not on the outside, but on the inside. There is no fuel tank, there is no motor and no exhaust. The vehicle is pedal powered, runs on "milk and cornflakes" and produces nothing but

Chicago, IL?", *Research in Transportation Economics* (in press, corrected proof available online 24 September 2016): [1].

exhalation and perspiration: it is the two passengers who propel it, keeping up with the motorized traffic by using nothing but human power.

This image of a human-powered vehicle on the freeway is not science fiction nor a vision of the future. It is a fact of history, taking us back to May 30th, 1980.[65] The location was the Californian Interstate 5 freeway, 80 miles northeast of San Francisco. The occasion was the first Energy and Transportation Fair organized by the California Department of Transportation (CalTrans) in Sacramento. To attract media attention for the Fair, Caltrans turned to the fastest Human-Powered Vehicle (HPV) around at that moment, undoubtedly to provide a telling example of alternatives to car transportation:[66] earlier in May, the Vector tandem (as the vehicle was called) had set a new world speed record on a Californian racetrack, breaking for the first time ever the 100 km/h limit (62,92 m/h) for unpaced and non-motorized vehicles over a distance of 200 m.[67] Bridging the 67 km between Stockton and Sacramento with an average speed of 80 km/h in real life circumstances was arguably an even more stunning accomplishment for a human-powered vehicle.

The Vector tandem was a record-breaking machine—but it was not unique. Similar human-powered vehicles or "HPVs"—tricycles and bicycles without power assist, but with an aerodynamic body and/or a recumbent design—were being built and raced in California since the first "International Human Powered Speed Championships" were held there in 1975. One year later, in 1976, the International Human Powered Vehicle Association (IHPVA) was established—in first place to regulate the records and organize future races.[68] The Caltrans employee who had suggested the "human power on the freeway" stunt, was a member of the IHPVA.[69] With a firm local basis in California, the IHPVA from the beginning attracted members from all over the Unites States and even abroad. The Board of Directors of the IHPVA in 1977 included members

65 F. Dan. Fernandes, "Human Power on the Freeway." *Human Power* 1, no. 5 (1980): 1–2.

66 According to the Caltrans-website "the 1970s were a time of austerity. The then-current political philosophy urged alternatives to highway building, a trend that would continue into the 1980s. Such thinking led [in 1973] to a new name for the department, Caltrans, short for the California Department of Transportation. The name change was emblematic of new thinking, and a rise in the concept that while highways have long been vital to the state, other forms of transportation were emerging to complement roadways", quoted from: www.dot.ca.gov/aboutcaltrans.htm.

67 See the recordlist available from www.ihpva.org.

68 See [anonymus], "History of IHPVA": 10, 19, and several publications by main organizer Chester Kyle. "A History of Human-Powered Land Vehicles and Competition." In *Human-Powered Vehicles*, edited by Abbott and Wilson, 95–111; Kyle, "A Brief History of the International Human-Powered Vehicle Association".

69 Fernandes, "Human power on the freeway", 2.

from several European countries, Australia and Japan.[70] In the early and mid-eighties, national HPV clubs were set up in the United Kingdom (1983), the Netherlands (informally 1984, formally 1985), Switzerland (informally 1984, formally 1985), West Germany (informally 1985, formally in 1986), and, followed in the nineties by France (1991), Denmark (1993), Belgium (1993) and some other countries. Even in the former Eastern Bloc, engineers with a similar interest in the development of innovative human-powered transport, though not formally associated with the IHPVA, apparently were encouraged by its activities to become active in HPV experiments, design-competitions and publications in the 1970s and 80s.[71] Although spread all over the west, the HPV movement is not—and never was—very popular. Available numbers suggest a membership-peak in the 1990s of somewhere between 5,000 and 10,000 members of HPV clubs worldwide.[72]

Although the rise of the Human-Powered Vehicle movement was encouraged by increasing concerns about the natural environment, as well as by the market trend towards product differentiation, it added a distinctive modernist perspective of its own. Far from inducing nostalgia, its innovative bicycles were meant to inspire visions of the future. The HPV movement was driven by a belief in technological progress. Most conspicuously, it challenged the monopoly on high speed transportation held by motorized vehicles. It consisted (and still consists) of a number of associations or clubs that were established in many

70 *Human Power* 1, no. 1 (1977), 11.

71 Documented by the Lithuanian engineer Vytas Dovydėnas, *Velomobile* (Berlin: Verlag Technik, 1990), see esp. pp. 5–8, 19–33 and the bibliography. Already David G. Wilson's HPV design contest of 1967 had invited two successful Polish entries, see David G. Wilson, "Man-powered land transport", *Engineering* 207, no. 5372 (11 April 1969): 567–573.

72 This rough estimate is based on incomplete data; also, it is not clear how many double memberships of both IHPVA and national clubs exist. According to founding member Chester Kyle, the IHPVA had its highest membership number in the beginning of the 1990s, with just over 2000 members (cf. Chester R. Kyle, "A Brief History of the International Human-Powered Vehicle Association 1976–1998", *Cycle History* 12 (2001): 134–145, here 143); in 1994 only about 1700 seem to have been left (*HPV News* march 1995, p. 5). Membership of the biggest European HPV club, the Dutch NVHPV increased rapidly from 60 in 1985 to more than 1500 in 1997 and has stabilized since then at a level of about 1750 (statistics in *Ligfiets&* 26, no. 5 (2010): 9). The German HPV club membership increased from about 60 in 1985 to about 800 members in 1995 (statistics in: *HPV Chronik. 10 Jahre HPV—Eine Chronik und mehr!*, ed. Michael Pohl (Erlangen: HPV Deutschland e.V., (1995), 71) and it had in 2009 together with the Swiss club about 1,400 members (according to the circulation of their joint newsletter *InfoBull*), the Swiss accounting for about 250 members (see "Über uns" on www.futurebike.ch). The Danish HPV Klub had about 200 members in 2008 (according to the membership-list in their club magazine)—other clubs probably even less. I have no information on the oldest European HPV club, the British BHPC.

western countries from the mid-1970s onwards, to promote the development and use of "human-powered vehicles". In the following sections, I will analyze the main activities and ideas of the HPV movement, addressing three major themes: the movement's attitude towards speed, towards politics and towards technology. Finally, I will evaluate its significance for the modern renaissance of the bicycle. The focus is on the HPV clubs in the United States, Great Britain, the Netherlands and Germany, my main sources being newsletters, magazines and other publications issuing from the HPV subculture.[73]

## 4 Speed

In its fascination with speed, the human-powered vehicle movement was distinctly modern or modernist. Instead of appreciating the bicycle as a slow vehicle and deliberately stressing the differences with cars, many HPV builders were driven by an ambition to minimize these differences and build a bicycle that could compete with cars when it came to comfort, reliability and speed. That is, they reconciled modernity and the bicycle by stressing and strengthening its modern aspects. In fact, the wish to establish speed records was the main reason to found the IHPVA. As their deviating designs—streamlined and/or recumbent bicycles and tricycles—were excluded from races organized by the world governing body of bicycle racing UCI, the HPV builders had to create an organization of their own to formalize and recognize speed records.[74] The records themselves were remarkable from the very beginning indeed, and

73 Reference will be made to the following HPV magazines: *Human Power* and *HPV News* (US), *BHPC Newsletter, Bike Culture Quarterly* and *Velo Vision* (UK), *HPV Nachrichten* (in *Provelo*) and *InfoBull* (Germany/Switzerland), *HPV Nieuws/Ligfiets&* (NL). The available histories of the HPV movement are mostly produced by members of the movement: "The history of IHPVA", *Human Power* 1, no. 1 (1977), 10+19; Dovydėnas, *Velomobile*; Gunnar Fehlau, *Das Liegerad* (Kiel: Moby Dick Verlag, 1996); Kyle, "A Brief History"; Pohl, *HPV Chronik*; David Gordon Wilson, Richard Forrestall, and Dereck Henden, "Evolution of Recumbent Bicycles and the Design of the Avatar Bluebell", in *Second International Human Powered Vehicle Scientific Symposium. Proceedings*, ed. Allan V. Abbott (Indianapolis: IHPVA, 1983), 92–103; David Gordon Wilson, "The development of modern recumbent bicycles", in *Human-powered vehicles*, ed. Allan V. Abbott and David G. Wilson (Champaign, Ill., [etc.]: Human Kinetics, 1995), 113–127; Arnfried Schmitz and Tony Hadland, *Human Power - the Forgotten Energy 1913—1992* (Coventry: Le Thor, 2000); Tony Hadland and Hans-Erhard Lessing, *Bicycle design: an illustrated history* (Cambridge, Mass.: the MIT Press, 2014), 473–492.

74 See Kyle, "A Brief History", 137. The UCI had banned streamlined and recumbent bicycles already before WW II, see: Arnfried Schmitz, "Why your bicycle hasn't changed for 106 years", *Human Power* 11 (1994): 4–9; Otto Beaujon, "Was Bicycle Development Stunted by

did much to create publicity for the movement and make its efforts visible to a wide audience. The best non-professional HPV athletes were from the start able to beat the top athletes complying with the UCI rules. For instance, at present (2016), the hour record for HPVs (currently 92.4 km) is approaching twice the maximum achieved on a UCI-approved racing bicycle (54.5 km).

Even more spectacular than the land speed competitions were the IHPVA's efforts to improve human-powered water- and aircraft. In 1977 Californian IHPVA member Paul MacCready won the Kremer prize for meeting the 18-year-old challenge to fly a one mile figure 8 using nothing but human power, followed two years later by a human-powered record-flight across the English Channel.[75] Stunts like this one, or the 100-km-ride of a Vector HPV through the German Mosel Valley in just over 1.5 hours in 1987,[76] fuelled the imagination as to what human-powered vehicles could achieve. Especially in the first years, the HPV races and records attracted a lot of media attention and this media effect was essential in spreading the movement internationally, through articles in magazines such as *Scientific American* and *National Geographic*.[77] New magazines such as *Bicycle magazine* (UK, 1982–), *Fiets* (NL, 1982-), or *ProVelo* (Germany, 1984–) reported enthusiastically and repeatedly about HPVs, bicycle fairs like the Amsterdam RAI of 1982 or the Cologne IFMA of 1982 showed the first commercial products and prototypes.[78] In many European countries, before HPV clubs were established there, HPV races and competitions were held and the first commercial builders had started their small-scale businesses.[79] In fact, in Europe, just as in the US, the HPV clubs originated out of organizing competitions. The movement's focus on speed records, in itself in line with

Organized Racing?" *Cycle History* 10 (1999): 67–76; Chester R. Kyle, "Bicycle Aerodynamics and the Union Cycliste Internationale", *Cycle History* 11 (2000): 118–131.

75 See MacCready's own report in *Human Power* 1, no. 1 (1977): 4–5.

76 See Wolfgang Gronen, "Die Vector-story", in *HPV Chronik*, ed. Pohl, 57.

77 See *Human Power* 1, no. 1 (1977): 7+19; Kyle, "A Brief History", 140.

78 See e.g. the reports in *Fiets* 1, no. 1 (1982): 6 and 1, no. 5 (1982): 30. Indicating the American influence on new cycling trends in Europe, "Holland's first recreative cycling magazine" *Fiets* (1982) was created by an American publisher Bob Rubinstein, living in Amsterdam. *Fiets* organised the first Dutch race for HPVs in Zandvoort (1983) and one of the Americans working for the magazine (Cary Peterson) became a founding member of the Dutch HPV club (see *Ligfiets&* (2005), no. 2: 6–7).

79 See Schmitz and Hadland, *Human Power*, 43ff, for an impression of the atmosphere at the early HPV races in Europe. Among the "pre-club" races were the ones in Brighton (1980), Den Haag & Zandvoort (Netherlands, 1983) and Nümbrecht (Germany, 1984). The first HPV/recumbents on the market were in the US the Avatar 2000 (1979), and in Europe the Windcheetah (UK, 1982), the Vélérique (Belgium, 1982), the Roulandt (NL, 1983), the Leitra (DK, 1983) and the Radius Peer Gynt (Germany, 1985).

established views of modernity, helped to make the bicycle visible again in the public domain.

## 5 Politics

Thus, racing and record attempts were an important public manifestation of the new HPVs and a main motive for organizing HPV clubs. But as the first issue of the IHPVA magazine *Human Power* stated in 1977, the construction of rapid HPVs was not an end in itself:

> It was hoped that the stimulus provided by the Speed Championships and by other events sponsored by the IHPVA would encourage radical changes and improvements in human powered transportation throughout the world.[80]

The magazine-title *Human Power* already indicated its wider program. In the first volume, the editor suggested:

> In broader terms, *Human Power* perhaps ought to even promote stationary power production in methods such as stairs (versus elevators), electricity generation (i.e., the pedal-powered television), the treadle sewing machine, and such reactionary devices as the manual toothbrush. Personally, I even have a great interest in using human power to overcome government mass and friction, and social inertia.[81]

And although this was written down half-jokingly, the underlying message was serious. In a contemporaneous Rodale publication devoted to *Pedal power in work, leisure, and transportation* (1977), prominent IHPVA member David Gordon Wilson (1928–) elaborately advocated the use of human-powered devices similar to the ones mentioned by the *Human Power*-editor. Thinking about "the future potential for muscle power", Wilson claimed that "intelligent application of muscle power could reverse historical trends and roll back the use of external power" modern man was increasingly and unfortunately dependent upon: "decades of mechanization and pollution have led us to reconsider human muscle potential ... Hopefully, all countries will reconsider the

80 *Human Power* 1, no. 1 (1977): 10.
81 *Human Power* 1, no. 1 (1977): 3.

enlightened use of pedal power. The high cost of energy and the failure of our transportation systems make such reconsiderations imperative".[82]

Wilson became involved in the IHPVA from the very beginning as a member of the initial Board of Directors, and quickly became a leading figure within the IHPVA, first as its President (1983/84) and then as the longtime editor of *Human Power* (1994–2002). Already before the establishment of the IHPVA this MIT professor had been actively encouraging efforts to innovate the bicycle and these efforts may well have set the scene from which the IHPVA emerged. As a professor of mechanical engineering he had organized a widely publicized international design competition to create an innovative human-powered land transport vehicle back in 1967. In 1974, he became co-author of the MIT publication *Bicycling Science* which gave currency to the term "human-powered" transport and concluded with the vision of a "hopeful future" in which more widespread use of "improved bicycles" would counter threatening prospects of ongoing pollution and climate change.[83] Now in its third edition, it is still the only available English textbook on the physics of the bicycle, in which the bicycle is presented as subject of science and innovation.

Unlike many HPV builders from the Californian scene, Wilson was less interested in top speeds than in improving the safety and comfort of the bicycle. Instead of designing fast but impractical streamlined HPVs, as other HPV enthusiasts did, Wilson made the first modern case for the recumbent bicycle as the ultimate HPV for everyday use. Born and educated in Great Britain, Wilson himself had experienced (and in the 1970s deliberately and repeatedly commemorated) the massive use of the bicycle as means of transport in the late 1940s and he once reported how he was considered a "very queer fish" when he first rode his bicycle in the States in 1953.[84] From the seventies onwards, he propagated the recumbent bicycle as the commuter-bike of the future. His experiments with different recumbent designs, one of which was given the significant name "Green Planet", led in 1979 to one of the first modern

82 David Gordon Wilson, "Human muscle power in history" / "The future potential for muscle power", in *Pedal Power - in Work, Leisure, and Transportation*, ed. James C. McCullagh (Emmaus, Penns.: Rodale Press, 1977), 1–36 / 106–123, quotes on 106–107 and 36 respectively.

83 Frank Rowland Whitt and David Gordon Wilson, *Bicycling Science: ergonomics and mechanics* (Cambridge, Mass.: MIT Press, 1974), 222–231 (including a report on the 1967 bicycle design competition).

84 See David Gordon Wilson, "Getting in gear: human-powered transportation", *Technology Review* 81, oct. (1979): 42–54, here 53 and 42. The anecdote about his UK experiences was reiterated several times by Wilson.

commercially available recumbent bicycles, with an equally significant name: the Avatar 2000—a "reincarnation" of the bicycle suitable for the year 2000.[85]

For Wilson innovating the bicycle was one of three main things that were needed to bring about, what he called, an "HPV revolution":[86]

> For there to be a major increase in the use of human-powered vehicles (HPVs) for commuting, shopping and recreation there needs to be a further development of vehicles to make them easier, safer and more convenient to use.

The other things needed to stop the negative effects of the massive car-mobility for society and the environment, was improving the infrastructure for cycling and a governmental policy that would make car-drivers pay for the true costs of using their vehicle, including the environmental costs—a plea that Wilson has repeated ever since the 1970s.[87] The combination of bicycle innovation and heavy taxation of car-use was intended to contribute to "a substantial proportion of the population freely choosing HPVs as their means of commuting".[88] "Virtually every city in the 'western' world is facing massive problems of traffic congestion ... Human-powered vehicles are not the complete solution to these problems, but they can play a vital role in a 'kinder, gentler transportation pattern'".[89]

Within the IHPVA similar considerations soon led to the organization of the first of a series of "practical vehicle design competitions", that would contribute to the development of HPVs suitable for "everyday, year-round use".[90] The (improved) bicycle was considered to have a "powerful future" as the "vehicle for a small planet", as World Watch researcher Marcia Lowe stated in *HPV*

85 See Wilson, "Development of modern recumbent bicycles".

86 David Gordon Wilson, "A Blueprint for an HPV Revolution", in *Third International Human Powered Vehicle Scientific Symposium. August 28 and 29, 1986. Proceedings*, ed. Allan V. Abbott (Indianapolis: IHPVA, 1986), 79–83, here 79.

87 Already in McCullagh, *Pedal Power* (p. 123) there is a short reference to it, repeated in Wilson, "Getting in gear", and again in David Gordon Wilson, "Human power in the future", *Human Power* 54 (2003): 13–14; see for one of his latest proposals: David Gordon Wilson, "White paper on proposed tax-and-rebate bill to reduce emissions and nonrenewable energy use [etc.]", *lessgovletsgo.org*. January 3, 2011. http://lessgovletsgo.org/wp-content/uploads/2013/10/White_paper_on_energy_3Jan2011.pdf.

88 Wilson, "Blueprint", 79.

89 David G. Wilson, "Editoral", *Human Power* 7, no. 2 (1988): 2.

90 See *Human Power* 2, no. 1 (1982): 2.

*News*.[91] The HPV builders didn't limit themselves to designing innovative bicycles for those who could afford an expensive hand-built machine. In *Human Power* an article appeared on the design of a human-powered "mobile home" for homeless people.[92] The magazine also published a series of articles on possibilities to improve rickshaws and HPVs for heavy load carrying in developing countries.[93] "We all share the goal of a desirable, sustainable world", human-powered flight pioneer Paul MacCready stated in 1995, "Human-powered vehicles will help move us in that direction".[94]

Not all HPV clubs were as explicit as the IHPVA publications on the "politics of human power". There were national differences in stressing the role of politics, some clubs focusing more on the technical or even the athletic competition than on politics. In some European HPV clubs the environmentalist inspiration for bicycle innovation was made explicit in the club's statutes (which was not even the case with the IHPVA).[95] The Swiss HPV club FutureBike, for instance, formulated as its main objective: "to promote environmentally friendly propulsion technology, in particular by muscular driven lightweight vehicles free of exhaust gases".[96] Similarly, the statutes of German HPV club stated that it wanted to "promote the protection of landscape and environment through organizing public manifestations and campaigns in favour of environmentally friendly means of transport".[97] In the German statutes, organizing races was only listed as the very last goal of the German HPV club, and some of its members expressed their criticisms of the "delusion of speed" that

91 Marcia Lowe, "Bicycles have a powerful future", *HPV News* 5, no. 3 (1988); Lowe, *The Bicycle, Vehicle for a Small Planet.*

92 See *Human Power* 7, no. 1 (1988): 10–11.

93 See *Human Power* 3, no 3 (1985): 18–20; *Human Power* 4, no. 4 (1985): 27–28; *Human Power* 5, no. 3 (1986): 19–20; *Human Power* 5, no. 4 (1986/7): 5–6; *Human Power* 48 (1999): 22–24; *Human Power* 49 (1999–2000): 15–18; see also McCullagh, *Pedal Power*, 37–56.

94 Paul B. MacCready, "The value and future of Human-Powered Vehicles", in *Human-powered vehicles*, ed. Abbott & Wilson, 265–267, here 267.

95 In 1980 the IHPVA statutes read: "The specific purposes of this corporation are to: [1] Encourage and support unrestrictive, innovative and creative design and construction of vehicles operated solely by human power for transportation on land, in the air and on the water. [2] Sanction and otherwise promote at minimum a yearly meet at which the vehicles may be showcased and tested. [3] Keep historical records (photographs, speeds, distances) of the results of these meets. [4] Provide focus for technical data regarding the design, construction and operation of these vehicles. [5] Publish a newsletter for dissemination of data and information regarding human powered vehicles of any type." See: www.ihpva.org/about.htm.

96 See: www.futurebike.ch > Über uns > Vereinstatuten.

97 See: http://www.humanpoweredvehicles.de/satzung.html.

HPV racing seemed to foster.[98] Founding member Wolfgang Gronen, who himself had a bicycle racing background and was clearly in the first place interested in speed competitions, is reported to have remarked once about his German club-members: "I couldn't co-operate with the majority of recumbent enthusiasts. They're not politically neutral, but either firmly red or rigidly green".[99]

For many human power enthusiasts in Switzerland, Germany and the United States it was not at all contradicting their focus on human power to investigate the possibilities of power assist systems on HPVs, because such systems would only strengthen the versatility of these vehicles without compromising (too much) the concern for "environmentally friendly propulsion technology".[100] In the same way, IHPVA members such as Paul MacCready became involved in *solar* powered vehicles. When it came to the rules for racing and competitions, however, such technical innovations were bound to lead to discussion and controversy—especially as the number of club-members who were more interested in *athletic* than technical competition grew, together with the availability of commercial racing recumbents.[101]

Compared with the American, Swiss and German HPV clubs, the British and the Dutch HPV-clubs were less pronounced in voicing environmentalist ideals. The *British Human Power Club*, the first European HPV association, founded in 1983, put most emphasis on organizing races—and the official club's publications sometimes downplayed any more wide ranging mission statements.[102]

---

98 See e.g. the section on "HPV Nachrichten" in *ProVelo* 35 (1993): 34; *ProVelo* 46 (1996): II; *ProVelo* 47 (1996): II–III; connected with the polarisation between environmentalists and racers within the German HPV club was the creation in 1995 of a separate German HPV club solely devoted to racing, see "HPV Nachrichten", *ProVelo* 44 (1996): VII.

99 Quoted in Schmitz & Hadland, *Human Power*, 93.

100 See e.g. John Tetz, "AHPV: assisted human powered vehicle", in *Fourth International Human Powered Vehicle Scientific Symposium August 6, 1992*, ed. Chester Kyle, Jean A. Seay, and Joyce S. Kyle (Indianapolis: IHPVA, 1994), 21–26; Andreas Fuchs & Theo Schmidt, eds., *Assisted Human Powered Vehicles and Velomobiles—Wettergeschützte Elektrofahrräder. Proceedings of the 4th Velomobile Seminar.* (Liestal: Future Bike CH, 1999); and several contributions in *Human Power* 10, no. 3 (1993): 19–23; *Human Power* 11, no. 2 (1994): 14–16; *Human Power* 11, no. 4 (1994/5): 13–14; *Human Power* 12, no. 2 (1995): 21–22; *Human Power* 12, no. 3 (1996): 13–14.

101 See e.g. several contributions by Peter Sharp in *Human Power* 10, no. 1 (1992), *Human Power* 10, no. 3 (1993), *Human Power* 11, no. 3 (1994) and *Human Power* 12, no. 2 (1995).

102 *BHPC Newsletter* 1 (1983), 1. Compare also the BHPC website (www.bhpc.org.uk/about-the-club.aspx): "The British Human Power Club was formed in 1983 to foster all aspects of human-powered vehicles—air, land and water—for competitive, recreational and utility activities, to stimulate innovation in design and development in all spheres of HPV's, and to promote and to advertise the use of HPV's in a wide range of activities. If you think this seems rather like a mission statement, you would be right: in reality, the BHPC was

Still, even here practical vehicle competitions were soon organized,[103] and one of its founders and presidents, the American born bestselling cycling author Richard Ballantine, published articles on "the politics of HPVs" and the governments obligation to stimulate cycling.[104] It was also Britain that became home to the only two commercial magazines propagating the central HPV message of "more bicycle use through bicycle innovation": *Bike Culture Quarterly* (BCQ, 1993–2000, published in an English and German version) and its successor *Velo Vision* (2000–). In the editorial with which he launched BCQ in 1993, Jim McGurn described the intended audience of BCQ as those who "see cycling as a way of life rather than an occasional leisure activity. *They wish to see cycle design cutting free from conventions, so that pedal-power can be used in more varied and meaningful areas of our lives*".[105] Both magazines publish(ed) many articles on HPVs and recumbent bikes and contributions by prominent BHPC members such as Mike Burrows and Richard Ballantine.

Remarkably, the Dutch NVHPV, that in the 1990s would become the biggest HPV club in Europe and the world, was initially quite silent about the politics of human power, and concentrated on organizing races—although its first newsletter casually remarked that "our recumbents [are] one of the best alternatives to the car".[106] Its 1985 statutes just mentioned that the club wanted "to promote the use and development of vehicles, vessels and aircraft that are powered by human muscle power".[107] According to the environmentalist organization *De Kleine Aarde* (Little Earth), that looked for co-operation with the NVHPV in the mid-1980s, the Dutch HPV club at that moment was a self-centered club of technicians and racers, (too) little concerned with problems of social acceptance.[108] This changed however from the end of the 1980s, when a growing number of its members became interested in the development of practical HPVs that could act as substitutes for cars. Articles on this topic

---

formed to allow people to race recumbent trikes and bikes and the mission statement was added to broaden the club's outlook."

103 See e.g. *BHPC Newsletter* 3 (1984), 4.

104 Richard Ballantine, "Go for the Money", *Bike Culture Quarterly* 1 (1993): 19; Richard Ballantine, "The politics of human powered vehicles", *Human Power* 54 (2003): 3.

105 Jim McGurn, "From editor Jim McGurn", *Bike Culture Quarterly* 1(1993): [2] (emphasis mine, MS). McGurn also published a series of seven buyer's guides on alternative and innovative bicycle designs: *Encycleopedia* (1993–2003). According to McGurn (*Bike Culture Quarterly* 13 (1997), 2) the number of subscriptions to BCQ had increased from 2000 to 8000 in three years.

106 *HPV Nieuws* 1, no. 1 (1984), 1. See for a detailed comparison of the IHPVA and the Dutch NVHPV: Stoffers, "The Politics of Bicycle Innovation."

107 See http://oud.ligfiets.net/nvhpv/statuten.php3.

108 See *HPV Nieuws* (1985), no. 4.

appeared from 1989 onwards in the club's newsletter (including a proposal by onetime Provo-member and bicycle activist Luud Schimmelpennink),[109] a sub-committee of the club was installed to stimulate this idea, and within years a successful design competition was organized, supported by government subsidies and with the cooperation of the Technical University of Eindhoven. The "365-days-bicycle" contest, that was concluded in 1993, would firmly establish the "velomobile" (or streamlined tricycle) as ultimate commuter bike within the HPV community in the Netherlands. From the 1990s onwards the Dutch club also started to argue for infrastructural adaptations that would facilitate long distance and high speed bicycle commuter traffic.[110]

Last but not least, in the East European version of the HPV movement, the politics of human power was again a prominent feature. The book on HPVs or "velomobiles", as he called them, by the leading Eastern Bloc proponent of HPV thinking, Lithuanian engineer and university professor Vytas Dovydėnas, concluded with a lengthy chapter on "velomobile transport systems" that would solve pressing urban environmental and congestion problems from above and, eventually, would lead to an ideal "ecopolis".[111] After the re-won independence of Lithuania, Dovydėnas, invited by government officials, undertook efforts to give what he called "biotransport" a more prominent position within Lithuanian traffic law, arguing that "governments should popularize biotransport" to stop the "degradation" of European towns brought about by the massive use of "such irrational transport" as the car.[112]

To conclude, the politics of human power played a role in all HPV clubs, though not always or in all of them equally pronounced. Since 2000, with a growing number of commercially available recumbent bicycles, the majority of members of the HPV clubs has changed from being HPV builders and hobby-technicians to consumers, a fair amount of whom appear to be predominantly interested in product information, experience sharing, sociability and racing. The fact that some clubs—i.e. in Denmark and the Netherlands—have recently considered to change their names into "recumbent cycling club", seems to

---

109 See Luud Schimmelpennink, "Solo: de 365-dagen fiets", *HPV Nieuws* (1990), no. 2: 15–17 (on Schimmelpennink's HPV project see also *Bike Culture Quarterly* 6 (1995), 21).

110 See *HPV Nieuws* (1993) no. 6; *HPV Nieuws* (1997), no. 5; *HPV Nieuws* (2005), no. 3; *HPV Nieuws* (2006), no. 1.

111 Dovydėnas, *Velomobile*, 95–121; the German edition was a translation of the Russian version of 1986, the book was originally published in Lithuanian in 1979.

112 See: Vytas Dovydėnas, "The draft of the biotransport law of the Lithuanian Republic", in *Safety and design. Proceedings of the second European seminar on velomobiles*, ed. Peter Zeller (Liestal: Future Bike CH, 1994), 163–166; Vytas Dovydėnas, "Biotransport and 'Green Towns' of Europe", in *Proceedings of the third European seminar on velomobile design*, ed. Carl Georg Rasmussen (Velomobile Forum, 1998), 13–15.

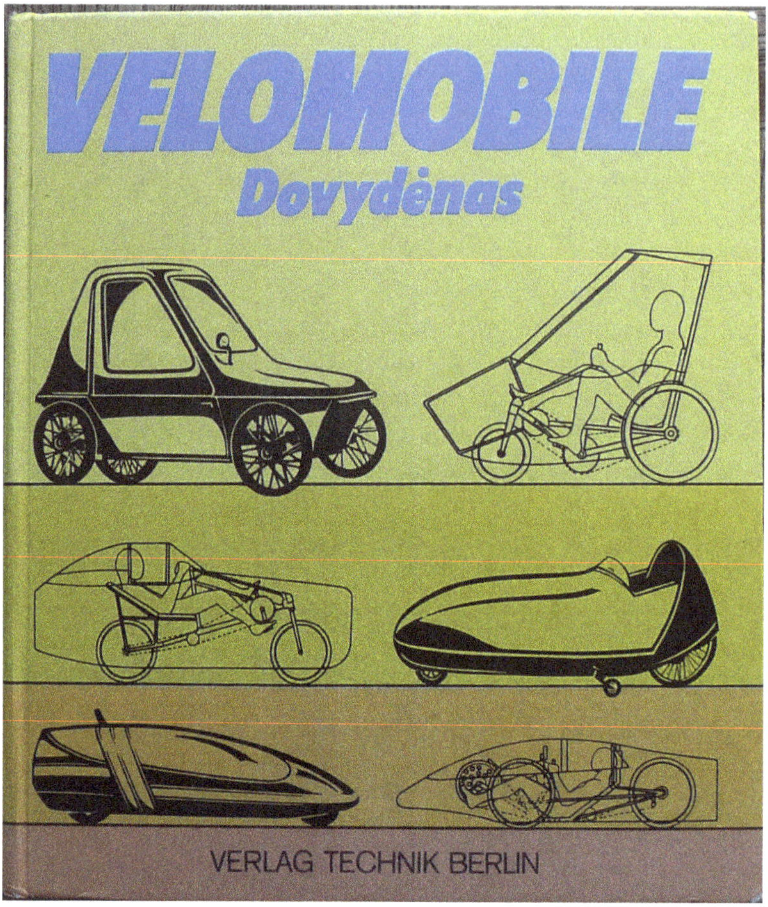


ILLUSTRATION 7.1 Cover of Vytas Dovydėnas book *Velomobile* (Berlin: Verlag Technik, 1990), showing several recent types of HPVs

indicate a growing indifference about the "politics of human power" within these clubs.

## 6 Technology

Recognizing the political inspiration of the HPV movement (next to its preoccupation with races and records), one may ask what exactly its position was in the general rise of bicycle advocacy and environmentalist groups since the

1960s. First of all, there were (and are), several interconnections between the HPV clubs and the (other) bicycle advocacy groups. For instance, David Gordon Wilson's Avatar 2000 recumbent was first presented in Europe, when Wilson was invited at the Velo-City Conference that the German ADFC had organized in Bremen in 1980.[113] A minority of the founding members of the German HPV club initially argued for integration in the ADFC.[114] The first European seminar on velomobile design was sponsored by the Danish Cycling Federation. At least in the Netherlands, many HPV club members seem to be members of the national bicycle advocacy organization as well.[115]

Still, the differences should not be overlooked. Whereas other bicycle advocacy organizations directed their efforts mainly toward changing government policies and improving the bicycle infrastructure, the specific focus of the HPV movement was on changing bicycle *technology*.[116] The characteristic that made HPV clubs stand out from other bicycle advocacy groups, was their belief in the value of technical innovation to lure people out of their cars and thus contribute to the pressing problems of fossil energy over-consumption. Their main technical targets were 1) to make the bicycle more comfortable (through a better seat and weather protection) and 2) more energy-efficient (through improving its aerodynamical qualities). One of the first Dutch recumbent producers succinctly called his firm M5, which stood for: Meer Meters Met Minder Moeite ("more meters with less effort").

The emphasis on innovative technology also gave the HPV clubs their specific place within the environmentalist movements of the seventies and eighties. The HPV clubs did *not* belong to the radical anti-modernist, technophobic wing within environmentalism. Instead, many HPV enthusiasts were optimistic adherents of the "appropriate technology" philosophy, believing in the real progress that scientific research and sensible technology could bring—for

113 See, Pohl, *HPV Chronik*, 36.

114 See, Pohl, *HPV Chronik*, 64.

115 *Ligfiets&* 26, no. 5 (2010), 25.

116 E.g. in 2009 the *VeloVision*-website explained that the magazine was not looking for contributions on "Car-bashing/preaching/politics: We feel the best way we can promote cycling is to offer practical, positive examples of bikes in action for transport and touring, and showcasing the new technology on offer and the possibilities it opens up for a greener, happier and community-centred lifestyle. [...] So [we] do NOT want anti-car arguments, environmental evangelism or politics, or accounts of injustices perpetrated against cyclists. If you feel the need to express such materials (and it may well be justified) then another forum is more appropriate." (see: http://www.velovision.com/archive, go to 'Magazine' > 'Notes for contributors').

instance in energy production and communications technology.[117] One of David Gordon Wilson's first pleas for the recumbent as commuter-bike of the future, was typically published in a volume devoted to several examples of appropriate technology.[118] In an intriguing combination of innovation-mindedness and anti-consumerist spirit, the German HPV scene issued in 1986 a popular DIY-guide on building recumbent bikes *from scrap*, in a series significantly called "Einfälle statt Abfälle" ("ideas rather than waste").[119] From the 1990s onwards, however, the DIY-HPV builders gave way to professionals, using sophisticated materials and production methods to produce slick looking, commercially available, high quality recumbent bicycles and tricycles.

Many HPV enthusiasts shared the environmentalists' appreciation of the bicycle as a "vehicle of sustainability".[120] In their fascination for an *innovated* bicycle, however, they added a different flavour to the positive evaluation of the bicycle. While many environmentalists preferred "sturdy" and "sensible" bicycles and appreciated the bicycle as a deliberately *slow* means of transport, for many HPV enthusiasts increasing the speed or (which comes down to the same point) increasing the energy-efficiency of cycling was a major target of their efforts to innovate the bicycle. Not only in its potential to contribute to sustainability, but also in its looks and construction, the bicycle for the HPV movement was a quintessentially *modern* means of transport.

Characteristic of the modernist appreciation of the bicycle were two popular types of images within the HPV subculture. On the one hand, there are the utopian pictures of the role of HPVs in future traffic, that can be found in publications from the HPV subculture, such as *BCQ*, *VeloVision* or the book on velomobiles by Dovydėnas. Such pictures typically show a diverse collection of bicycles and HPVs cruising along in sunny weather, in the middle of clean, modernistic architecture and surrounded by beautiful nature. These images celebrate the diversity of HPV designs for all possible purposes. On the other hand, there are popular cartoons from the HPV subculture, depicting (in

117 Compare Paul Rosen, "Up the Vélorution: Appropriating the Bicycle and the Politics of Technology", in *Appropriating Technology: Vernacular Science and Social Power*, ed. Ron Eglash et al. (Minneapolis: University of Minnesota Press, 2004), 365–389. See on the AT-movement: Andrew Kirk, "Machines of loving grace. Alternative technology, environment and the counterculture", in Peter Braunstein and Michael William Doyle,). *Image Nation. The American counterculture of the 1960s and '70s* (New York, London: Routledge, 2002), 353–378.

118 McCullagh, *Pedal Power*.

119 Christian Kuhtz and Werner Stiffel, *Chopper-Fahrräder* [Einfälle statt Abfälle, Reihe Fahrrad, Heft 5] (Kiel: Verlag Christian Kuhtz, 1986).

120 Also Illich's views on cycling were appreciated by HPV fans, see e.g. the reviews in *Bike Culture Quarterly* 14 (1997): 14–15 and *Human Power* 13, no. 2 (1998): 10–11.

several variants) an evolutionary line of development, from the fish or monkey via the racing cyclist to, ultimately, the HPV rider as the final stage of evolution. In fact, the linear, evolutionary perspective on the development of transport technology, that is typical for many transport planners,[121] also seems to be characteristic of the many HPV fans who consider their innovative vehicles to be superior to any other, whether ordinary bicycle or car.

Both the image of happy HPV diversity and that of HPV superiority confirm the belief in technological progress within the HPV movement. This belief came naturally to many HPV club members, many of whom had a professional background or were educated in science or engineering. Of the twelve founding members of the IHPVA, for instance, more than half were engineers, three had finished a doctoral dissertation, and two were working at a university.[122] Arguably the two most prominent members of the IHPVA, Chester Kyle (initiator of the first HPV speed championships and the IHPVA) and David Gordon Wilson, were both university professors who became leading experts in the field of bicycling science. Without a trace of irony Wilson once remarked in *Human power*: "we are not read by the masses, but by a small, highly intelligent, far-sighted group of individuals".[123]

Both American and European HPV clubs were active in organizing scientific symposia on HPV development and related issues.[124] The statutes of the German HPV club even put "stimulating science and research" in the very first place of its club-goals. If one looks at the contributors to the European seminars, the German academics and university teachers dominate. Involved from the very beginning was Professor Schöndorf of the *Fachhochschule* Cologne, who was a member of the initial Board of Directors of the IHPVA and a founding member of the German HPV club.[125] Also Professor Falk Riess, from the University of

121 See for an analysis and criticism: Frederick Van De Walle, *The Velomobile as a Vehicle for more Sustainable Transportation. Reshaping the Social Construction of Cycling Technology* [MSc thesis] (Stockholm: Kungliga Tekniska Högskolan, 2004), esp. 37–39. Also: Peter Cox and Frederick Van De Walle, "Bicycles don't evolve—Velomobiles and the Modelling of Transport Technologies", in *Cycling and Society*, ed. Dave Horton, Paul Rosen and Peter Cox (Aldershot: Ashgate, 2007), 113–131.

122 See the list in Kyle, "A brief history", 137.

123 *Human Power* 9, no. 1 (1991), 2.

124 Between 1981 and 1992 four scientific symposia were organized by the IHPVA in America. The "European seminars on velomobile design", which started in 1993, have been held eight times until 2015, alternatedly in Denmark (3×), Switzerland (2×), Germany (1×), the Netherlands (1×) and Austria (1×).

125 See: Paul Schöndorf, "20 Jahre Fahrradforschung an der FH Köln", in Pohl, *HPV Chronik*, 31–33; Schöndorf contributed to the first (1981) and second (1983) IHPV Scientific Symposium.

ILLUSTRATION 7.2 An HPV-utopia? Poster by Dutch HPV-designer Bauke Muntz, 2002
REPRODUCED WITH KIND PERMISSION OF BAUKE MUNTZ

Oldenburg, was already in 1980s involved in the movement and contributed to the development of an HPV concept model (OLF) at his university.[126] The same applied to Walter Rohmert, a renowned ergonomics professor at the University of Darmstadt (who also supervised a doctoral dissertation by Stefan Gloger on the subject),[127] and several other academics teaching at *Fachhochschulen* or universities.[128] Outside Germany, one may note that Dovydėnas, the major proponent of the East-European HPV movement, was teaching at Vilnius Technical University, and that the prominent Swiss FutureBike member Dr. Andreas Fuchs taught physics at the *Fachhochschule* Bern. In Great Britain

126 See e.g. Falk Rieß, "Leichtfahrzeuge für den Verkehr der Zukunft: Vom Fahrrad zum Alltags-Vector", *Provelo* 10 (1987), 23–32.

127 Rohmert (and Gloger) published several papers in the proceedings of the first and second "European seminar on velomobile design" (1993 and 1994).

128 See e.g. the following contributors to the European seminars on velomobile design: Prof. dr. H-P. Barbey (*Fachhochschule* (FHS = university of applied sciences) Bielefeld), Prof. Jürgen Eick (FHS Wiesbaden), Prof. G. Rinne and Prof. P. Wollschläger (FHS Braunschweig/ Wolfenbüttel).

and the Netherlands, on the other hand, it seems that involvement of universities in the HPV movement was mainly restricted to students.

## 7 The Human-Powered Vehicle Movement and the Modern Image of the Bicycle

What was the significance of this small, international movement, dominated by engineers and scientists? How did it contribute to the renaissance of the bicycle since the 1970s? First of all, unlike other, much bigger bicycle advocacy organizations that in some countries developed quickly from street activist clubs into lobby organizations with access to policy makers,[129] the HPV clubs cannot claim a huge political significance—although efforts to influence politics were sometimes undertaken, for instance by Wilson in the United Staes (proposing a new car tax) and by Dovydėnas in Lithuania (drafting a new traffic law). In the Netherlands, the Department of Transport showed some sensitivity to the ideas of the HPV movement at the time of the ambitious Bicycle Master Plan (1990–1997). For instance, public money was spent on Luud Schimmelpennink's *Solo* HPV project from the early 1990s, on the 365-days-bicycle contest in 1993 and (indirectly) on the (failed) development of the MITKA (1996–2003), a concept pedelec-trike that, along the lines of HPV thinking, was meant to seduce car-drivers to choose a more environmentally friendly way of commuting.[130] But these were exceptions. In general, the Dutch government took the line that bicycle innovation and "the development of HPVs" is not the government's business, but should be left to the market.[131] The fact that in August 2009 the Dutch government announced that it would spend €25 million on building "fietssnelwegen" (bicycle highways) between several big cities, was certainly not directly the result of the lobbying by the Dutch HPV

129 For the Dutch Fietsersbond, see Duizer, *"in Het Nut Van Actie Moet Je Geloven"*; for the German ADFC see: Ulrich von Staszewski, "Die ADFC-Chronik. Mehr als 25 Jahre Engagement für das Fahrrad", *www.adfc.de* (accessed 19 April 2010, not available anymore).

130 See on the development of the MITKA: Luca Berchicci, "The Green Entrepreneur's Challenge. The Influence of Environmental Ambition in New Product Development" (PhD thesis, Delft University of Technology, 2005); cf. my comments in *Ligfiets&* (2006), no. 1: 36–37.

131 See the report on the speech by Ton Welleman, head of the "Master Plan Fiets" at the occasion of the 365-days-bicycle contest, in: *HPV Nieuws* no. 3 (1993): 8–9. Ten years later, Welleman, then head of the *Fietsberaad* (a government funded bicycle think tank), explained that the *Fietsberaad* was not interested anymore in receiving the national HPV magazine.

club, although these plans were in line with their recurrent pleas for such high speed, long distance bicycle paths.[132]

The impact of the HPV movement on the bicycle market also seems to have been rather small, at first sight at least. Production and sales numbers are not very high and recumbents and velomobiles have only a very modest share of the bicycle market. The recumbent bicycle has evidently not become the next sales hit after the introduction of the mountain bike in the early 1980s—and major bicycle producers have been reserved (though not completely absent) in developing these different types of bicycles and tricycles. In the Netherlands, a recent estimate suggests that there are no more than 50,000 individuals using a recumbent bicycle or velomobile.[133]

Still, the significance of this movement for the bicycle market should not be underestimated either. The HPV movement managed to take the step from developing unique proto-types to producing a huge diversity of commercially available models, from recumbent low racers and back-to-back tandems to rowing bikes, transport bikes and streamlined all-weather velomobiles. Many early HPV experimenters and builders tried their luck as producers, and a fair amount of them were able to make a living out of it. In all western countries, there are recumbent- or velomobile-builders active. Since the mid-nineties a fair is organized in Germersheim (Germany) for all "special" bicycles, claiming in its 2016 edition more than one hundred exhibitors and about 10,000 visitors.[134] This commercial production serves real needs, for instance the needs of cyclists who are not able to ride an ordinary bicycle or who have a longer-range commuter distance to cover every day, and it provides a non-motorized solution to these transport needs. The expectations, among those HPV enthusiasts who are convinced of the superiority of their machines, of an "HPV breakthrough" have proven to be idle, but the other HPV ideal of design-diversity definitely has been realized. At the moment, there are probably more different bicycle designs around than ever before, certainly many more than in the 1970s, all adapted to specific uses. Apart from the flexibilization of bicycle production and the market trend toward differentiation of the bicycle as a lifestyle product, the HPV movement played a role in creating room for all these alternative designs and presenting the bicycle as a vehicle that could be innovated. Somewhat ironically, HPV enthusiasts like Chester Kyle and Mike Burrows, or more recently Gerard Vroomen of the Cervélo Racing Team, also

132 See *Ligfiets&* (2006), no. 1: 10–11 and 32–33.

133 See *Ligfiets&* 26, no. 5 (2010), 16.

134 See: http://www.spezialradmesse.de/pressemitteilungen.html.

became leading experts in the innovation of professional racing bicycles since the 1970s.[135]

Arguably the most important contribution of the HPV movement to the western revival of cycling since the 1970s was its role in breaking up the idea of the bicycle as an obsolete technical artifact and propagating the idea of the bicycle as a pre-eminently *modern* vehicle. Especially in the beginning, the remarkable distance and speed records of HPVs created a lot of publicity and reached a wide audience. This was also the case with two publications from the HPV scene, that appeared for the first time in the 1970s and have been reprinted since. In the first place, David Gordon Wilson's textbook *Bicycling Science*, that presented the bicycle to a restricted but relevant international audience of engineers and scientists, as a vehicle worthy of research, innovation and promotion as utility vehicle. For a more general public, prominent IHPVA and BHPC member Richard Ballantine's bestselling *Richard's Bicycle Book* did the same. First published in 1972 and continuously reprinted, revised and translated, by 1992 it was claimed to have sold over 1,000,000 copies worldwide. In addition to the maintenance and repair sections that formed the core of it, the book tried to convince its readers of the need for more cycling in modern cities, casually encouraged them to become bicycle activists, and kept them informed about the newest developments in bicycle design and innovation, especially those that provided increased utility, whether in transport bicycles, folding bicycles or HPVs.

In the background of this chapter stood the question of how exactly the trend towards neglecting the bicycle as a means of transport was reversed. One aspect of this reversal, I have argued, was the change in the image of the bicycle from an outdated to a modern means of transportation. Borrowing a distinction made by the sociologist Hartmut Rosa, I suggest that the western revival of cycling can only partly be seen as an ideologically motivated, *intentional* slowdown of "modern acceleration and its effects", that has inspired many environmentalists and bicycle advocates. The revival of cycling was also partly a way to deal with traffic jams and congestion, thus counteracting the *unintentional* slowdown of our "high speed society" and strengthening the image of

135 At the Olympics of 1984, United States cyclists won nine medals, the first since 1912, partly because of the aerodynamic improvements made under the direction of professor Chester Kyle, founding member of the IHPVA. Almost ten years later, BHPC-founding member Mike Burrows was involved in designing the Lotus, an innovative streamlined, monocoque racing bicycle which won Chris Boardman a gold medal at the Olympics of 1992, the first British cycling gold medal at the Olympics in 72 years. Compare Hadland and Lessing, *Bicycle design*, 398–411.

cycling as a "rational" choice in contemporary urban traffic.[136] Furthermore, the marketing of the bicycle as an adult lifestyle product, especially via its incorporation in the increasingly important outdoor recreation market, has made the bicycle a trendy vehicle that has become strongly associated with the undisputed modern values of youth, health and fitness. Last but not least, the engineers and scientists involved in the HPV movement have successfully challenged the conception of the bicycle as an outdated utility vehicle and reinstated it as a modern means of transportation: the subject of science, innovation and functional differentiation.

136 Rosa, *High-Speed Society*, 94–95.

CHAPTER 8

# History, Tweed and the Invisible Bicycle

*Nicholas Oddy*

## 1 Introduction

There is an important catch in the theme of this book and this chapter with in it. It is necessary to differentiate between the bicycle and the bicyclist, as one can be more invisible than the other. At present in the United Kingdom the bicycle remains largely invisible, but the cyclist does not. To illustrate this, we need go no further than the nearest city street. One sees many cars, but how many motorists? As almost all motor cars are now made fully enclosed with tinted windscreens and windows, their drivers and occupants become invisible. In day to day language we acknowledge this difference of visibility by tending to give agency to the car: "that car nearly ran me down"; yet the chances are, when faced with a bicycle, we will attribute agency to the cyclist: "that cyclist went straight through those red lights".

With this in mind, "History, Tweed and the Invisible Bicycle" develops the themes of a number of short papers delivered by the author to the International Cycling History Conference, and another delivered as part of the "Invisible Bicycle" strand at the ICHSMT Conference in 2013.[1] It also touches on some of the themes raised in "This Hill Is Dangerous", an article in *Technology and Culture* discussing road signage and politics in the late nineteenth and early twentieth centuries.[2] The chapter considers the way in which cycling has been perceived in the context of increasing "invisibility" during much of the twentieth century; then in the context of increasing visibility and assertiveness in recent years.

History has had a very significant role in cycling. The first written histories of the machine were published in the late 1860s, shortly before the high wheeled machine of the 1870s, the starting point of this chapter, was invented. They

1 Nicholas Oddy, "Let Bygones be Bygones", Ch. 6.3 in *Cycle History 21*, ed. Andrew Ritchie (Birmingham: Cyclin History (Publishing) Ltd, 2012), 11–16; "The Roadster—Iconic Bygone of the Future", Ch. 2.4 in *Cycle History 24*, ed. Andrew Ritchie and Gary Sanderson (Birmingham, Cycle History (Publishing) Ltd, 2014), 82–87; and "A Rather Tweedy History" delivered at Manchester University 2013 (unpublished).

2 Nicholas Oddy, "This Hill is Dangerous", *Technology and Culture* 56/2 (2015): 335–369.

© KONINKLIJKE BRILL NV, LEIDEN, 2019 | DOI:10.1163/9789004289970_009

were the first ever histories that focused on a modern consumer durable.[3] They set a pattern in which historic machines were presented mainly to demonstrate the superiority of newer ones, a model that remained dominant for well over a century and is still common. It was not long before actual demonstrations of "ancient" machines were employed to do the same as a type of living history.[4]

In this chapter, such historical presentation is seen to become significant only during the first half of the twentieth century, when both bicycles and their riders became steadily less visible, even though there were far more of them than there had been previously. In contrast, the "antique" bicycle was often very visible as part of fairs and processions, this will be explored in some depth.

The growing invisibility of bicycles and their riders on the road was not only cultural; bicycles were seen as presenting a hazard to motorists who might not notice them. During the 1920s, against the advice of the Cyclists' Touring Club (CTC), the most significant voice of cycling at government level, the United Kingdom government took steps to make *bicycles* more visible but not cycling, a key difference that subtly discouraged the activity. Attention waned as cycling diminished in the post-war era, and with it what visibility it had.

However, in the last three decades, with an increase in the activity, cycling has become far more visible and, in turn, attention has turned to make *cyclists* more visible, something that has been reflected in wider culture. I conclude the chapter in looking at the "Tweed Run", where once again "antique" (or now "retro and vintage") bicycles are presented for public display, to explore the present state of visibility of bicycles and their riders.

## 2 Red Lights for the Hazardous Bicycle

Undoubtedly, the bicycle enjoyed its moment of greatest visibility in the 1870s and 1880s when the "ordinary" bicycle was one with a high front wheel fitted to the inside leg of its rider, giving him (almost invariably him) the elevated riding position of an equestrian. The bicycle was the fastest vehicle on the road and it was the fantastic nature of the machine that captured the public imagination, as in this reminiscence by Flora Thompson in *Over to Candleford*:

3 For instance—Velox, (pseud.): *Velocipedes. Bicycles and Tricycles: How to Make and How to Use Them. With a Sketch of Their History, Invention and Progress* (London: G. Routledge, 1869).

4 One of the pioneers of cycle design, Pierre Lallement, rode his machine in such a demonstration in 1885: See David Herlihy, *Bicycle* (Yale University Press: New Haven CT, USA, 2004), 232.

> How fast those new bicycles travelled and how dangerous they looked! Pedestrians backed almost into the hedges when they met one of them ... it was thrilling to see a man hurtling through space on one high wheel with another tiny wheel wobbling helplessly behind.[5]

The bicycle's visibility was compromised when that style of machine was superseded by the chain-driven-rear-wheel "safety" bicycle, a process largely complete by the mid-1890s. The 'safety' brought the rider down to the level that we are familiar with today, making the bicycle less conspicuous and adjusting downwards both the cyclist's view of his (and now her) view of the road and the visibility of the bicyclist to other road users. The bicyclist's head was now at the same level and sometimes slightly below that of pedestrians of much the same height, giving the possibility of direct eye contact; no longer did the cyclist take the elevated position of "king of the road".[6]

The process of invisibility was slowed by cycling reaching a fashionable peak during the "bicycle boom", which lasted from 1894–1897. Manufacturers were keen to sustain the bicycle's visibility by constant (if minor) changes to fittings and details to differentiate the most recent from the previous; but the boom was a time when public attention was more focused on the bicyclists. Many were wealthy, powerful and/or famous and their cycling in the parks and boulevards of major cities, in full view of idle spectators, was the stuff of press reports. It could be argued that visibility had transferred from the machines to their riders.

Meanwhile, a much more visible machine had begun to appear on the road, the motor car. This would not have much effect on the number of cyclists in the United Kingdom until after the Second World War, but it did have serious effect on their status as road users. After the "boom", the bicycle became adopted as a form of transport, mainly by the middle classes. For the first four decades of the twentieth century, bicycle use grew as it became more demotic and, as a consequence, was increasingly unremarkable. Both bicycles and their riders entered a period of increasing "invisibility" as part of day-to-day road traffic, while throughout a steady stream of wealthier cyclists abandoned their bicycles in favour of motor vehicles when they became within realistic financial reach.

---

5 Flora Thompson, *Over to Candleford* (London: Oxford University Press, 1941), 18.

6 This term was commonly used by riders of high bicycles to describe the experience of cycling. It was used as the trade name of a famous series of Lucas cycle headlamps and, in the 1970s, Andrew Ritchie used it as the title of his seminal history of cycling: *King of the Road* (London: Wildwood House, 1975).

The invisibility of both bicycles and bicyclists in the twentieth century is one that is entirely connected with the development of motoring and later road planning. Indeed, in the United Kingdom it can be almost pinpointed to a particular event in 1906, the case mounted in the High Court to test whether or not the CTC could expand its constitution to include motorists. The court decided that it could not and the die was cast—motoring and cycling clubs were placed in opposition to one another.[7]

This was significant. In the United Kingdom, motoring had not existed to any extent prior to 1896 when the Locomotives on Highways Act of that year removed many restrictions to the use of "light locomotives of under three tons unladen", which it defined as "motor cars", on public roads.[8] Until then, the "road improvement" lobby had been more or less entirely composed of middle and upper-class cyclists devoted to asserting their right to the highway. On the introduction of motor cars, many wealthy and influential cyclists began to turn their interests to motoring, taking their campaigning skills with them. The 1906 ruling effectively excluded cycling from being directly associated with motoring, placing it in what was to become the category of "other road users", alongside foot and animal traffic. It ensured that many early motorists would become *former* cyclists, rather than *also* cyclists.

Anywhere there were motor cars there was heavy pressure placed on authorities to give them primacy over other road users; their expense restricted their market to the most powerful socio-economic groups and their size, power and potential speed was largely incompatible with normal traffic, which was steadily squeezed off the road. In most industrially developed nations the second paradigm of automobility, in which the motor car was accepted as the definitive road vehicle, had been achieved by the early to mid-1930s.[9]

To illustrate how quickly cycling was downgraded and written out of road transport history in the United Kingdom, one only needs to turn to the seminal work by Sydney and Beatrice Webb, *English Local Government—The Story of the King's Highway*, published in 1913, which devotes only a couple of pages to cycling.[10] Had this book been written only a decade earlier, its content would have been very different.

7 James T. Lightwood, *The Cyclists' Touring Club: Being the Romance of Fifty Years'*. (CTC: London, 1928) 84–92, 232–235; and William Oakley, *Winged Wheel* (CTC: Godalming, 1977), 19–21.

8 John A. Williamson, *The Motor Car Acts 1896 & 1903*. (*Autocar*: London, 1903).

9 Peter Norton, "Four Paradigms of Traffic Safety", *Technology and Culture* 56/2 (April 2015): 321–331.

10 Sydney and Beatrice Webb, *English Local Government—The Story of the King's Highway* (London: Longmans Green, 1913), 239, 240. (Reprinted by ULAN n.d.)

This invisibility was paradoxical given that during the early twentieth century cycling was ever increasing and, during the inter-war years had never been so popular. However, its spread was simultaneous with steadily reducing social status. In 1903, cycling was still largely a middle-class leisure pursuit, with practical application. By 1933 a brand-new bicycle could be bought for between £3 and £4, and cycling had become a working class means of utilitarian transport. That it was more economic than public transport by train, tram or bus only served to emphasize that the bicycle was the transport of the poor.[11] On the other hand, motor cars were still high luxury items, with makers such as Ford and Morris aspiring to bring the retail price of their cheapest models down to £100.[12]

Clearly the socio-political context of both forms of transport was significant in the amount of political sway their lobbyists had at the government level, but it could be reasonably argued that the bicycle became "invisible" because of its lack of "presence" against a much larger, more powerful vehicle. Moreover, as mentioned above, the motor lobby was keen to create a paradigm in which "road traffic" became synonymous with motor vehicles. Both factors were to play out in what became the last serious attempt by the United Kingdom's cycling lobby, represented by the CTC, to assert the visibility of the bicycle as a legitimate form of traffic, rather than an obstruction to it, in its opposition to proposals for a "Red Light Act". This would require cyclists to carry tail lamps at night to allow motorists to see them at a distance beyond the scope of their headlamps. The cycle lobby argued that motorists should not drive faster than their lamps allowed; it was the responsibility of motorists to assume that there would be unmarked traffic on the road travelling at low speeds and they should drive in accordance with it.[13] The result of the debate was a compromise in 1927 and another in 1934. First; that a reflector should be fitted to a bicycle if no lamp was carried; then, that the reflector be augmented with eight square

11 The Hercules Cycle and Motor Company developed a sales policy based on the weekly price of tram and bus fares for an urban worker, with easy hire purchase for machines costing about £4. During this period Hercules was the largest cycle producer in the world, with an annual output of approximately 500,000 machines per annum between 1933 and 1939. Andrew Millward, *The Founding of the Hercules Cycle & Motor Co.* in *Cycle History, Proceedings of the 5th International Cycle History Conference*, ed. Rob van der Plas. (San Francisco: Bicycle Books, 1995), 99.

12 In June 1935, the two-door "Tudor" 8hp Ford model Y was introduced at a retail price of £100, a price it held until July 1937. It became known as the "£100 saloon"; the price was never bettered by any other maker. See Ronald Baker, "*Henry's Old Pop: comparison with Youthful Escort*", *Autocar. Vol. 130 no. 3806. (23 January 1969), 45.*

13 *Why Cyclists Object To Compulsory Rear Lights* (CTC, London, 1927).

inches of white surface.[14] While the CTC had fended off the requirement for tail lamps, it was a Pyrrhic victory that still demanded that bicycles be made "visible" to the drivers of motor vehicles driving beyond the distance of their night vision, throwing the onus of responsibility on to the cyclist.

It is important to differentiate this sort of constructed conspicuousness, in which today many riders bedeck themselves in fluorescent and reflective clothing, fearful of otherwise being overlooked by motorists, from the sort of visibility cycling had enjoyed in the late nineteenth century. Constructed conspicuousness is largely an imposition with the aim to aid the free passage of motorists; but, it is generally presented under a camouflage of "safety", which acts as a tool of coercion. Whatever its value in preventing cyclists being run down by motor vehicles (or rather, their "invisible" drivers), it is tantamount to an acknowledgement of cycling's invisibility as a legitimate form of traffic. Rather, it treats cycling as an obstruction. It would be reasonable to argue that the more visible cycling is in cultural terms, the less the requirement to make cyclists conspicuous.

In the context of road legislation in the 1930s, cycling tended to be allied to foot and animal traffic, the representatives of the past. Foot traffic was inevitable, but it was hoped could be restricted to sidewalks and "official" road crossing points.[15] Cycles and horse drawn vehicles were often perceived as more problematic because they shared the road with motor vehicles. Typical of this view was a three-fold pamphlet put out by the British Roads Federation in the late 1930s, presented as a conversation between *The Railwayman and The Motor Driver*. Here cycling and horse traffic become allied by default as enemies of the modern transport systems. The text reads, "Look at the layout of your railways. They're designed to serve horse transport instead of motor transport", illustrated by two cartoons, one of a congested goods yard full of horses and wagons along with another captioned, "Horses and cyclists cause congestion and pay no tax".[16] In this model it does not matter how modern the cycle might be (the cartoon shows a lightweight type machine) it is representative of something that is rapidly becoming an obstruction to progress and will be displaced.

It is in this context that we return to the presentation of the antique machine. After the almost total transition from the high wheel bicycle to the

---

14 Gilbert Woodward, *Woodward's Road Traffic Acts* (London: Eyre & Spottiswoode, 1934), 45, 283.

15 Woodward, *Woodward's Road Traffic Acts*, 44, 45.

16 British Roads Federation, *The Railwayman* (London: British Roads Federation, 6174a, n.d. c1935), 1i–1iii.

safety bicycle, the former soon became derisorily known as the "penny-farthing" and, not long after, became an iconic bygone that could be wheeled out to demonstrate progress. Moreover, it was eminently display-worthy because of its obvious visual difference from more modern machines, its height and its riders' seeming acrobatic ability just to ride it. To emphasize the impractical antiquity of the machine, the rider would often dress in "Victorian" costume of top hat and tailcoat, very different from the sort of clothing actually worn by bicyclists in the 1880s. The fact that the machines had been made in large numbers made them readily available, they became an expected part of processions at fairs during the period of increasing automobilism.

## 3 From the Ancient Machine to the Mountain Bike

The idea of the "ancient" machine being used as part of public exhibitions to illustrate notions of western progress is a fairly obvious visual trope that could be compared to the public display of motorists driving veteran cars, typified by the London–Brighton veteran car rally (first run in 1927 for pre-1905 motor vehicles).[17] But, there was a significant difference. The veteran car was the predecessor of the motor car, the vehicle of the future, making its place honourable. Veteran motor cars were rather the same as modern ones, allied to the twentieth century rather than the nineteenth. The London–Brighton run was (and remains) on the public road, in the midst of "normal" traffic, over a distance that would have been substantial when the vehicles were in their currency.

In contrast, the penny-farthing's symbolic values are more complex and demeaning. The penny-farthing was the predecessor of a transport form that was simultaneously losing a visible place on the road and seen as an obstruction to progress. As if to emphasize its uselessness, and unlike the veteran car, it tended only to be demonstrated over a short distance on roads closed to normal traffic. In overall effect, the penny-farthing became representative of obsolete pre-motor traffic, even though the bicycle itself was far from obsolete.

The linear nature of modernist ideas of history added to this. Commonly, bicycle enthusiasts point out that without the technologies developed by the bicycle industry, the motor car could not have been built, but this does not validate cycles built after those technologies have been adopted by the more

17 Piers Brendon, *Motoring Century—The Story of the Royal Automobile Club* (London: Bloomsbury, 1997), 18.

advanced and modern transport form.[18] The seemingly ludicrous nature of the penny-farthing as a bicycle type given the problems of height and balance, compared to more modern forms of road transport, further demeaned it and cycling as a whole as the product of Victorian eccentricity, unusable in modern road conditions.

As such, the penny-farthing was seen by many to be an object of ridicule that represented the whole of cycling in a period of mass motorization, a form of transport for the poor that would eventually be replaced by universal motoring. This was a very different reading from that of the 1890s and first years of the twentieth century when the historic significance of the high bicycle was merely that it had been superseded by another bicycle.

In the United Kingdom, there was little to challenge this perception of cycling until the early 1980s. A notable attempt was made by Alex Moulton in the early 1960s to offer a new type of bicycle for city riding that looked very different from established bicycle forms and was aimed at a middle-class, car-owning market. Small wheeled, ungendered and clearly modern, it seemed to have the possibility of repositioning the popular perception of cycling. But, although it was enthusiastically taken up by the cultural historian Rayner Banham amongst others, and widely publicized, generating significant initial sales figures, it coincided with the high-tide of motorization in the United Kingdom and remained an eccentricity.[19]

It could be argued that Moulton's bicycle failed by being intended to fit into transport infrastructure, which, by the 1960s, largely ignored cycling. It can be compared to the most advanced designs of steam locomotives that were contemporary with it. They attracted the attention of enthusiasts, but in the larger scheme of things the steam locomotive was obsolete, no matter how modern. To become visible cycling needed to position itself outside of transport infrastructure and therefore outside of public expectations of a cheap form of short-distance travel that would be better by car. Yet somehow, this position would need to avoid falling into the "other" public perception of cycling provided by racing and overt competition against the clock. These placed the activity in the category of "sport" limited to enthusiasts and with as much relevance to transport as Formula-One motor racing had to day-to-day motor traffic.

---

18 Carlton Reid, *Roads Were Not Built For Cars–How cyclists were the first to push for good roads & became the pioneers of motoring* (Newcastle on Tyne: Front Page Creations, 2014), xi–xv, 3–36.

19 Bruce Epperson, "A New Class of Cyclist", in *Cycle History* 22, ed. Andrew Ritchie and Gary Sanderson (Cycling History (Publishing) Ltd: Birmingham), 68–72.

The introduction of the commercial mountain bike to the United Kingdom market in 1981–1982 did just that.[20] Is it just a convenient coincidence that this type of bicycle emerged as postmodernism became established as the dominant cultural discourse in western nations? Whatever the case may be, there is a good case to claim it as a postmodern bicycle, not only in its production and consumption, but even in its style, which to some extent reflected the revival of a design discredited by modernist history, such as art deco and "50s contemporary". Just as it was almost impossible to fit popular design styles into established modernist understandings of design history, the mountain bike's large diameter tube construction, fat tyres, wide saddle and level bars could neither be easily related to modernist understanding of the machine beautiful as utilitarian transport, nor of cycle racing and time trialling.

Indeed, there was no intention to lock the mountain bike into transport infrastructure or established cycle sport. Its development was as a plaything in a context of Bay-area hippy youth, appealingly romantic to middle-class young adults elsewhere. The mountain bike was designed to be ridden down steep hillsides over rough terrain. It had its origins in American "balloon tire clunkers" from the 1930s–60s.[21] These machines were despised by "serious" cyclists as triumphs of style over practicality, aimed largely at children and teenagers; completely outside of modernist expectations of technological development.[22]

The significance of the mountain bike was that it re-established an understanding of cycling that had been lost with the rise of motoring, that cycling could be conducted for pleasure, by young (in particular male) adults, as a fashion statement. Indeed, to many non-cycling motorists, mountain-bikes were made most visible not by being seen on the road, but rather strapped to rear carriers on relatively expensive cars, en-route to the out-of-town, off-road sites of their intended use, emphasising their *pointlessness*. They seemed superfluous, both in the context of established transport and established forms of cycle sport, and, being in a position of being extra to them, were therefore clearly luxuries. This was significant because it was a clear indication that the bicycle

---

20 Tony Hadland and Hans Lessing, *Bicycle Design* (Cambridge (Mass): MIT Press, 2014), 434–444. The Specialized "Stump-Jumper" was the first commercially made mountain bike to be introduced to the United Kingdom market. It was soon followed by brands such as Muddy Fox. The names chosen reflect the style of riding. Alex Newson, *Fifty Bicycles That Changed The World* (London: Design Museum & Conran/Octopus, 2013), 58, 59.

21 Frank Berto, *The Birth of Dirt* (San Francisco: Van Der Plas/Cycle Publishing, 2008), 34–54.

22 The first fully developed Balloon Tire machine was the Schwinn "Aero-cycle" of 1933. Its styling was derived from stream-form motor cycles being developed by the Indian and Harley Davidson companies. Berto considers the near contemporary Schwinn "Excelsior" to have set the pattern for early mountain bikes. *Birth of Dirt*, 35.

could be a personal choice of vehicle for those who owned motor cars. This undermined the still commonly held belief that cycling was merely a cheap form of transport for those did not, in turn undermining the lingering assumption that bicycles were the poor-person's substitute for motor cars. [23]

The mountain biker *chose* to ride a bicycle for no good reason other than pleasure, and in doing so recaptured the high point of cycling in the nineteenth century when it was the stuff of the leisured classes. Then, even in sporting competition, the most influential proponents of the activity were keen to establish it as superfluous to necessity, jealously promoting its "amateur" status against those who might use it to make a living.[24] Although manuals of the time sometimes pay lip service to utility, few cyclists seem actually to have used their machines for day-to-day travel; indeed, their major lobbying organization was specifically for "touring", distancing the activity from utility.[25] All this was important in terms of reflecting the social class of the cyclists. Cycling involved a fair bit of physical effort, the stuff of the working class, therefore cycling tended to be treated as a leisure activity that one would *choose* to do. It is not difficult to see that cycling's invisibility in the twentieth century in the United Kingdom was closely allied to how necessary it became as a cheap form of utility transport. The mountain biker repositioned the bicycle as a lifestyle and fashion accessory.

An important aspect of the mountain bike was its relationship to overt consumerism in the context of the then dominant belief in an entrepreneurial free-market economy that permeated United States and United Kingdom politics. Mountain bikes tended to be made by small and medium-sized newcomers to the industry, trading on an international basis across the United

---

23 It is now commonplace to find 4 × 4 car advertisements in cycling magazines.

24 The issue of amateur status in competition was particularly marked in cycling, with both the CTC and National Cyclists' Union jealously guarding the concept that members be 'amateur' to protect the idea that cycle racing was merely done for honour. This model was copied by other cycling organizations, notably the League of American Wheelmen (LAW). As cycle sport developed in the late 1880s and 1890s, it became increasingly difficult for cycling organizations to control the amateur and professional issue, with the LAW facing the added problem of race. Andrew Richie, *Major Taylor* (San Francisco: Bicycle Books, 1988), 56–58, 101–107.

25 The CTC was founded in a context of middle and upper-class men riding high bicycles that had little practical application other than for leisure riding. Through the first part of the twentieth century, bicycling became commonplace as a means of commuting, but the CTC persisted in focusing mainly on "touring". After c. 1930 this was characterized by long distance riding on light-weight, drop-handlebar machines, very different from the kind of working class, short to medium distance travel to work on full roadsters that characterized most bicycle travel.

States, the United Kingdom, mainland Europe and Japan. Rosen proposed that the manufacture of the machine moved into a post-modern framework of flexibility within globalization, which allowed makers to respond rapidly to developments in technologies and styles.[26] This was very different from the then established model of mass manufacturers such as Raleigh making utility machines and tending to operate on Fordist principles of standardization, with the specialist market being supplied finely crafted, lightweight machines turned out by small workshops, with distribution usually contained within national or even local boundaries.

An essential feature of the mountain bike was its potential hybridity. Designed outside of any established expectations, its market had few preconceptions as to what it "should" be. The mountain bike was subject to variation in detail not seen since the bicycle boom with (often costly) novelties in suspension, gearing and frame design which often bore uncanny similarities to that earlier period, for instance the "biometric" chain ring.[27] Bicycles again started to be able to be differentiated on a seasonal basis with such novelties quickly becoming dated. An observation here is that there is a clear link to the bicycle's most "visible" period in terms of fashion and status before it began to slip into invisibility against the motor car; it was almost as if the bicycle, if not the cyclist, had regained something of that time.

The commercial mountain bike might first have been designed with off-road conditions in mind; but, laden with gears and with an upright riding position, it could easily be ridden on-road to make an urban fashion statement. Its rugged design lent itself to kerb-hopping and "traffic busting" on road surfaces cut up by motor vehicles. As mountain bikes began to move from extreme leisure pursuit to a commonplace in city streets, they had the potential of making cycling a viable, fashion-conscious alternative to motoring in the urban context, at least to younger adults. But overall, whether ridden on road, or strapped to the back of a motor car on the way to some distant hillside, the mountain bike was very visible to all as something very different in cycling culture.

The mountain bike entered a streetscape already populated by "invisible" cyclists representing the remnants of utility riding. However, despite their relative rarity and the fact that they were inconspicuous on the road, they were certainly not an easily defined group in socio-economic class, values or beliefs. To a great extent the huge decrease in working class utility cycling had

---

26 Paul Rosen, *Framing Technologies* (Cambridge MA and London: MIT Press, 2002), 133–154.

27 The biometric chain ring used computer technology to develop a nineteenth-century concept that an ovoid chain ring would more exactly follow the strengths and weaknesses of leg muscles in the circular motion of pedalling.

destroyed the previously most visible cycling culture, leaving an inhomogeneous mix of the remnants of utility riding on everything from small-wheelers to full roadsters and "club" style touring and time trialling on lightweight "racing" machines. The addition of the mountain bike to this could be seen to begin a process of multiplication of discourses of cycling practice and culture. The mountain bike itself diversified, the "ATB" (all-terrain bicycle) aimed at general off-road use and "hybrids" (mountain bike frames with higher gearing and narrower rims carrying lighter tyres) for the road.

While the mountain bike began a process of altering the perception of cycling in the 1980s, one should look at rising concern for health and increasing awareness of the environment and sustainability for what was to give it legitimacy during the 1990s. These were to give a moral imperative for using bicycles, but crucially, they were framed by an assumption that cycling was a preferable alternative to driving a motor vehicle for the same journey. It might be noted that few post-1980 exhortations to use cycling as an alternative to motoring focus on economy, which could feed lingering impressions of it being the transport of the poor.[28]

Whatever the mountain bike might have achieved in terms of reviving a fashionability akin to the bicycle boom, its underlying values were more those of the 1870s and 80s; young, masculine and athletic. Of these, athleticism seems by far to have been the most significant in setting cycle fashion. An important factor here was rising awareness of competition cycling, fuelled by British successes on the track by the likes of Chris Boardman, using new and visible technologies. Boardman's carbon-fibre monocoque, mono-blade machine was complemented by its rider clad in a skin of Lycra and plastic, all proven by wind-tunnel testing.[29] Truly, it seemed, rider and machine were one, operating in perfect mechanical harmony. The aesthetic was taken up by cutting edge fashion that locked into gymnastics and dance in clothing ranges put out by edgy labels such as BOY London, DKNY, Bodymap and Pineapple Studios.[30] Lycra lent itself to the psychedelia of rave dance culture of the early 1990s.

Meantime, an equal influence came from the do-or-die riding style of bicycle couriers. Couriers became noticeable during the 1980s as document delivery in financial centres could be achieved far more quickly by bicycle than

28 Typical is a recent campaign in Edinburgh using bus-mounted posters showing a large beer belly entitled "used a car" against a much leaner one "used a bike".

29 Newson, *Fifty Bicycles*, 80, 81.

30 Nick Clements, "This is Cycle Style Today", in Horst Friedrich, *Cycle Style* (Munich: Prestel, 2012), 1.

any other way in a period before digital file transfers. This was largely because of the lax legislation applying to cyclists, who could "traffic-bust" with relative impunity; pavement hopping, breaking traffic signals, squeezing through narrow gaps and alleys. Most used ATB style machines, which lent themselves to mounting kerbs and running down steps. Couriering was a real rarity in that it was a truly utilitarian use for bicycles that made obvious commercial sense, but it was made far more extraordinary by the fact that it looked and was urban, rebelliously daring and exciting, with many of the characteristics of a youth sub-culture.[31] This could hardly be said of any previous utilitarian cycle use.

Was all this positive in terms of cycling culture? While the machines were more and more variegated, it might be noted that they might not have been the most visible part of the equation. Lycra at raves, and the publicity surrounding the riding practices of couriers seems more a response to the visibility of cyclists, rather than bicycles. Although both couriers and lycra-wearing "hard" cyclists were small groups within a whole, their high visibility lent itself to stereotyping.

All these newly visible cycling forms were closely allied to overtly visible fitness and, in the United Kingdom, were dominant at the time that environmentalists and transport planners began to forward the idea that cycling should begin to replace motoring for short-haul journeys. Two imperatives underlay its uptake at government level. The first was long-standing but ever-increasing: namely concerns regarding global warming and carbon emissions by motor vehicles. The second was new: the cost to the National Health Service in treating rising levels of obesity in an increasingly sedentary population.

Responding to these new imperatives, many United Kingdom cities' local authorities began to reassess the place of cycling in motor-dominated road provision. Their aim is invariably to encourage a general uptake of cycling amongst all age groups and both sexes, and to achieve this laudable aim many of their policies have been based round increasing public awareness of cycling. Rather appropriately, many have adopted a technique of visibly marking the roads, a tactic first developed by cycling clubs on country roads by erecting "danger boards", the first modern road signs, in the 1880s.[32] Today, local authorities favour road markings more than signs alone, particularly in urban areas,

---

31 Ben Fincham, "Generally Speaking, People are in it for the Cycling and the Beer", *Sociological Review*, 55:2 (2007): 189–202: and 'Bicycle Messengers—Image, Identity and Community', Ch.9 in *Cycling and Society*, ed. David Horton et al (Aldershot: Ashgate, 2007) 180–193. The attractive nature of couriering to a youth audience was reflected in the television series *Streetwise* aimed at "tweens", screened on ITV in 1989–92.

32 Nicholas Oddy, "Signing Off", Ch.4.4 in *Cycle History 25*, ed. Andrew Ritchie and Gary Sanderson (Birmingham: Cycle History (Publishing) Ltd, 2015), 166.

in an attempt to encourage utility cycling, most obviously commuting. While the design and implication of these remains controversial amongst cycling lobbying groups, their effect has been to make the invisible far more visible by asserting a presence even when there are no cyclists on the road.[33] Although such infrastructure is often accused of being tokenistic, its importance as an indicator of changing attitudes to cycling as day-to-day transport should not be underestimated. It serves to legitimize the activity on streets largely given over to motor vehicles.

The overall effect of all this is that in the United Kingdom, the first two decades of the twenty-first century have seen a marked shift in which cycling has been made visible in a way it had not been since the early 1930s. Moreover, for those who care to look, a multiplicity of *bicycles* has become visible, along with a multiplicity of riding styles and cultures to suit the bicycles. However, it could be argued that far more visible than the bicycles are the bicyclists who dress themselves up in the most conspicuous manner. Thus, it is that public perceptions of what a cyclist should look like seemingly remain dominated by modernist beliefs in technological progress, largely informed by cycle sport and enhanced by "safety" features. Indeed, in the United Kingdom, much of cycling's utility is concealed beneath a quasi-competitive aesthetic, with many commuters presenting themselves in full sporting gear. Fashionable it may be, but a long way from the sort of everyday practice that is associated with high percentages of utility cycling that can be seen in, say, Amsterdam or Copenhagen. [34]

It is here that many of the values that are displayed in the Tweed Run come into play. What makes the Tweed Run significant is that it represents an overtly historicist view of cycling based on ideas of tradition and longevity. But this should not be confused with the rarefied, established collector/enthusiast groups devoted to veteran cycles that tend toward antiquarianism. Rather, the Tweed Run attracts those who see cycling as a part of retro fashion. It is the most overt manifestation of "vintage" in cycling, but it has a serious political undercurrent in terms of what type of cycling should be visible in the urban context.

---

33 In particular the extent to which they might be seen to be mandatory, rather than advisory. Nicholas Oddy 'Signing Off', fig3.

34 Although 'Cycle Revolution' an exhibition presented at the Design Museum, London 18 November 2015–30 June 2016 proposed that there are four "tribes" of cyclists on United Kingdom roads, the exhibition was dominated by competition style "High Performers" and "Thrill Seekers". See Nicholas Oddy "Bicycle Histories, they have a past, but do they have a future?" in *West86* 23(1) (2016), 129–130.

## 4 Tweed Runs

The first Tweed Run was organized in London in 2009. It has since grown into a larger franchised event that has been taken up world-wide.[35] Its success has spawned copyists under slightly adjusted names, for instance the "Harris Tweed Ride" in Scotland. With this has come diversity; the scale and commercial management of the events range from intense to non-existent. Not surprisingly, the more commercial, the more the sponsorship. This comes from fashion houses: The Tweed Run itself has enjoyed sponsorship by Rugby Ralph Lauren.[36] In all of these cases the cycling is leisured: the opposite of those devoted to "hard riding" or distance achievement. As one might expect, riders are not expected to dress as "cyclists" as the term would be understood today. Unlike previous manifestations of cycling fashion which were informed by competition riding, the tweed run is informed by early-to-mid-twentieth-century images of bourgeois leisure and utility cycling found in commercial art, in which riders are respectably dressed in practical tweeds or city suits, able to go about their business without looking out of place, in the same way a motorist might.[37] To emphasize this aspect one only needs to look at the itinerary of many Tweed Runs. A "Harris Tweed Ride" recently attended by the author as part of the research for this chapter involved meeting at a boutique hotel for coffee, then visiting a range of up-market city centre clothes shops, a cocktail bar and a fashionable eatery.[38]

Importantly, although the concept of the Tweed Run is informed by popular understandings of history, there is no demand for a historic machine. The choice of bicycle reflects the dress, encapsulating an attitude to cycling informed by an idea of "traditional" cycling employing a bricolage of historic and contemporary references that are loosely termed "retro". Thus, while some riders ride older machines, many appear on updated takes on 'traditional' machines generally taking from the roadster (the upright "black bicycle"); others use machines that in some way encapsulate the nature of non-competitive

35 The official website of the Tweed Run is www.tweedrun.com.

36 This was a brand of Polo Ralph Lauren aimed at the young adult market, it ran from 2004–2013. In February 2018, its website http://www.rugby.com/ led direct to the parent company site.

37 Oddy, "Cycling, A Game for all Players", 13–15.

38 http://www.harristweed.org/blog/2013/06/the-edinburgh-harris-tweed-ride/, accessed July 2015; this source has now been deleted but see: http://www.campaignforwool.org/2014/10/08/the-harris-tweed-ride-edinburgh/. Arguably Tweed Runs follow current trends for social activities and experiential cultural participation often related to visual media, such as ones in the manner of BBC television programmes Bake Off and Strictly Come Dancing on Tour.

bourgeois riding, but have little connection to traditional cycle design, for instance the folding, small-wheeled "Brompton".[39] In such a context, a rider on an Edwardian roadster, now some 110 years old, hardly stands out from the other bicycles present. This places it in a very different historical understanding from the penny-farthing ridden as iconic bygone. The veteran bicycle is notable in being so little different from its modern counterparts, the Dutch-style heavy roadsters and Pashley-style safety roadsters, that it demonstrates a type of timelessness largely immune from technological or stylistic change.[40] It is, ironically, almost invisible; as at home on the city streets of 2018 as it was on those of 1908.

This is reflected in the pictures of the various Tweed Runs freely available on the web. Most of the riders, though dressed for the event, would rarely look out of place, were they cycling outside of it.[41] Indeed, an obvious observation is that they would not look out of place if they were *not* cycling. Rather, they look like many of the day-to-day riders that Colville Andersen photographs to illustrate *Cycle Chic*.[42] Moreover, the sort of cycling culture promoted by tweed runs tends to be gender equal, unlike the dominant masculinity of ATB and Tour de France-fetishising, sports riding that inform much visible cycling on United Kingdom streets. A likely reason for this is its close relationship to vintage and retro, which have become main-stream in feminine fashion idioms. Moreover, there is no referencing to athleticism, which gives a further distance from the idea that cycling is inherently about masculine competitive and/or extreme exercise (summed up by the current acronym MAMIL[43]), rather than transport. However, to the larger public, which is more visible, the rider in "normal" clothes on some sort of utilitarian city-bike, or the MAMIL?[44]

---

39 Hadland & Lessing, *Bicycle Design* 463, 464. The Brompton rises to a "Brompton Championship" where "The 800 plus competitors have to wear jackets and ties". Walsh, "No Froome". In *The Times* 29 June 2013: http://www.thetimes.co.uk/tto/business/industries/consumer/article3827943.ece; also: https://www.google.co.uk/search?q=brompton+championship.

40 Dutch, heavy, or full roadsters, follow the pattern of early twentieth-century "popular" models, typically they have high frames with 28 inch wheels, full chain case and mudguards. The "safety roadster" is a similar machine built on 26 inch wheels and with a lower bottom bracket. It was introduced in the late 1920s to facilitate interaction with the stop-start of motorized traffic, police and light-controlled junctions.

41 https://www.google.co.uk/search?q=tweed+run

42 Mikael Colville-Andersen, *Cycle Chic* (London: Thames & Hudson, 2012). See also Friedrich, *Cycle Style* for a similar photographic survey, but based in London and with a more overtly fashion driven interest.

43 Middle Aged Men In Lycra.

44 It might be noted that the lycra/crash helmet/high visibility of the MAMIL is such that the acronym is easily misapplied as the aesthetic conceals both age and gender; by so doing it plays into the acronym's stereotyping.

In spite of its cultural positioning moving away from established exclusive tropes, this does not mean that the concept of the Tweed Run is not without its own issues. Like historic bourgeois cycling that lies at its roots, it could be accused of being socially and racially exclusive. While open to all, the runs seem to tend to appeal to those who feel confident in matters of history and/or style, buying-in to the grand narrative of British history when "Britain was still great". A comparison could be drawn to "reclaim the streets", a cycle campaign that involves groups of riders taking up the road-space of motor vehicles to reduce traffic speeds to that of cyclists. It is open to all and is promoting general access through road-share, but is inherently exclusive to those who wish to be visible as activists and campaigners.

It could be argued that the Tweed-Run provides a visual display of what *should* be invisible bicycles and their riders, whose presence relies on being accepted as an integral part of the street, as they were a century ago. This is something very different from the idea of their having to be *made* visible to prevent being mown down by motorists who are expected not to notice them otherwise, a practice that has tended to inform United Kingdom traffic planning and official attitudes since the "Red Light Act" controversy in the late 1920s and 30s.

## 5 Conclusion

In conclusion, rather than invisibility being the issue for the bicycle's place as part of road traffic, it seems that the opposite might be the case. Maybe the problem is to be found in the attempts to make the bicycle highly visible, particularly to motorists, to make it stand out as a potential hazard and obstruction? Looking at the way cycling was marginalized during and after the inter war years it is reasonable to argue that the CTC were right to be suspicious of "visibility" regulations in the inter war years, but perhaps not just for the reasons it gave at the time.

Rather, reflectors and white paint in 1934 have led to fluorescent reflective coatings, flashing LEDs and "high vis—hi tech" clothing today, effectively moving the visible bicycle to the visible cyclist, but one hidden beneath a shell of equipment that fits to assumptions regarding athleticism and road safety. The more visible does this style of cyclist become, the less visible and "other" are those cyclists who do not mark themselves out in such a way. To an extent, the Tweed Run does something to demonstrate to a larger public that there are others outside of the stereotype; but as yet, the visibility of one particular type of cyclist and the invisibility of others makes it still too easy for those outside of the activity (and indeed many within it) to fall into the trap of being able to define what cyclists *should* look like, whatever the diversity of cyclists themselves.[45] Perhaps for cycling to take a fully integrated place in a shared road space, invisibility might be the best option?

---

45 Elaine Mullan, and Margriet Groot, "Obsessing on safety: teenage girls attitudes to cycling" (Paper presented to Cycling and Society Symposium, Newcastle, September, 2014). https://www.researchgate.net/publication/266698205_Obsessing_on_safety_teenage_girls_attitudes_to_cycling . See also: *Obsessing on safety: Teenage girls' attitudes to cycling* (second version). http://newcycling.org/wp-content/uploads/CSS2014_MullanEtAl_comb.pdf and *Drivers, teenagers, danger and tolerance: Views from Ireland* (Waterford: Department of Health, Sport & Exercise Sciences, 2014). http://repository.wit.ie/2994/1/C%26S%20symposium%202014%20-%20Drivers,%20teens%20%26%20cycling%20presentation%202%20-%20E.%20Mullan.pdf.

# Bibliography


*40 Jaar - uit het archief der Kampioen.* ANWB, 1923.

Advisory Council for Transport, Public Works and Water Management, The Netherlands. *Wie ik ben en waar ik ga: Advies over de effecten van veranderingen in demografie en leefstijlen op mobiliteit.* The Hague: Advisory Council for Transport, Public Works and Water Management, 2010.

Agentschap NL. *Een terugblik in vogelvlucht, als het gaat om octrooien.* Den Haag: Ministerie van Economische Zaken, 2010.

Agerwig Carstensen, Trine, and Anne-Katrin Ebert. "Cycling Cultures in Northern Europe: From 'Golden Age' to 'Renaissance'". In *Cycling and Sustainability*, edited by John Parkin, 23–58. Bingley: Emerald, 2012.

Agudo, Shirley. *Bicycle Mania.* Den Haag: XPat Media, 2009.

Akar, Gulsa, and Kelly J. Clifton. "The Influence of Individual Perceptions and Bicycle Infrastructure on the Decision to Bike." *Transportation Research Record* 2140 (2009): 165–72.

Albert de la Bruhèze, Adri A., and Frank C.A. Veraart. *Fietsverkeer in praktijk en beleid in de twintigste eeuw. Overeenkomsten en verschillen in het fietsgebruik te Amsterdam, Eindhoven, Enschede, Zuid-Oost Limburg, Antwerpen, Manchester, Kopenhagen, Hannover en Basel.* Den Haag: Ministry of Transport, Public Works and Water Management/Stichting Historie der Techniek, 1999.

Albert de la Bruhèze, Adri A., and Frank C.A. Veraart. "Fietsen en verkeersbeleid. Het fietsgebruik in negen West-Europese steden in de twintigste eeuw". *NEHA-jaarboek voor economische, bedrijfs- en techniekgeschiedenis* 62 (1999): 138–70.

Aldred, Rachel. "On the Outside: Constructing Cycling Citizenship". *Social and Cultural Geography* 11, no. 1 (2010): 35–52.

Aldred, Rachel. "Incompetent or Too Competent? Negotiating Everyday Cycling Identities in a Motor Dominated Society." *Mobilities* 8, no. 2 (2012): 252–71.

Aldred, Rachel. "The Role of Advocacy and Activism". In *Cycling and Sustainability*, Transport and Sustainability, Volume 1, edited by John Parkin, 83–100. Bingley: Emerald, 2012.

Aldred, Rachel. "Governing Transport from Welfare State to Hollow State: The Case of Cycling in the UK". *Transport Policy* 23 (2012): 95–102.

Aldred, Rachel. "Who are Londoners on Bikes and What Do They Want? Negotiating Identity and Issue Definition in a 'Pop-up' Cycle Campaign." *Journal of Transport Geography* 30 (2013): 194–201.

Aldred, Rachel, and Katrina Jungnickel. "Constructing Mobile Places between 'Leisure' and 'Transport': A Case Study of Two Group Cycle Rides". *Sociology* 46, no. 3 (2012): 523–39.

Aldred, Rachel, and Katrina Jungnickel. "Why Culture Matters for Transport Policy: The Case of Cycling in the UK." *Journal of Transport Geography* 34 (2014): 78–87.

Amenda, Lars. "Mit dem Fahrrad um die Welt—Radfernreisen vor hundert Jahren." In *Das Fahrrad: Kultur, Technik, Mobilität*, edited by Mario Bäumer, 112–16. Hamburg: Junius, 2014.

Anable, Jillian, and Birgitta Gatersleben. "All Work and No Play? The Role of Instrumental and Affective Factors in Work and Leisure Journeys by Different Travel Modes". *Transportation Research Part A: Policy and Practice* 39, nos. 2–3 (2005): 163–81.

Anaya, Ester, and Santiago Gorostiza. "The Historiography of Cycling Mobility in Spain in the Twentieth Century". *Mobility in History* 5, no. 1 (2014): 37–42.

Antonakos, Cathy L. "Environmental and Travel Preferences of Cyclists." *Transport Research Record* 1438 (1994): 25–33.

Arnold, David. *Everyday Technology: Machines and the Making of India's Modernity.* Chicago: University of Chicago Press, 2013.

Atsma, Joop. *Fietsen in Nederland ... een tandje erbij. Initiatiefnota met voorstellen voor actief fietsbeleid in Nederland.* The Hague: CDA Tweede Kamerfractie, 2008.

"Att cykla i Stockholm: Så tycker stockholmarna". Stockholm: Utrednings- och statistikkontoret, Stockholms stad, 2006.

*Australia Cycling 1999–2004: The National Strategy.* Sydney: Austroads, 1999.

Australian Bicycle Council and Austroads. *Gearing up for Active and Sustainable Communities: The Australian National Cycling Strategy 2011–2016.* Sydney: Austroads, 2010.

Austroads Incorporated. *The Australian National Cycling Strategy 2005–2010.* Sydney: Austroads Incorporated, 2005.

Ballantine, Richard. "Go for the Money". *Bike Culture Quarterly* 1 (1993): 19.

Ballantine, Richard. "The Politics of Human Powered Vehicles". *Human Power* 54 (2003): 3.

Baltes, Michael R. "Factors Influencing Nondiscretionary Work Trips by Bicycle Determined from 1990 US Census Metropolitan Statistical Area Data." *Transportation Research Record* 1538 (1996): 96–101.

Bamberg, Sebastian. "Understanding and Promoting Bicycle Use: Insights from Psychological Research". In *Cycling and Sustainability*, Transport and Sustainability, Volume 1, edited by John Parkin, 219–46. Bingley: Emerald, 2012.

Bamberg, Sebastian, Izek Ajzen, and Peter Schmidt. "Choice of Travel Mode in the Theory of Planned Behavior: The Roles of Past Behavior, Habit, and Reasoned Action". *Basic and Applied Social Psychology* 25, no. 3 (2003): 175–87.

Banister, Chris. "Planning for Cycling." *Planning Practice & Research* 5, no. 1 (1990): 27–29.

Banister, Chris, and Nick Gallent. "Trends in Commuting in England and Wales—Becoming Less Sustainable?" *Area* 30, no. 4 (1998): 331–41.

Barker, Ronald. "*Henry's Old Pop: Comparison with Youthful Escort*", *Autocar* 130, no. 3806 (1969), 45–47.

Barnes, Gary and Kevin J. Krizek. "Estimating Bicycling Demand". *Transportation Research Record* 1939 (2005): 45–51.

Basford, L., S. Reid, T. Lester, J. Thomson, and A. Tolmie. *Drivers' Perceptions of Cyclists*. London: Department of Transport, and Crowthorne: Transport Research Limited, 2002.

Bassett, David R. Jr., John Pucher, Ralph Buehler, Dixie L. Thompson, and Scott E. Crouter. "Walking, Cycling, and Obesity Rates in Europe, North America, and Australia". *Journal Physical Activity and Health* 5, no. 6 (2008): 795–814.

Bäumer, Mario. "Vom aud und ab des Fahrrads—Fahrradmobilität im Wandel". In *Das Fahrrad: Kultur, Technik, Mobilität*, edited by Mario Bäumer, 156–59. Hamburg: Junius, 2014.

Bax, Florine M.W. "Bicycle Use in the Netherlands versus the United States". Master's thesis, Rijksuniversiteit Groningen, 2004.

Beaujon, Otto. "Was Bicycle Development Stunted by Organized Racing?" *Cycle History 10: Proceedings of the 10th International Cycling History Conference*, edited by Rob van der Plas, (1999): 67–76.

Beduhn, Ralf. *Die Roten Radler. Illustrierte Geschichte des Arbeiterradfahrerbundes 'Solidarität'.* Münster: Lit Verlag, 1982.

Beduhn Ralf and Jens Klocksin, eds. *Rad-Kultur-Bewegung. 100 Jahre rund ums Rad: Rad- und Kraftfahrerbund Solidarität. Illustrierte Geschichte 1896–1996*. Essen: Klartext, 1995.

Bell, David Christopher. "6 Technologies Conspicuously Absent from Sci-Fi Movies". *Cracked.com*, 27 July 2011. http://www.cracked.com/article_19325_6-technologies-conspicuously-absent-from-sci-fi-movies.html.

Benoist, Christian. "Une Expérience francaise et ses difficultés: Rennes". In *The Bicycle: Global Perspectives: Papers Presented at the Conference vélo mondiale Pro Bike–Velo-city, Montreal, Canada, 13–17 September 1992*, edited by Robert Boivin and Jean-François Pronovost,, 128–31. Québec: Vélo Québec, 1992.

Benoît, Lambert. *Cyclopolis, ville nouvelle. Contribution à l'histoire de l'écologie politique*. Genève: Editions Médecine & Hygiène, 2004.

Berchicci, Luca. "The Green Entrepreneur's Challenge: The Influence of Environmental Ambition in New Product Development". Dissertation, Delft University of Technology, 2005.

Bergman, Martin. *Effektivare cykeltrafik: Planeringsmässiga förutsättningar för ökad cykelanvändning*. Stockholm: Royal Institute of Technology, 1994.

Berlage, Titia, ed. *De fiets*. Rotterdam: Museum Boymans-Van Beuningen, 1977.

Bernhoft, Inger Mary, and Gitte Carstensen. "Preferences and Behaviour of Pedestrians and Cyclists by Age and Gender". *Transportation Research Part F: Traffic Psychology and Behavior* 11, no. 2 (2008): 83–95.

Bertho-Lavenir, Catherine. "Normes de comportement et contrôle de l'espace. Le Touring Club de Belgique avant 1914". *Le Mouvement Social* 178 (1997): 69–87.

Bertho-Lavenir, Catherine. *La Roue et la Stylo. Comme nous sommes devenus touristes*. Paris: Odile Jacob, 1999.

Bertho-Lavenir, Catherine. *Voyages à vélo, du vélociped au Vélib*. Paris: Paris Bibliotheques, 2011.

Berto, Frank J. "The Great American Bicycle Boom". *Cycle History 10: Proceedings of the 10th International Cycling History Conferences* (1999): 133–41.

Berto, Frank J., ed. *The Dancing Chain: History and Development of the Derailleur Bicycle*. 3rd rev. & exp. ed. San Francisco: Cycle Publishing/Van der Plas Publications, 2009.

Berto, Frank J. *The Birth of Dirt: Origins of Mountain Biking*. 2nd ed. San Francisco: Cycle Publishing/Van der Plas Publications, 2009.

Besse, Nadine, ed. *The Velocipede, Object of Modernity 1860–1870/Le velocipede, objet de modernité 1860–1870*. Saint-Étienne: Musée d'Art et d'Industrie, 2008.

"Bicycle Statistics: Usage, Production, Sales, Import, Export", International Bicycle Fund, accessed 28 December 2017, http://www.ibike.org/library/statistics-data.htm.

Bijker, Wiebe. *Of Bicycles, Bakelites, and Bulbs: Toward a Theory of Sociotechnical Change*. Cambridge, MA: MIT Press, 1995.

Birkholz, Tim. "Die stille Revolution—das Fahrrad kommt zurück". In *Das Fahrrad: Kultur/Technik/Mobilität*, edited by Mario Bäumer and Hamburg Museum der Arbeit, 160–67. Hamburg: Junius, 2014.

Blickstein, Susan G. and S. Hanson. "Critical Mass: Forging a Politics of Sustainable Mobility in the Information Age". *Transportation* 28, no. 4 (2001): 347–62.

Blickstein, Susan G. "Automobility and the Politics of Bicycling in New York City". *International Journal of Urban and Regional Research* 34, no. 4 (2010): 886–905.

Blue, Elly, ed. *Bikes in Space: A Feminist Science Fiction Anthology*. Portland, OR: Microcosm, 2013.

Blue, Elly, ed. *Bikes in Space: Volume 2: More Feminist Science Fiction*. Portland, OR: Microcosm, 2014.

Blue, Elly, ed. *Pedal Zombies: Thirteen Feminist Bicycle Science Fiction Stories*. Portland, OR: Microcosm, 2015.

Boal, Ian A. "Towards a World History of Cycling". In *Cycle History 11: Proceedings of the 11th International Cycling History Conference*, edited by Andrew Ritchie and Rob van der Plas, 16–22. San Francisco: Van der Plas Publications, 2001.

Boivin, Robert. "The Bicycle: Constraint, Duty or Choice? As Seen by the Devil's Advocate". In *The Bicycle: Global Perspectives: Papers Presented at the Conference vélo mondiale Pro Bike–Velo-city, Montreal, Canada, 13–17 September 1992*, edited by Robert Boivin and Jean-François Pronovost, 14–19. Québec: Vélo Québec, 1992.

Bond van Rijwiel- en Motorhandelaren in Nederland. *Een kwart eeuw organisatie. Gedenkboek. Uitgegeven door den Bond van Rijwiel- en Motorhandelaren bij zijn vijf-en-twintigste jaarfeest.* Amsterdam: Bond van Rijwiel- en Motorhandelaren in Nederland, 1928.

Bonham, Jennifer, and Peter Cox. "The Disruptive Traveller? A Foucauldian Analysis of Cycleways". *Road and Transport Research* 19, no. 2 (2010): 42–53.

Bonham, Jennifer, and Marilyn Johnson, eds. *Cycling Futures*. Adelaide: University of Adelaide Press, 2015.

Bonham, Jennifer, and Anne Wilson. "Women Cycling through the Life Course: An Australian Case Study". In *Cycling and Sustainability*, Transport and Sustainability, Volume 1, edited by John Parkin, 59–81. Bingley: Emerald, 2012. Börjesson, Maria, and Jonas Eliasson. "The Benefits of Cycling: Viewing Cyclists as Travelers Rather than Non-Motorists". In *Cycling and Sustainability*, Transport and Sustainability, Volume 1, edited by John Parkin, 247–68. Bingley: Emerald, 2012.

Bosworth, R.J.B. "The Touring Club Italiano and the Nationalization of the Italian Bourgeoisie". *European History Quarterly* 27, no. 3 (1997): 371–410.

Bowerman, Les. "Clubs: Their Part in the Study of Cycles and Cycling History". In *Cycle History 5: Proceedings of the 5th International Cycling History Conference*, edited by Rob van der Plas, 133–38. San Francisco: Bicycle Books, 1995.

Brandenburg, Christiane, Andreas Matzakaris and Arne Arnberger. "The Effects of Weather on Frequencies of Use by Commuting and Recreation Bicyclists". In *Advances in Tourism Climatology*, edited by Andreas Matzarakis, C.R. De Freitas, and Daniel Scott, 189–97. Freiburg: Berichte des Meteorologischen Instituts der Universität Freiburg, 2004.

Brandenburg, Christiane, Andreas Matzakaris, and Arne Arnberger. "Weather and Cycling—A First Approach to the Effects of Weather Conditions on Cycling". *Meteorological Applications* 14, no. 1 (2007): 61–67.

Brendon, Piers. *Motoring Century: The Story of the Royal Automobile Club*. London: Bloomsbury, 1997.

Briese, Volker. "From Cycling Lanes to Compulsory Bike Path: Bicycle Path Construction in Germany, 1897–1940". In *Cycle History 5: Proceedings of the 5th International Cycling History Conference*, edited by Rob van der Plas, 123–28. San Francisco: Bicycle Books, 1995.

British Roads Federation. *The Railwayman and the Motor Driver.* London, British Roads Federation, 6174a, n.d. (c1935).

Brooke, Stephen. "Gender and Working Class Identity in Britain during the 1950s". *Journal of Social History* 34, no. 4 (2001): 773–96.

Brown, Gerard, and Graeme Fife. *The Elite Bicycle: Portraits of Great Marques, Makers and Designers*. London: Bloomsbury Sport, 2013.

Brown, Jeffrey, Eric A. Morris, and Brian D. Taylor. "Planning for Cars in Cities: Planners, Engineers, and Freeways in the 20th Century". *Journal of the American Planning Association* 75, no. 2 (2009): 161–77.

Brusse, W.J. *De Groninger rijwielenfabriek A. Fongers*. Vol. 63, *Neerlands Welvaart*, 1920.

Buehler, Theodore J. "Fifty Years of Bicycle Policy in Davis, CA". Davis: University of California, 2007. *http://www.davishistoryresearch.org/3-authors/BuehlerThesisFinalDraft.pdf.*

Buenstorf, Guido. "Designing Clunkers: Demand-Side Innovation and the Early History of the Mountain Bike". In *Change, Transformation and Development*, edited by J. Stanley Metcalfe and Uwe Cantner, 53–70. Heidelberg: Physica, 2003.

Buis, Jeroen, and Roefof Wittink. *The Economic Significance of Cycling: A Study to Illustrate the Costs and Benefits of Cycling Policy*. The Hague: VNG uitgeverij, 2000.

Burr, Thomas. "Building Community, Legitimating Consumption: Creating the US Bicycle Market, 1876–1884". *Socio-Economic Review* 4, no. 3 (2006): 417–46.

Burr, Thomas. "National Cycling Organizations in Britain, France, and the United States, 1875–1905". In *Cycle History 18: Proceedings of the 18th International Cycling History Conference*, edited by Rob van der Plas, 34–42. San Francisco: Van Der Plas Publications, 2009.

Burton, Beryl. *Personal Best*. Horsham: Springfield Press, 1986.

Butler, Tim, and Chris Hamnett. "Social and Geographic Interpretations of Housing Spaces". In *The Sage Handbook of Housing Studies*, edited by David Clapham, William Clark, and Kenneth Gibb, 147–62. London: Sage, 2012.

Byron, Mark. "Modernist Wheelmen". In *Flann O'Brien & Modernism*, edited by Julian Murphet, Rónán McDonald, and Sascha Morrell. New York: Bloomsbury, 2014.

Cao, Xinyu (Jason), Patricia L. Mokhtarian, and Susan L. Handy. "Examining the Impacts of Residential Self-Selection on Travel Behaviour: A Focus on Empirical Findings". *Transport Reviews* 29, no. 3 (2009): 359–95.

Carleheden, Mikael, Carl-Göran Heidegren, and Rasmus Willig. "Recognition, Social Invisibility, and Disrespect". *Distinktion: Scandinavian Journal of Social Theory* 13, no. 1 (2012): 1–3.

Cardoza, Anthony. "Making Italians? Cycling and National Identity in Italy: 1900–1950". *Journal of Modern Italian Studies* 15, no. 3 (2010): 354–77.

Carlsson, Chris, ed. *Critical Mass: Bicycling's Defiant Celebration*. Oakland: AK Press, 2002.

Carré, Jean-René. "La situation de la bicyclette en France". In *The Bicycle: Global Perspectives: Papers Presented at the Conference vélo mondiale Pro Bike–Velo-city,*

*Montreal, Canada, 13–17 September 1992*, edited by Robert Boivin and Jean-François Pronovost, 49–55. Québec: Vélo Québec, 1992.

Carstensen, Trine Agervig, and Anne-Katrin Ebert. "Cycling Cultures in Northern Europe: From 'Golden Age' to 'Renaissance'". In *Cycling and Sustainability*, Transport and Sustainability, Volume 1, edited by John Parkin, 23–58. Bingley: Emerald, 2012.

Cavill, Nick, Sonja Kahlmeier, Harry Rutter, Francesca Racioppi, and Pekka Oja. "Economic Analyses of Transport Infrastructure and Policies Including Health Effects Related to Cycling and Walking: A Systematic Review". *Transport Policy* 15, no. 5 (2008): 291–304.

Centraal Bureau voor de Statistiek. *Bedrijfstelling 31 december 1930*. Den Haag: Centraal Bureau voor de Statistiek, 1935–1937.

Centraal Bureau voor de Statistiek, "De Nederlandsche rijwielmarkt." *De Nederlandsche Conjunctuur*, February (1937): 20–27.

Centraal Bureau voor de Statistiek. *Tweehonderd jaar statistiek in tijdreeksen 1800–1999*. Voorburg/Heerlen: Centraal Bureau voor de Statistiek, 2001.

Cervero, Robert. *The Transit Metropolis: A Global Enquiry*. Washington, DC: Island Press, 1998.

Cervero, Robert. "The Built Environment and Travel: Evidence from the United States". *European Journal of Transport and Infrastructure Research* 3, no. 2 (2003): 119–37.

Cervero, Robert, and Michael Duncan. "Walking, Bicycling, and Urban Landscapes: Evidence from the San Francisco Bay Area". *American Journal of Public Health* 93, no. 9 (2003): 1478–83.

Chakrabarty, Dipesh. *Provincializing Europe: Postcolonial Thought and Historical Difference*. New edition. Princeton: Princeton University Press, 2007. First published 2000.

Christmas, Simon, Shaun Helman, Su Buttress, Cecilia Newman, and Rebecca Hutchins. *Cycling, Safety and Sharing the Road: Qualitative Research with Cyclists and Other Road Users*. Road Safety Web Publication No. 17. London: Department for Transport, 2010.

City of Copenhagen. *Cycle Policy 2002–2012*. Copenhagen: City of Copenhagen, Roads & Parks Department, 2002.

Clarsen, Georgine. "Pedaling Power: Bicycles, Subjectivities and Landscapes in a Settler Colonial Society". *Mobilities* 10, no. 5 (2015): 706–25.

Claudy, Marius C., and Mark Peterson. "Understanding the Underutilization of Urban Bicycle Commuting: A Behavioral Reasoning Perspective". *Journal of Public Policy & Marketing* 33, no. 2 (2014): 173–87.

Claxton, E.C. "The Future of the Bicycle in Modern Society". *Journal of Royal Society of Arts* 116, no. 5138 (1968): 114–34.

Clayton, Nick. "The Quest for Safety: What Took So Long?" In *Cycle History 8: Proceedings of the 8th International Cycling History Conferences*, edited by Nicholas Oddy and Rob van der Plas, 15–20. San Francisco: Van der Plas Publications, 1998.

Cleaveland, Fay, and Frank Douma. "The Impact of Bicycling Facilities on Commute Mode Share". Paper presented at the Transportation Research Board 88th Annual Meeting, Washington, DC, 11–15 January 2009.

Clements, Nick. "This Is Cycle Style Today". Introduction to Friedrich, Horst A, *Cycle Style*. Munich: Prestel, 2012.

Cohen, Deborah, Amber Sehgal, Stephanie Williamson, Daniela Golinelli, and Nicole Lurie. "Impact of New Bicycle Path on Physical Activity". *Preventive Medicine* 46 (2008): 80–81.

Colebrook, Claire. *Gilles Deleuze*. London: Routledge, 2002.

Colville-Andersen, Mikael. *Cycle Chic*. London: Thames & Hudson, 2012.

Cordell, H. Ken, et al. "Outdoor Recreation Participation Trends". In *Outdoor Recreation in American Life: A National Assessment of Demand and Supply Trends,* edited H. Ken Cordell and Susan M. McKinney, 219–321. Champaign: Sagamore, 1999.

Cornelissen de Beer, G.D. "Invoering en gebruik van het rijwiel bij de Europese legers in de 19e eeuw". *Armamentaria. Jaarboek Legermuseum* 19 (1984–1985): 60–87.

Coulouma, Flore. "Cycling in Circles: Flann O'Brien's Free-wheeling Stories in The Third Policeman". *The International Journal of the History of Sport* 29, no. 12 (2012): 1715–28.

Cox, Peter, and Frederick van de Walle. "Bicycles Don't Evolve: Velomobiles and the Modelling of Transport Technologies". In *Cycling and Society*, edited by Dave Horton, Paul Rosen, and Peter Cox, 113–31. Aldershot: Ashgate, 2007.

Cox, Peter. "Ideas in Motion: Cycling: Image and Imagery in the Cultural Turn: Review Essay". *Transfers* 2, no. 1 (2012): 159–64.

Cox, Peter. "'A Denial of Our Boasted Civilisation': Cyclists' Views on Conflicts over Road Use in Britain, 1926–1935". *Transfers* 2, no. 3 (2012): 4–30.

Cox, Peter. "Strategies Promoting Cycle Tourism in Belgium: Practices and Implications". *Tourism Planning and Development* 9, no. 1 (2012): 25–39.

Cox, Peter, ed. *Cycling Cultures*. Chester: University of Chester Press, 2015.

Cox, Peter. "Cycling Cultures and Social Theory". In *Cycling Cultures*, edited by Peter Cox, 14–42. Chester: University of Chester Press, 2015.

Cox, Peter. "Cycling, Environmentalism and Change in 1970s Britain". Paper presented at the Conference on Mobility and Environment, Kerschensteiner Kolleg, Deutsches Museum, Munich, 13–14 February 2015.

Cronon, William. "A Place for Stories: Nature, History, and Narrative". *Journal of American History* 78, no. 4 (1992): 1347–76.

Crowther, Geoffrey. Preface to *Traffic in Towns: The Specially Shortened Edition of the Buchanan Report*. Harmondsworth: Penguin, 1964.

"Cykelplan 1978 för Stockholms kommun". Stockholm: Gatukontoret, Stockholms stad, 1979.

"Cykelplan 2006 för Stockholms innerstad". Stockholm: Trafikkontoret, 2006.

"Cykelplan 2012". Stockholm: Trafikkontoret, Stockholms stad, 2012.

"Cykelräkningar 2009". Stockholm: Trafikkontoret, Stockholms stad, 2010.

"Cykelstockholm i siffror 2010". Stockholm: Trafikkontoret, Stockholms stad, 2010.

*Cycling Manual*. 2nd ed. 1948.

Daley, Michelle, and Chris Rissel. "Perspectives and Images of Cycling as a Barrier or Facilitator of Cycling". *Transport Policy* 18, no. 1 (2011): 211–16.

Dalmulder, J.J.J. "De vraag naar rijwielen". *Maandschrift Centraal Bureau voor de Statistiek* 36, no. 3/4 (1941): 360–62.

Dauncey, Hugh, and Geoff Hare, eds. *The Tour de France: A Century of Sporting Structures, Meanings and Values*. London: Routledge, 2003.

Dauncey, Hugh. *French Cycling: A Social and Cultural History*. Liverpool: Liverpool University Press, 2012.

Davids, Mila, Harry Lintsen, and Arjan van Rooij. *Innovatie en kennisinfrastructuur. Vele wegen naar vernieuwing*. Vol. 5, *Bedrijfsleven in Nederland in de twintigste eeuw*. Amsterdam: Boom, 2013.

Davidson, Avram. "Or All the Seas with Oysters". *Galaxy Science Fiction*, May 1958.

Deegan, Brian. "Mapping Everyday Cycling in London". In *Cycling Cultures*, edited by Peter Cox, 106–29. Chester: University of Chester Press, 2015.

de Certeau, Michel. *The Practice of Everyday Life*. Translated by Steven Rendall. Oakland, CA: University of California Press, 1984.

Dekoster, J., and U. Schollaert. *Cycling: The Way Ahead for Towns and Cities*. Luxembourg: Office for the Official Publications of the European Communities, 1999.

Department of Trade and Industry, the United Kingdom. *Annual Statement of the Trade of the United Kingdom with Foreign Countries and British Possessions*. London: HMSO, 1892–1937.

Department of Transport, the United Kingdom. *National Cycling Strategy*. London: Department of Transport, 1996.

Department of Transport, the United Kingdom. *Ten Year Transport Plan*. London: Department of Transport, 2000.

Department for Transport, the United Kingdom. *Walking and Cycling: An Action Plan*. London: Department for Transport, 2004.

Department of Transport, the United Kingdom. *A Sustainable Future for Cycling*. London: Department of Transport, 2008.

Department of Transport, Ireland. *Ireland's First National Cycle Policy Framework*. Dublin: Department of Transport, 2009.

Department of Transportation, Federal Highway Administration, the United States. *The National Bicycling and Walking Study: Transportation Choices for a Changing America*. Washington, DC: Federal Highway Administration, 1994.

Department of Transportation, the United States. *The National Bicycling and Walking Study: A Ten-Year Status Report*. Washington, DC: Department of Transportation, 2004.

Department of Transportation, the United States. *The National Bicycling and Walking Study: 15-Year Status Report*. Washington, DC: Department of Transportation, 2010.

Department of Transportation, the United States. *Policy Statement on Bicycle and Pedestrian Accommodation Regulations and Recommendations*. Washington, DC: Department of Transportation, 2010.

Di Donna, Giuseppe. "Bicycle World in Italy". In *The Bicycle: Global Perspectives: Papers Presented at the Conference vélo mondiale Pro Bike–Velo-city, Montreal, Canada, 13–17 September 1992*, edited by Robert Boivin and Jean-François Pronovost, 430–31. Québec: Vélo Québec, 1992.

Dickinson, Janet E., Simon Kingham, Scott Copsey, and Deborah J. Pearlman Hougie. "Employer Travel Plans, Cycling and Gender: Will Travel Plan Measures Improve the Outlook for Cycling to Work in the UK?" *Transportation Research Part D: Transport and Environment* 8, no. 1 (2003): 53–67.

Di Falco, Daniel. "Prometheus fährt Rad: Die Mythen der industriellen Moderne in der frühen Fahrradwerbung". In *Bilder vom besseren Leben: Wie Werbung Geschichte erzählt*, edited by Daniel di Falco, Peter Bär, and Christian Pfister, 83–96. Bern: Haupt, 2002.

Dill, Jennifer and Theresa Carr. "Bicycle Commuting and Facilities in Major U.S. Cities: If You Build Them, Commuters Will Use Them—Another Look". *Transportation Research Record* 1828 (2003): 116–23.

Dill, Jennifer. "Bicycling for Transportation and Health: The Role of Infrastructure". *Journal of Public Health Policy* 30, no. 1 (2009): 95–110.

Directorate-General for Internal Policies, Policy Department B, Structural and Cohesion Policies, Transport and Tourism. *The Promotion of Cycling*. Brussels: European Union, 2010.

Divall, Colin, and George Revill. "Cultures of Transport: Representation, Practice and Technology". *Journal of Transport History* 26, no. 1 (2005): 99–111.

Divall, Colin, ed. *Cultural Histories of Sociabilities, Spaces and Mobilities*. London: Pickering & Chatto, 2015.

Dodge, Pryor. *The Bicycle*. Paris and New York: Flammarion, 1996.

Dovydėnas, Vytas. *Velomobile*. Berlin: Verlag Technik, 1990.

Dovydėnas, Vytas. "The Draft of the Biotransport Law of the Lithuanian Republic". In *Safety and Design: Proceedings of the Second European Seminar on Velomobiles*, edited by Peter Zeller, 163–66. Liestal: Future Bike CH, 1994.

Dovydėnas, Vytas. "Biotransport and 'Green Towns' of Europe". In *Proceedings of the Third European Seminar on Velomobile Design*, edited by Carl Georg Rasmussen, 13–15. Velomobile Forum, 1998.

Dower, John. *Embracing Defeat: Japan in the Wake of World War II*. New York: W.W. Norton, 2000.

Duizer, Bram. *'In Het Nut Van Actie Moet Je Geloven': Dertig Jaar Actievoeren Door De Fietsersbond*. Utrecht: Wetenschapswinkel Letteren Universiteit Letteren, 2005.

Ebert, Anne-Katrin. "Cycling towards the Nation: The Use of the Bicycle in Germany and the Netherlands, 1880–1940". *European Review of History* 11, no. 3 (2004): 347–64.

Ebert, Anne-Katrin. "Zwischen 'Ratreiten' und 'Kraftmaschinen'. Der bürgerliche Radsport am Ende des 19. Jahrhundert". *Werkstatt Geschichte* 44 (2007): 27–45.

Ebert, Anne-Katrin. "Het 'paard der democratie'. Fatsoenlijk fietsen in Nederland, 1900–1920". In *Fatsoenlijk vertier. Deugdzame ontspanning voor arbeiders na 1870*, edited by Christianne Smit, 209–37. Amsterdam: Bert Bakker, 2008.

Ebert, Anne-Katrin. *Radelnde Nationen. Die Geschichte des Fahrrads in Deutschland und der Niederlanden bis 1940*. Frankfurt am Main: Campus, 2010.

Ebert, Anne-Katrin. "Liberating Technologies? Of Bicycles, Balance and the 'New Woman' in the 1890s". *ICON Journal of the International Committee for the History of Technology* 16 (2010): 25–52.

Ebert, Anne-Katrin. "When Cycling Gets Political: Building Cycling Paths in Germany and the Netherlands, 1910–40". *Journal Of Transport History* 33, no. 1 (2012): 115–37.

Edgerton, David. *The Shock of the Old: Technology and Global History since 1900*. London: Profile Books, 2006.

Eickmans, Luuk, and Imelda Nasei. *Sustainable Mobility for African Cities*. Nairobi: UN-HABITAT, 2011.

Ekman, Sven. "Cykelplanering på 1980-talet i Stockholm". In *Ger mer cykeltrafik bättre städer? Rapport från ett seminarium anordnat i Linköping 22–23 augusti 1988 av Väg- och trafikinstitutet (VTI), Cykelfrämjandet och Föreningen för oskyddade trafikanter (FOT)*, 54–59. Linköping: Statens väg- och transportforskningsinstitut, 1988.

Embacher, Michael. *Cyclepedia: A Tour of Iconic Bicycle Designs*. London: Thames and Hudson, 2011.

Emanuel, Martin. "Understanding Conditions for Bicycle Traffic Through Historical Inquiry: The Case of Stockholm". *Journal of the Institute of Urban Transport of India*, December (2010): 1–16.

Emanuel, Martin. "Constructing the Cyclist: Ideology and Representations in Urban Traffic Planning in Stockholm, 1930–70". *Journal of Transport History* 33, no. 1 (2011): 67–91.

Emanuel, Martin. *Trafikslag på undantag. Cykeltrafiken i Stockholm 1930–1980*. Stockholm: Stockholmia, 2012.

Emanuel, Martin. "Stockholm: Where Public Transit Eclipses Cycling". In *Cycling Cities: The European Experience: Hundred Years of Policy and Practice*, edited by Ruth Oldenziel, Martin Emanuel, Adri Albert de la Bruhèze, and Frank Veraart. Eindhoven: Foundation of the History of Technology, 2016.

Emmerson, P., T. Ryley, and D.G. Davies. "The Impact of Weather on Cycle Flows". *Traffic Engineering and Control* 39, no. 4 (1998): 238–43.

Emmet, Robert, and Frank Zelko, eds. *Minding the Gap: Working across Disciplines in Environmental Studies*. Rachel Carson Center Perspectives 2014/2. Munich: Rachel Carson Center, 2014.

Emond, Catherine, Wei Tang, and Susan L. Handy. "Explaining Gender Difference in Bicycling Behavior". *Transportation Research Record* 2125 (2009): 16–25.

Engbers, Luuk H., and Ingrid J.M. Hendriksen. "Characteristics of a Population of Commuter Cyclists in the Netherlands: Perceived Barriers and Facilitators in the Personal, Social and Physical Environment". *International Journal of Behavioral Nutrition and Physical Activity* 7, no. 1 (2011): 89–94.

Epperson, Bruce. "After Pope: The Pope Manufacturing Company and the American Bicycle Industry, 1899–1990". *Cycle History 10: Proceedings of the 10th International Cycling History Conference*, edited by Rob van der Plas, 32–42. San Francisco: Van der Plas Publications, 2000.

Epperson, Bruce. "Failed Colossus: Strategic Error at the Pope Manufacturing Company, 1878–1900". *Technology and Culture* 41, no. 2 (2000): 300–20.

Epperson, Bruce. "Chasing the 'isms': Fordism, Taylorism, Popeism, and the Search for Meaning in the History of the American Bicycle Industry". In *Cycle History 20: Proceedings of the 20th International Cycling History Conference*, edited by Gary Sanderson, 195–205. Chesham: John Pinkerton Memorial Publishing Fund, 2010.

Epperson, Bruce D. "The Great Schism: Federal Bicycle Safety Regulation and the Unraveling of American Bicycle Planning". *Transportation Law Journal* 37, no. 2 (2010): 73–118.

Epperson, Bruce. "A New Class of Cyclists: Reyner Banham's Bicycle and the Pop Culture Revolution in Cycling That Never Happened". In *Cycle History 22: Proceedings of the 22nd International Cycling History Conference*, edited by Andrew Ritchie and Gary Sanderson, 68–75. San Francisco: Cycle History, 2012.

Epperson, Bruce. "A New Class of Cyclists: Banham's Bicycle and the Two-Wheeled World It Didn't Create". *Mobilities* 8, no. 2 (2013): 238–51.

Epperson, Bruce D. *Bicycles in American Highway Planning: The Critical Years of Policy-Making 1969–1991*. Jefferson: McFarland, 2014.

Esfehani, Amir Moghaddas. "The Bicycle's Long Way to China: The Appropriation of Cycling as a Foreign Cultural Technique". In *Cycle History 13: Proceedings of the 13th International Cycling History Conferences*, edited by Nick Clayton and Andrew Ritchie, 94–102. San Francisco: Van der Plas Publications, 2003.

European Commission. *Policy and Provision for Cyclists in Europe*. Brussels: Commission of the European Community, Directorate General for Transport, 1989.

European Commission. *Promotion of Measures for Vulnerable Road Users: Measures to Promote Cyclist Safety and Mobility*. Brussels: European Commission, 2001.

European Conference of the Ministers of Transport. *Implementing Sustainable Urban Travel Policies: National Policies to Promote Cycling*. Paris: Organisation for Economic Co-operation and Development, 2004.

Ewing, Reid, and Robert Cervero. "Travel and the Built Environment: A Meta-Analysis". *Journal of the American Planning Association* 76, no. 3 (2010): 265–94.

Fallan, Kjetil. "Kombi-Nation: Mini Bicycles as Moving Memories". *Journal of Design History* 26, no. 1 (2013): 65–85.

Federal Highway Administration, the United States. *The FHWA National Bicycling and Walking Study Case Study No. 3: What Needs to Be Done to Promote Bicycling and Walking?* Washington, DC: Federal Highway Administration, 1992.

Federal Highway Administration, the United States. *The FHWA National Bicycling and Walking Study Case Study No. 4: Measures to Overcome Impediments to Bicycling and Walking*. Washington, DC: Federal Highway Administration, 1993.

Federal Highway Administration, the United States. *The National Bicycling and Walking Study: Five-Year Status Report by the US Department of Transportation*. Washington, DC: Federal Highway Administration, 1999.

Federal Highway Administration. *The National Bicycling and Walking Study*. Washington, DC: US Department of Transportation/Federal Highway Administration/Pedestrian and Bicycle Information Center 2004.

Federal Ministry of Agriculture and Forestry, the Environment and Water Resources, Austria. *Masterplan Radfahren, Strategie zur Förderung des Radverkehrs in Österreich*. Vienna: Federal Ministry of Agriculture and Forestry, the Environment and Water Resources, 2006.

Federal Ministry of Transport, Building and Urban Development, Germany. *Erster Bericht der Bundesregierung über die Situation des Fahrradverkehrs in der Bundesrepublik Deutschland*. Bonn: Federal Ministry of Transport, Building and Urban Development, 1998.

Federal Ministry of Transport, Building and Urban Development, Germany. *Erster Bericht der Bundesregierung über die Situation des Fahrradverkehrs in die Bundesrepublik Deutschland*. Bonn: Federal Ministry of Transport, Building and Urban Development, 1999.

Federal Ministry of Transport, Building and Urban Development, Germany. *Nationaler Radverkehrsplan 2002–2012, FahrRad! Massnahmen zur Förderung des Radverkehrs in Deutschland*. Berlin: Federal Ministry of Transport, Building and Urban Development, 2002.

Federal Ministry of Transport, Building and Urban Development, Germany. *Zweiter Bericht der Bundesregierung über die Situation des Fahrradverkehrs in der Bundesrepublik Deutschland 2007*. Berlin: Federal Ministry of Transport, Building and Urban Development, 2007.

Fehlau, Gunnar. *Das Liegerad*. Kiel: Moby Dick Verlag, 1996.

Fehlau, Gunnar. *Das Modul-Bike. Faltbare Fahrräder*. Kiel: Moby Dick Verlag, 1997.

Fernandes, F. Dan. "Human Power on the Freeway". *Human Power* 1, no. 5 (1980): 1–2.

Filarski, Ruud. *The Rise and Decline of Transport Systems: Changes in a Historical Context*. Rotterdam: 2004.

Filarski, Ruud (in co-operation with Gijs Mom). *Shaping Transport Policy*. Den Haag: SDU Uitgevers, 2011.

Fincham, Ben. "Bicycle Messengers and the Road to Freedom". In *Against Automobility*, edited by Steffan Böhm, Campbell Jones, Chris Land, and Matthew Peterson, 208–22. Malden: Blackwell, 2006.

Fincham, Ben. "'Generally Speaking, People are in it for the Cycling and the Beer': Bicycle Couriers, Subculture and Enjoyment". *Sociological Review*, 55, no. 2 (2007) 189–202.

Fincham, Ben. "Bicycle Messengers: Image, Identity and Community". In *Cycling and Society*, edited by David Horton, Paul Rosen, and Peter Cox. Aldershot: Ashgate, 2007.

Firth, Daniel. "Framkomlighetsstrategin". Stockholm: Stockholms stad, Trafikkontoret, 2012.

Fischer, Edward L., et al. *Pedestrian and Bicyclist Safety and Mobility in Europe*. Washington, DC: US Department of Transportation, 2010.

Fishman, Elliot, Simon Washington, and Narelle L. Haworth. "Understanding the Fear of Bicycle Riding in Australia". *Journal of the Australasian College of Road Safety* 23, no. 3 (2012): 19–27.

Fitzpatrick, Jim. *The Bicycle and the Bush: Man and Machine in Rural Australia*. Oxford: Oxford University Press, 1980.

Fitzpatrick, Jim. *The Bicycle in Wartime*. Washington, DC: Brassey's, 1998.

Fitzpatrick, Jim. *Wheeling Matilda: The Story of Australian Cycling*. Kilcoy: Star Hill Studio, 2013.

Florida, Richard. *The Rise of the Creative Class: And How It's Transforming Work, Leisure, Community, and Everyday Life*. New York: Basic Books, 2002.

Flanagan, Elizabeth, Ugo Lachapelle, and Ahmed El-Geneidy. "Riding Tandem: Does Cycling Infrastructure Investment Mirror Gentrification and Privilege in Portland, OR, and Chicago, IL?" *Research in Transportation Economics* 60 (2016), 14–24.

Fläschner, Thomas. "Out of Date, Out of Mind: Public Awareness of the Bicycle during the 1950s in Germany's Saarland State". In *Cycle History 13: Proceedings of the 13th International Cycling History Conference*, edited by Nick Clayton and Andrew Ritchie, 46–50. San Francisco: Cycle Publishing/Van der Plas Publications, 2003.

Fläschner, Thomas. "Stahlroß auf dem Aussterbe-Etat: Zur Geschichte des Fahrrads und seiner Verdrängung in den 50er Jahren". *Eckstein. Journal für Geschichte* 9 (2000): 4–22.

*Flere cykler på sikker vej i staten. Transport- og Energiministeriets cykelstrategi.* Copenhagen: Transport- og Energiministeriet, 2007.

Forester, John. "American Cycling History from the 1940s; As I Remember It". John Forester's Home Page, accessed 28 December 2017. http://www.johnforester.com/Articles/Social/My%20History.htm.

Forester, John. "Fight for Your Right to Cycle Properly!" John Forester's Home Page, accessed 28 December 2017. http://www.johnforester.com/.

Forester, John. "Bicycling, Transportation and the Problem of Evil". American Dream Coalition, accessed 28 December 2017. http://www.americandreamcoalition.org/safety/CTEvil.pdf.

Forester, John. *Bicycle Transportation.* Cambridge, MA: MIT Press, 1983.

Forester, John. "Objective and Psychological Explanations for Differences in the Bicycling Programs of Different Nations". In *The Bicycle: Global Perspectives: Papers Presented at the Conference vélo mondiale Pro Bike–Velo-city, Montreal, Canada, 13–17 September 1992*, edited by Robert Boivin and Jean-François Pronovost, 434–37. Québec: Vélo Québec, 1992.

Forester, John. "Ideas in Motion: The Bicycle Transportation Controversy". *Transportation Quarterly* 55, no. 2 (2001): 7–18.

Forester, John. "The Place of Bicycle Transportation in Modern Industrialized Societies". American Dream Coalition, accessed 28 December 2017. http://americandreamcoalition.org/safety/forester.pdf.

"Förslag till trafikprogram 93". Stockholm: Stadsbyggnadskontoret, Stockholms stad, 1993.

Forsyth, Ann, and Kevin J. Krizek. "Promoting Walking and Bicycling: Assessing the Evidence to Assist Planners". *Built Environment* 36, no. 4 (2010): 429–46.

Forsyth, Ann, Kevin Krizek, and Daniel Rodríguez. "Non-Motorized Travel Research and Contemporary Planning Initiatives". *Progress in Planning* 71, no. 4 (2009): 170–83.

Fotheringham, William. *Cyclopedia: It's All about the Bike.* Chicago: Chicago Review Press, 2011.

Fraser, Simon, and Karen Lock. "Cycling for Transport and Public Health: A Systematic Review of the Effect of the Environment on Cycling". *European Journal of Public Health* 21, no. 6 (2010): 738–43.

Friedrich, Horst A. *Cycle Style*. Munich: Prestel, 2012.

Friss, Evan. "The Path Not Taken: The Rise of America's Cycle Paths and the Fall of Urban Cycling". In *Cycle History 20: Proceedings of the 20th International Cycling History Conference*, edited by Gary Sanderson, 67–72. Chesham: John Pinkerton Memorial Publishing Fund, 2010.

Friss, Evan. "Writing Bicycles: The Historiography of Cycling in the United States". *Mobility in History* 6 (2015): 127–33.

Fuchs, Andreas, and Theo Schmidt, eds. *Assisted Human Powered Vehicles and Velomobiles—Wettergeschützte Elektrofahrräder. Proceedings of the 4th Velomobile Seminar*. Liestal: Future Bike CH, 1999.

Furness, Zachary Mooradian. "'Put the Fun between Your Legs!' The Politics and Counterculture of the Bicycle". Dissertation, University of Pittsburgh, 2005.

Furness, Zack. "Critical Mass, Urban Space and Vélomobility". *Mobilities* 2, no. 2 (2007): 299–319.

Furness, Zack. *One Less Car: Bicycling and the Politics of Automobility*. Philadelphia: Temple University Press, 2010.

Gaboriau, Philippe. "Les trois ages du vélo en France". *Vingtième Siècle. Revue d'histoire* 29 (1991): 17–33.

Gaboriau, Philippe. *The Tour de France et le vélo. Histoire sociale d'une epopee contemporaine*. Paris: L'Harmattan, 1995.

Galtung, Johan. *Launching Peace Studies: The First PRIO Years: Strategies Findings Implications*. Grenzach-Wyhlen: TRANSCEND University Press, 2008.

Gardiner, Chris, and Rosalie Hill. "Cycling on the Journey to Work: Analysis of Socioeconomic Variables from the UK 1991 Population Census Samples of Anonymised Records". *Planning Practice & Research* 12, no. 3 (1997): 251–61.

Garrard, Jan, Geoffrey Rose, and Sing Kai Lo. "Promoting Transportation Cycling for Women: The Role of Bicycle Infrastructure". *Preventive Medicine* 46 (2008): 55–59.

Gatersleben, Birgitta, and Katherine M. Appleton. "Contemplating Cycling to Work: Attitudes and Perceptions in Different Stages of Change". *Transportation Research Part A: Policy and Practice* 41, no. 4 (2007): 302–12.

Gatersleben, Bigitta, and Hebba Haddad. "Who Is the Typical Bicyclist?" *Transportation Research Part F: Traffic Psychology and Behaviour* 13, no. 1 (2010): 41–48.

Gatersleben, Birgitta, and David Uzzell. "Affective Appraisals of the Daily Commute: Comparing Perceptions of Drivers, Cyclists, Walkers, and Users of Public Transport". *Environment and Behavior*, 39, no. 5 (2007): 416–31.

Gerwen, Jacques van, and Ferry de Goey. *Ondernemers in Nederland. Variaties in ondernemen*. Vol. 1, *Bedrijfsleven in Nederland in de twintigste eeuw*. Amsterdam: Boom, 2008.

Gerzon, Frits. *Nederland, een volk van struikrovers? De herinvoering van de Nederlandse octrooiwet (1869–1912)*. Den Haag: Orde van Octrooigemachtigden, 1986.

Geus, Bas de. "Cycling to Work: Psychosocial and Environmental Factors Associated with Cycling and the Effect of Cycling on Fitness and Health Indexes in an Untrained Working Population". Dissertation, Free University Brussels, 2007.

Geus, Bas de, Ilse De Bourdeaudhuij, Caroline Jannes, and Romain Meeusen. "Psychosocial and Environmental Factors Associated with Cycling for Transport among a Working Population". *Health Education Research* 23, no. 4 (2007): 697–708.

Gibson, Timothy A. "The Rise and Fall of Adrian Fenty, Mayor-Triathlete: Cycling, Gentrification and Class Politics in Washington, DC". *Leisure Studies* 34, no. 2 (2015): 230–49.

Gill, Tom. "Those Restless Little Boats: On the Uneasiness of Japanese Power-Boat Gambling". *The Asia-Pacific Journal* 11, issue 42, no. 2 (2013), http://www.japanfocus.org/-Tom-Gill/4012.

Goetzke, Frank, and Tilmann Rave. "Bicycle Use in Germany: Explaining Differences between Municipalities with Social Network Effects". *Urban Studies* 48, no. 2 (2011): 427–37.

Goeverden, Kees van and Tom Godefrooij. "Ontwikkeling van het fietsbeleid en—gebruik in Nederland". Bijdrage aan het Colloquium Vervoersplanologisch Speurwerk 25 en 26 november 2010, accessed 28 December 2017. https://www.cvs-congres.nl/cvspdfdocs/cvs10_062.pdf.

Golbuff, Laura, and Rachel Aldred. *Cycling Policy in the UK: A Thematic and Historical Overview*. London: University of East London, 2011.

Goodman, Anna. "Walking, Cycling and Driving to Work in the English and Welsh 2011 Census: Trends, Socio-Economic Patterning and Relevance to Travel Behaviour in General". *Plos One* 8 no. 8: e7190 (2013). http://journals.plos.org/plosone/article?id=10.1371/journal.pone.0071790.

Gordon, P., and H. Richardson. "Bicycling in the United States: A Fringe Mode?" *Transportation Quarterly*, 52, no. 1 (1998): 9–11.

Gorostiza, Saniago and Esther Anaya. "Under the Shade of Bahamontes: Bicycle Decline in Spain between 1950 and 1975". In *Cycle History 23: Proceedings of the 23rd International Cycling History Conference*, edited by Andrew Ritchie, 63–70. San Francisco: Cycling History, 2013.

Gorz, André. "L'idéologie sociale de la bagnole". *Le Sauvage*, September–October (1973).

Grieco, Margaret, Jeff Turner, and E.A. Kwakye. "A Tale of Two Cultures: Ethnicity and Cycling Behaviour in Urban Ghana". *Transportation Research Record* 1441 (1994): 101–07.

Gronen, Wolfgang. "Die Vector-story", In *HPV Chronik. 10 Jahre HPV—Eine Chronik und mehr!* edited by Michael Pohl, 57. Erlangen: HPV Deutschland e.V., 1995.

Gullberg, Anders, and Karolina Isaksson, eds. *Congestion Taxes in City Traffic: Lessons Learnt from the Stockholm Trial.* Lund: Nordic Academic Press, 2009.

Guo, Jessica Y., Chandra R. Bhat, and Rachel B. Copperman. "Effect of the Built Environment on Motorized and Non-Motorized Trip Making: Substitutive, Complementary, or Synergistic?" *Transportation Research Record* 2010 (2007): 1–11.

Guroff, Margaret. *The Mechanical Horse: How the Bicycle Reshaped American Life.* Austin: University of Texas Press, 2016.

Hachleitner, Bernhard, Matthias Marschik, Rudolf Müllner, and Michael Zappe, eds. *Motor bin ich selbst. 200 Jahre Radfahren in Wien.* Vienna: Metroverlag & Wienbibliotek im Rathaus, 2013.

Hadland, Tony. *The Moulton Bicycle.* 2nd, rev. ed. Coventry: Hadland Books, 2000.

Hadland, Tony. *Raleigh: Past and Presence of an Iconic Bicycle Brand.* San Francisco: Cycle Publishing, 2011.

Hadland, Tony, and Hans-Erhard Lessing. *Bicycle Design: An Illustrated History.* Cambridge, MA: MIT Press, 2014.

Hahn, Hans Peter. "The Appropriation of Bicycles in West Africa: Pragmatic Approaches to Sustainability". *Transfer* 2, no. 2 (2012): 31–48.

Hahn, Hans Peter. "Use and Cycling in West Africa", In *Cycling and Recycling. Histories of Sustainable Practices,* edited by Ruth Oldenziel and Helmuth Trischler, 15–32. New York and Oxford: Berghahn, 2016.

Haigh, Thomas. "Technology's Other Storytellers: Science Fiction as History of Technology". In *Science Fiction and Computing: Essays on Interlinked Domains,* edited by David L. Ferro and Eric G. Swedin, 13–37. Jefferson: McFarland, 2011.

Haixiao, Pan. "Evolution of Urban Bicycle Transport Policy in China", in *Cycling and Sustainability,* edited by John Parkin, 161–80. Bingley: Emerald, 2012.

Hall, Bronwyn H. "Innovation and Diffusion". Ch. 17 in *The Oxford Handbook of Innovation,* edited by Jan Fagerberg, David C. Mowery, and Richard R. Nelson, 459–84. Oxford: Oxford University Press, 2005.

Hall, Peter. *Cities in Civilization: Culture, Innovation, and Urban Order.* London: Weidenfeld & Nicolson, 1998.

Hallenbeck, Sarah. *Claiming the Bicycle: Women, Rhetoric, and Technology in Nineteenth-Century America.* Carbondale: Southern Illinois University Press, 2016.

Hamer, Mick. *Wheels within Wheels: A Study of the Road Lobby.* London: John Murray, 1987.

Hanenbergh, Kaspar, and Michiel Röben, eds. *Ons stalen ros. Nederland wordt een land van fietsers 1820 tot 1920.* Utrecht: Uitgeverij Ons Stalen Ros, 2015.

Hanna, Erika. "Seeing Like a Cyclist: Visibility and Mobility in Modern Dublin, c. 1930–1980". *Urban History* 42, no. 2 (2015): 273–89.

Hård, Mikael, and Andrew Jamison. *Hubris and Hybrids: A Cultural History of Technology and Science*. New York: Routledge, 2005.

Harrison, James. "Planning for More Cycling: The York Experience Bucks the Trend". *World Transport Policy & Practice* 7, no. 3 (2001): 21–27.

Hass-Klau, Carmen. *The Pedestrian and City Traffic*. London: Belhaven, 1990.

Haynes, Amanda. "Mass Media Re-Presentations of the Social World: Ethnicity and 'Race'", in *Media Studies: Key Issues and Debates*, edited by Eoin Devereux, 162–89. Los Angeles: Sage, 2007.

Hebdige, Dick. "Object as Image: The Italian Scooter Cycle". In *Material Culture: Critical Concepts in the Social Sciences*, Vol. 2, edited by Victor Buchli, 121–60. London: Routledge, 2004.

Heinen, Eva, Bert van Wee, and Kees Maat. "Commuting by Bicycle: An Overview of the Literature". *Transport Reviews* 30, no. 1 (2010): 59–96.

Heinen, Eva, Kees Maat, and Bert van Wee. "The Effect of Work-Related Factors on the Bicycle Commute Mode Choice in the Netherlands". *Transportation* 40, no. 1 (2013): 23–43.

Heinen, Eva, Kees Maat, and Bert van Wee. "The Role of Attitudes toward Characteristics of Bicycle Commuting on the Choice to Cycle to Work over Various Distances". *Transportation Research Part D: Transport and Environment* 16, no. 2 (2011): 102–09.

Hendriksen, I.J.M., M. Fekkes, M. Butter, and V.H. Hildebrandt. *Beleidsadvies. Stimuleren van fietsen naar het werk*. TNO-rapport. The Hague: Ministry of Health, Welfare and Sport, 2010.

Henricson, Daniel, and Cecilia Strömbom. *Cykelplan 1998 för Stockholms innerstad*. Stockholm: Gatu- och fastighetskontoret, Stockholms stad, 1998.

Henshaw, David. *The Great Railway Conspiracy*. Rev. ed. Dorchester, UK: A to B Books, 2013.

Henry, Raymond. "Origins and Brief History of the Fédération Française de Cyclotourisme". In *Cycle History 19: Proceedings of the 19th International Cycle History Conference*, edited by Anne Henry, 77–82. St. Etienne: Musée d'Arts et d'Industries, 2010.

Herlihy, David V. *Bicycle: The History*. New Haven: Yale University Press, 2004.

Hertel, Christof. "Fahrradnutzung ausserhalb von Europa—gegensätzliche Entwicklungen in den Metropolen der Welt". In *Das Fahrrad: Kultur, Technik, Mobilität*, edited by Mario Bäumer, 192–97. Hamburg: Junius, 2014.

Hinrichs, Peter, Ingo Kolboom, and Hans Joachim Neyer. "Zwischen Fahrrad und Fliessband. Culture technique in Frankreich zwischen Belle Epoque and Front Populaire". In *Absolut modern sein: Culture technique in Frankreich 1889*–1937, edited by Hans Joachim Neyer, 21–58. Berlin: Elefanten Press, Staatliche Kunsthalle, 1986.

"History of IHPVA". *Human Power* 1, no. 1 (1977): 10+19.

Hodges, Karl. "Did the Emergence of the Automobile End the Bicycle Boom?" In *Cycle History 4: Proceedings of the 4th International Cycle History Conference*, edited by Rob van der Plas, 39–42. San Francisco: Bicycle Books, 1995.

Hoepen, L. van, ed. *Van veertig zegenrijke melkjaren: 1908–1948. Gedenkboek uitgegeven ter gelegenheid van het veertig-jarig bestaan van de Vereeniging voor Zuivelindustrie en Melkhygiëne.* Bussum: Kruyt, 1948.

Hogenkamp, George J.M.G. *De geschiedenis van Burgers Deventer is de geschiedenis van de fiets.* Deventer: Burgers ENR, 1939.

Hogenkamp, George J.M.G. *Een halve eeuw wielersport.* Amsterdam: 1916.

Holmstedt, Stig. *Ett halvsekel i Stockholmstrafiken: Politik, planing och utbyggnader.* Stockholm: Stockholmia, 2012.

Holt, Richard. *Sport and Society in Modern France.* London and Oxford: Palgrave Macmillan, 1981.

Holt, Richard. "The Bicycle, The Bourgeoisie and the Discovery of Rural France, 1880–1914". *British Journal of Sports History* 2, no. 2 (1984): 127–39.

Hoogstraten, S. van. "De rijwielindustrie in 1929." *Economisch Statistische Berichten* 15 (1930): 680–81.

Hopkinson, P. and Mark Wardman, "Evaluating the Demand for New Cycle Facilities". *Transport Policy* 3, no. 4 (1996): 241–49.

Horn, Burkhard. "Vom Niedergang eines Massenverkehrsmittels. Zur Geschichte der städtischen Radverkehrsplannung". Thesis, Gesamthochschule Kassel, 1990.

Horn, Burkhard. "Geschichte der städtischen Radverkehrsplanung." In *Handbuch der kommunalen Verkehrsplanung*, edited by Tillman Bracher et al., Chapter 2.1.1.2. Berlin: VDE Verlag, 2002.

Horton, Dave. "Environmentalism and the Bicycle". *Environmental Politics* 15, no. 1 (2006): 41–58.

Horton, Dave, Paul Rosen, and Peter Cox, eds. *Cycling and Society*. Aldershot: Ashgate, 2007.

Horton, Dave. "Fear of Cycling". In *Cycling and Society*, edited by Dave Horton, Paul Rosen and Peter Cox, 133–52. Aldershot: Ashgate, 2007.

Horton, Dave, and John Parkin. "Conclusion: Towards a Revolution in Cycling". In *Cycling and Sustainability*, Transport and Sustainability, Volume 1, edited by John Parkin, 303–25. Bingley: Emerald, 2012.

Horton, Dave, and Tim Jones. "Rhetoric and Reality: Understanding the English Cycling Situation". In *Cycling Cultures*, edited by Peter Cox, 63–77. Chester: University of Chester Press, 2015.

Hout, Kurt van. *Literature Search Bicycle Use and Influencing Factors in Europe*. Hasselt: Hasselt University/Transportation Research Institute, 2008.

Hudson, Mike. *The Bicycle Planning Book*. London: Open Books, 1978.

Hunt, John Douglas, and J.E. Abraham. "Influences on Bicycle Use". *Transportation* 34, no. 4 (2007): 453–70.

Huwer, Ulrike. "Let's Bike: The 10 Point Pedalling Action Programme to Support Cycling". *World Transport Policy and Practice* 6, no. 2 (2000): 40–44.

Illich, Ivan. *Tools for Conviviality*. New York: Harper and Row, 1973.

Illich, Ivan. "Energy & Equity", in Illich, *Toward a History of Needs* (Berkeley: Heyday Books, 1978).

Independent Commission on Transport. *Changing Directions: The Report of the Independent Commission on Transport.* London: Coronet, 1974.

Interministerial Co-ordinator for Bicycle Policy. *Plan pluriannuel d'actions de l'État en faveur du vélo proposé en 2007*. Paris: Ministry of Ecology, *Sustainable* Development and Energy/Ministère Délégué aux Transports à la Mer et à la Pêche, 2007.

Ishikawa, Yoshinori. "Japanese Publicly Managed Gaming (Sports Gambling) and Local Government". Papers on the Local Governance System and its Implementation in Selected Fields in Japan No. 16. Council of Local Authorities for International Relations, Institute for Comparative Studies in Local Governance of the National Graduate Institute for Policy Studies, Tokyo, 2010. http://www.clair.or.jp/j/forum/honyaku/hikaku/pdf/BunyabetsuNo16en.pdf.

Jacobsen, Peter L. "Safety in Numbers: More Walkers and Bicyclists, Safer Walking and Bicycling". *Injury Prevention* 9, no. 3 (2009): 205–09.

Jacobsen, Peter L., Francesca Racioppi, and Harry Rutter. "Who Owns the Roads? How Motorized Traffic Discourages Walking and Bicycling". *Injury Prevention* 15, no. 6 (2009): 369–73.

Jamieson, Duncan R. "Bicycle Touring in the Late Nineteenth Century". In *Cycle History 12: Proceedings of the 12th International Cycling History Conference*, edited by Andrew Ritchie and Rob van der Plas, 68–75. San Francisco: Cycle Publishing, 2002.

Jamison, Andrew. *The making of green knowledge: Environmental Politics and Cultural Transformation*. Cambridge, UK: Cambridge University Press, 2001.

Jitensha Sangyō Shinkō Kyōkai, ed. *Jitensha no isseki: Nihon jitensha sangyōshi* [The Bicycle Century: The History of Japan's Bicycle Industry]. Tokyo: Jitensha Sangyō Shinkō Kyōkai, 1973.

Jones, Marcus. "Promoting Cycling in the UK: Problems Experienced by the Practitioners". *World Transport Policy & Practice* 7, no. 3 (2001): 7–12.

Jones, Phil. "Performing the City: A Body and a Bicycle Take on Birmingham, UK". *Social & Cultural Geography* 6, no. 6 (2005): 813–30.

Jones, Tim. "Getting the British Back on Bicycles: The Effects of Urban Traffic-Free Paths on Everyday Cycling". *Transport Policy* 20 (2012): 138–49.

Jong, Sicco de. *Geschiedenis eener Nederlandsche vereeniging: RAI 1893–1968*. Bussum: C.A.J. van Dishoeck, 1968.

Jongsma, Martinus, and Bernard Pijper. *Wielrennen in Friesland door de eeuwen heen: Fan tsjiltsjeriden oant hurdfytsen.* Franeker/Surhuisterveen: Pijper and Jongsma, 1998.

Jordan, Pete. *De fietsrepubliek.* Amsterdam: Podium, 2013.

Jordan, Pete. *In the City of Bikes: The Story of the Amsterdam Cyclist.* New York: Harper Collins, 2013.

"Jūichi gatsu ichi nichi wa jitensha kinenbi—waga kuni te kotoshi ga saisho" [11 November is Bicycle Commemoration Day, Held for the First Time in Japan]. *Doyō yūkan,* 10 October 1927.

Kallioinen, Johanna. *Pyöräilyn institutionaalinen asema liikennesuunnittelussa.* VATT-keskustelualoitteita 267. Helsinki: VATT Institute for Economic Research, 2002.

Karlsson, Börje. "Stockholmspartiets uppgång och fall". In *Från almstrid till trängselskatt: De stora slagen om Stockholm,* edited by Börje Karlsson and Torbjörn Tenfält. Stockholm: Stockholmia, 2011.

Kastrup, Marie. "Hverdagens beskedne demokrati. Analyser af cyklen som symbol pa danskhed". Master's thesis, University of Copenhagen, 2007.

Kastrup, Marie. "Identity and Bicycle Culture: A Danish Perspective". (Presentation, Cycling and Society Symposium, Research Group CSRG, University of Bolton, UK, 7 September 2009.)

Kaufmann, Vincent. "Mobility: Trajectory of a Concept in the Social Sciences". In *Mobility in History: The State of the Art in the History of Transport, Traffic and Mobility,* edited by Gijs Mom, Gordon Pirie, and Laurent Tissot, 41–60. Neuchâtel: Éditions Alphil—Presses universitaires suisses, 2009.

Kingham, Simon, Janet Dickinson, and Scott Copsey. "Travelling to Work: Will People Move Out of Their Cars?" *Transport Policy* 8, no. 2 (2001): 151–60.

Kirk, Andrew. "Machines of Loving Grace: Alternative Technology, Environment and the Counterculture". In *Image Nation: The American Counterculture of the 1960s and '70s,* edited by Peter Braunstein and Michael William Doyle, 353–78. New York, London: Routledge, 2002.

Kloof, Angela van der. "Lessons Learned through Training Immigrant Women in the Netherlands to Cycle". In *Cycling Cultures,* edited by Peter Cox, 78–105. Chester: University of Chester Press, 2015.

Knudsen, Walther B., Thomas Krag, and Dansk Cyklistforbund. *På cykel i 100 år. Dansk Cyklistforbund 1905–2005.* Copenhagen: Dansk Cyklistforbund, 2005.

Knuts, Stijn. "Converging and Competing Courses of Identity Construction: Shaping and Imagining Society through Cycling and Bicycle Racing in Belgium before World War Two". Dissertation, KU Leuven, 2014.

Koetse, Mark J., and Piet Rietveld. "The Impact of Climate Change and Weather on Transport: An Overview of Empirical Findings". *Transportation Research Part D: Transport and Environment* 14, no. 3 May (2009), 205–21.

Koglin, Till. "Organisation Does Matter—Planning for Cycling in Stockholm and Copenhagen". *Transport Policy* 39 (2015): 55–62.

Koglin, Till. "Vélomobility—A Critical Analysis of Planning and Space". Dissertation, Lund University, 2013.

Koglin, Till. "Vélomobility and the Politics of Transport Planning". *GeoJournal* 80, no. 4 (2014): 569–86.

Koglin, Till, and Tom Rye. "The Marginalisation of Bicycling in Modernist Urban Transport Planning". *Journal of Transport & Health* 1, no. 4 (2014): 214–22.

Koorey, Glen F. "Why a Cycling Strategy on Its Own Will NOT Increase Cycling". Paper presented at the New Zealand Cycling Conference, North Shore, 10–11 October 2003.

Kouno, Hiroyuki. "Bicycle Racing in Japan and the History of Keirin". *Cycle History 11: Proceedings of the 11th International Cycling History Conference*, edited by Andrew Ritchie and Rob van der Plas, 132–36. San Francisco: Van der Plas, 2000.

Krag, Thomas. "Bicycle Promotion Strategies in Denmark". Polish Seminars on Bicycle Promotion, accessed 28 December 2017. http://www.Friefugle.dk/poland/promotion_tk_en.html.

Krag, Thomas. "Cycling, Environment, Exercise and Health". In *Cost Benefit Analysis of Cycling*, edited by Gunnar Lind, 64–68. Copenhagen: Tema Nord, Nordic Council, 2005.

Krausse, Joachim. "Versuch auf's Fahrrad zu kommen. Zur Technik und Ästhetik der Velo-Evolution." In *Absolut modern sein: Culture technique in Frankreich 1889–1937*, edited by Hans Joachim Neyer, 59–74. Berlin: Elefanten Press, Staatliche Kunsthalle, 1986.

Krizek, Kevin J., Pamela J. Johnson, and Nebiyou Tilahun. "Gender Differences in Bicycling Behavior and Facility Preferences". In *Research on Women's Issues in Transportation: Report of a Conference, Volume 2: Technical Papers*. Transportation Research Board Conference Proceedings 35, 31–40. Chicago: Transportation Research Board of the National Academies, 2004.

Krizek, Kevin J., and Rio W. Roland. "What Is at the End of the Road? Understanding Discontinuities of On-Street Bicycle Lanes in Urban Settings". *Transportation Research Part D: Transport and Environment* 10, no. 1 (2005): 55–68.

Krizek, Kevin J., Gavin Poindexter, Gary Barnes, and Paul Mogush. "Analysing the Benefits and Costs of Bicycle Facilities via Online Guidelines". *Planning, Practice & Research* 22, no. 2 (2007): 197–213.

Krizek, Kevin J., Gary Barnes, and Kristin Thompson. "Analyzing the Effect of Bicycle Facilities on Commute Mode Share over Time". *Journal of Urban Planning and Development* 135, no. 2 (2009): 66–73.

Krizek, Kevin J., Ann Forsyth, and L. Baum. *Walking and Cycling International Literature Review: Final Report*. Melbourne: Department of Transport, State of Victoria, 2009.

Krizek, Kevin J., Susan Handy, and Ann Forsyth. "Explaining Changes in Walking and Bicycling Behavior: Challenges for Transportation Research". *Environment and Planning B: Urban Analytics and City Science* 36, no. 4 (2009): 725–40.

Krizek, Kevin J. "Cycling, Urban Form and Cities: What Do We Know and How Should We Respond?" In *Cycling and Sustainability*, Transport and Sustainability, Volume 1, edited by John Parkin, 111–30. Bingley: Emerald, 2012.

Kuhtz, Christian, and Werner Stiffel. *Chopper-Fahrräder*. Einfälle statt Abfälle, Reihe Fahrrad, Heft 5. Kiel: Verlag Christian Kuhtz, 1986.

Kuipers, Giselinde. *De fiets van Hare Majesteit. Over nationale habitus en sociologische vergelijking*. Rotterdam: Erasmus Universiteit Rotterdam, 2010.

Kuipers, Giselinde. "De fiets van Hare Majesteit: Over nationale habitus en sociologische vergelijking". *Sociologie*, 6, no. 3 (2011): 3–25.

Kuipers, Giselinde. "The Rise and Decline of National Habitus: Dutch Cycling Culture and the Shaping of National Similarity". *European Journal of Social Theory* 16, no. 1 (2012): 17–35.

Kuner, Herbert. "Fahrräder und Fahrradindustrie in den Niederlanden (Teil 1)". *Knochenschüttler* 31 (2004): 2–5.

Kyle, Chester. "A History of Human-Powered Land Vehicles and Competition". In *Human-Powered Vehicles*, edited by Allan V. Abbott and David G. Wilson, 95–111. Champaign: Human Kinetics, 1995.

Kyle, Chester R. "Bicycle Aerodynamics and the Union Cycliste Internationale".*Cycle History 11: Proceedings of the 11th International Cycling History Conference*, edited by Andrew Ritchie and Rob van der Plas, 118–31. San Francisco: Van der Plas, 2000.

Kyle, Chester R. "A Brief History of the International Human-Powered Vehicle Association 1976–1998". In *Cycle History 12: Proceedings of the 12th International Cycling History Conference*, edited by Andrew Ritchie, 134–45. San Francisco: Cycle Publishing, 2002.

Kylliäinen, Mikko. "Cycling Towards Civil Society: Estonian Cycling History in the 19th Century". In *Cycle History 21: Proceedings of the 21st International Cycling History Conference*, edited by Andrew Ritchie, 71–79. Birmingham: Cycling History Publishing, 2012.

Kynaston, David. *Family Britain, 1951–1957*. London: Bloomsbury, 2009.

Laferrere, Geneviève. *Comparison of National Cycling Policy in European Countries*. Brussels: Association for European Transport, 2002.

Lagström, Per. *Trafikutvecklingen i Stockholms stad 1945–1993*. Stockholm: Stockholm Konsult, 1994.

Lal, Vinay, ed. *Dissenting Knowledges, Open Futures: The Multiple Selves and Strange Destinations of Ashis Nandy*. New Delhi: Oxford University Press, 2000.

Larsen, Jacob, and Ahmed El-Geneidy. "A Travel Behavior Analysis of Urban Cycling Facilities in Montréal, Canada". *Transportation Research Part D: Transport and Environment* 16, no. 2 (2011): 172–77.

Lefevre, Nicolas. "Popularité du Cyclisme et Cyclisme Populaire. Pour en Fin avec le Mythe et le Misérablilisme". In *Cycle History 23: Proceedings of the 23rd International Cycling History Conference*, edited by Andrew Ritchie, 91–109. San Francisco: Cycling History, 2013.

Lente, Dick van. *Techniek en ideologie. Opvattingen over de maatschappelijke betekenis van technische vernieuwingen in Nederland, 1850–1920*. Groningen: Wolters Noordhoff/Forsten, 1988.

Lightwood, James T. *The Cyclists' Touring Club: Being the Romance of Fifty Years*. London: CTC, 1928.

Lilja, Sven. "Miljöpartier i Stockholmspolitiken 1966–2006". In *Makten i stadshuset: Stockholms lokalpolitik under 1900-talet*, edited by Anders Gullberg and Sven Lilja, 152–90. Stockholm: Stockholmia, 2008.

Lind, Gunnar, ed. *Cost Benefit Analysis of Cycling*. Copenhagen: Tema Nord, Nordic Council, 2005.

Linders-Rooijendijk, Mathea F.A. "Gebaande wegen voor mobiliteit en vrijetijdsbesteding: De ANWB als vrijwillige associatie 1883–1937". Dissertation, Katholieke Universiteit, Brabant, 1989.

Lintsen, Harry. *Ingenieurs in Nederland in de negentiende eeuw: Een streven naar erkenning en macht*. 's-Gravenhage: Martinus Nijhoff, 1980.

Linusson, Svante. "De politiska förhandlingarna". In *1000 meter cykelfält som skakade Stockholm*, edited by Erik Beckman and Svante Linusson, 11–19. Bromma: Forum08, 2009.

Liu, Rongfang, and Jiahua Song. "Adversary bicycle policies and their impact on urban transportation in China". *TRB 85th Annual Meeting Compendium of Papers CD-ROM*, paper no. 06-1651. Washington, DC: Transportation Research Board, 2006.

Lloyd-Jones, Roger, and Myrddin John Lewis. *Raleigh and the British Bicycle Industry: An Economic and Business History, 1870–1960*. Burlington, VT: Ashgate, 2000.

Loft, Charles. *Last Trains: Dr Beeching and the Death of Rural England*. London: Biteback Publishing, 2013.

London, Daniel Hart. "Keeping a Respectable Distance: The Rise and Fall of the Bicycle as an Instrument of Gentility." In *Cycle History 20: Proceedings of the 20th International Cycling History Conference*, edited by Gary Sanderson, 90–101. Chesham: John Pinkerton Memorial Publishing Fund, 2010.

Longhurst, James. *Bike Battles: A History of Sharing the American Road*. Seattle: University of Washington Press, 2015.

López, Bernat. "The Failed Vuelta Ciclista a Espana of 1913 and the Launching of the Volta a Catalunya (1911–1913): Centre versus Periphery in the Struggle for the

Governance of Cycling in Early Twentieth-Century Spain". *Sport in History* 30, no. 4 (2010): 547–69.

López, Bernat. "Sport, Media, Politics and Nationalism on the Eve of the Spanish Civil War: The First Vuelta Ciclista a Espana". *The International Journal of the History of Sport* 27, no. 4 (2010): 635–57.

Lowe, Marcia. "Bicycles Have a Powerful Future". *HPV News* 5, no. 3 (1988).

Lowe, Marcia D. *The Bicycle: Vehicle for a Small Planet.* Worldwatch Paper 30. Washington, DC: Worldwatch Institute, 1989.

Lumsdon, Les, and Rodney Tolley. "The National Cycle Strategy in the UK: To What Extent Have Local Authorities Adopted Its Model Strategy Approach?" *Journal of Transport Geography* 9, no. 4 (2001): 293–301.

Lundin, Per. "Att tänka om staden med historia: En introduktion till Bilstaden". In *Bilstaden: USA visade vägen*, edited by Uno Åhrén and Per Lundin, 9–22. Stockholm: Stockholmia, 1960.

Ma, Liang. "The Objective versus the Perceived Environment: What Matters for Active Travel". Dissertation, Portland State University, 2014.

Ma, Liang, and Jennifer Dill. "Associations between the Objective and Perceived Built Environment and Bicycling for Transportation". *Journal of Transport and Health* 2, no. 2 (2015): 248–55.

Mackintosh, Phillip Gordon. "A Bourgeois Geography of Domestic Cycling: The Responsible Use of Public Space in Toronto and Niagara-on-the-Lake, 1890–1900". *Journal of Historical Sociology* 2, no. 1–2 (2007): 128–57.

Mackintosh, Philipp G., and Glen Norcliffe. "Men, Women and the Bicycle: Gender and Social Geography of Cycling in the Late 19th Century". In *Cycling and Society*, edited by Dave Horton, Paul Rosen, and Peter Cox, 153–77. Aldershot: Ashgate, 2007.

Maddox, Heath. "Another Look at Germany's Bicycle Boom: Implications for Local Transportation Policy & Planning Strategy in the USA." *World Transport Policy and Practice* 7, no. 3 (2001): 44–48.

Männistö-Funk, Tiina. "The Crossroads of Technology and Tradition: Vernacular Bicycles in Rural Finland 1880–1910". *Technology and Culture* 52, no. 4 (2011): 733–56.

Männistö-Funk, Tiina. "The Prime, Decline, and Recalling of Rural Cycling: Bicycle Practices in 1920s' and 1930s' Finland Remembered in 1971–1972". *Transfers* 2, no. 2 (2012): 49–69.

Mapes, Jeff. *Pedaling Revolution: How Cyclists are Changing American Cities*. Corvallis: Oregon State University Press, 2009.

Marchesini, Daniele. *L'Italia del Giro d'Italia*. Bologna: Il Mulino, 1996.

Mari, Carlo. "Putting the Italians on Bicycles: Marketing at Bianchi, 1885–1955". *Journal of Historical Research in Marketing* 7, no. 1 (2015): 133–58.

Martens, Karel. "The Bicycle as a Feedering Mode: Experiences from Three European Countries". *Transportation Research Part D: Transport and Environment* 9, no. 4 (2004): 281–94.

Martens, Karel. "Role of the Bicycle in the Limitation of Transport Poverty in the Netherlands". *Transport Research Record: Journal of the Transportation Research Board* 2387 (2013): 20–25.

Mauranen, Tapani. *Hopeasiipi. Sata vuotta Helkamaa*. Helsinki: Otava, 2005.

MacCready, Paul B. "The Value and Future of Human-Powered Vehicles". In *Human-Powered Vehicles*, edited by Allan V. Abbott and David G. Wilson, 265–67. Champaign: Human Kinetics, 1995.

McCarthy, Deborah. "'I'm a Normal Person': An Examination of How Utilitarian Cyclists in Charleston South Carolina Use an Insider/Outsider Framework to Make Sense of Risks". *Urban Studies* 48, no. 7 (2011): 1439–55.

McClintock, Hugh. *The Bicycle and City Traffic: Principles and Practice*. London: Belhaven, 1992.

McClintock, Hugh. *Planning for cycling: Principles, Practice and Solutions for Urban Planners*. Boca Raton: CRC Press, 2002.

McClintock, Hugh. "Post-War Planning and Provision for the Bicycle". In *The Bicycle and City Traffic: Principles and Practice*, edited by Hugh McClintock, 19–39. London: Belhaven, 1992.

McGurn, James. *On Your Bicycle: An Illustrated History of Cycling*. London: John Murray, 1987.

McGurn, Jim. *On Your Bicycle: The Illustrated Story of Cycling*. York: Open Road Publishers, 1999.

McGurn, Jim. "From Editor Jim McGurn". *Bike Culture Quarterly* 1 (1993): 2.

Merchant, Carolyn. Preface to *Ecological Revolutions: Nature, Gender, and Science in New England*. 2nd ed. Chapel Hill: University of North Carolina Press, 2010.

Menghini, Gianluca, Nelson Carrasco, Nadine Schüssler, and Kay W. Axhausen. "Route Choice of Cyclists in Zurich". *Transportation Research Part A: Policy and Practice* 44, no. 9 (2010): 754–65.

Merriman, Peter. "Mobilities, Crises, and Turns: Some Comments on Dissensus, Comparative Studies, and Spatial Histories". *Mobility In History* 6 (2015): 20–34.

Miller, Daniel. *Stuff*. Cambridge: Polity Press, 2010.

Miller, Tom, and Ralf Österberg. *Medborgarinflytande i kommunal planering: Försök till utvärdering av trafiksamrådet i Stockholm 1973–1974*. Stockholm: Statens råd för byggnadsforskning, 1977.

Millward, Andrew. "The Founding of the Hercules Cycle & Motor Co. Ltd." In *Cycle History 5: Proceedings of the 5th International Cycle History Conference*, edited by Rob van der Plas, 99–106. San Francisco: Bicycle Books, 1995.

Millward, Andrew. "Factors Contributing to the Sustained Success of the UK Cycle Industry 1870–1939". Dissertation, University of Birmingham, 1999.

Ministry of Transport, Public Works and Water Management, The Netherlands. *Nota Fietsverkeer 1983: Een volledig beeld*. The Hague: Ministry of Transport, Public Works and Water Management, 1983.

Ministry of Transport, Public Works and Water Management, The Netherlands. *Beleidsnotitie Masterplan Fiets*. The Hague: Ministry of Transport, Public Works and Water Management, 1991.

Ministry of Transport, Public Works and Water Management, The Netherlands. *Cities Make Room for Cyclists*. The Hague: Ministry of Transport, Public Works and Water Management, 1995.

Ministry of Transport, Public Works and Water Management, Directorate-General for Passenger Transport, The Netherlands. *Eindrapport Masterplan Fiets. Samenvatting, evaluatie en overzicht van de projecten in het kader van het Masterplan Fiets, 1990–1997*. The Hague: Ministry of Transport, Public Works and Water Management, 1998.

Ministry of Transport, Public Works and Water Management, Directorate-General for Passenger Transport, The Netherlands. *The Dutch Bicycle Master Plan: Description and Evaluation in a Historical Context*. The Hague: Ministry of Transport, Public Works and Water Management, 1999.

Ministry of Transport, Public Works and Water Management. *Cycling in the Netherlands*. The Hague: Ministry of Transport, Public Works and Water Management, 2006.

Ministry of the Flemish Community, Department of Environment and Infrastructure, Belgium. *Ontwerp Vlaams Totaalplan Fiets*. Brussels: Ministry of the Flemish Community, 2002.

Ministry of Transport, the United Kingdom. *Roads for the Future: The New Inter-Urban Plan for Transport*. Cmnd. 4369. London: HMSO, 1970.

Ministry of Transport, Denmark. *Promoting Safer Cycling: A Strategy*. Copenhagen: Ministry of Transport, 2000.

Ministry of Transport and Communications, Finland. *Towards Healthy, Sustainable Transportation: Implementation of the WHO London Charter in Finland*. Helsinki: Ministry of Transport and Communications, 2002.

Ministry of Transports and Communications, Finland. *Finland Moving on Two Wheels*. Helsinki: Ministry of Transports and Communications, 1993.

Mom, Gijs. "The Crisis of Transport History: A Critique, and a Vista". *Mobility In History* 6 (2015): 7–19.

Moore, Richard, and Daniel Benson. *Bike! A Tribute to the World's Greatest Cycling Designers*. London: Aurum Press, 2012.

Morckel, Victoria, and Katryn Terzano. "The Influence of Travel Attitudes, Commute Mode Choice, and Perceived Neighborhood Characteristics on Physical Activity". *Journal of Physical Activity and Health* 11, no. 1 (2014): 91–98.

Moritz, W.E. "Adult Bicyclists in the United States: Characteristics and Riding Experience in 1996". *Transportation Research Record* 1636 (1998): 1–8.

Moritz, W.E. "A Survey of North American Bicycle Commuters: Design and Aggregate Results". *Transport Research Record* 1578 (1997): 91–101.

Morris, Hazel. "Interesting facts". LHC Machine Outreach, compiled April 2004. http://cern.ch/lhc-machine-outreach/lhc-interesting-facts.htm.

Moudon, Anne Vernez, Chanam Lee, Allen D. Cheadle, Cheza W. Collier, Donna Johnson, Thomas L. Schmid, and Robert D. Weather. "Cycling and the Built Environment: A US Perspective". *Transportation Research Part D: Transport and Environment* 10, no. 3 (2005): 243–61.

Moulton, Dave. "Once upon a Time, Britain Had a Bike Culture". *Dave Moulton's Blog*, 29 July 2010. http://davesbikeblog.squarespace.com/blog/2010/7/29/once-upon-a-time-britain-had-a-bike-culture.html.

Mullan, Elaine, and Margriet Groot. "Obsessing on Safety: Teenage Girls' Attitudes to Cycling". Department of Health, Sport and Exercise Sciences, Waterford Institute of Technology. Paper presented at Cycling and Society Symposium, Newcastle, UK, 15–16 September 2014. https://www.researchgate.net/publication/266698205_Obsessing_on_safety_teenage_girls_attitudes_to_cycling.

Murphy, Kate. *Fears and Fantasies: Modernity, Gender, and the Rural–Urban Divide*. New York: Peter Lang, 2010.

Myllyntaus, Timo. "Methods in Environmental History". In *Thinking through the Environment: Green Approaches to Environmental History*, edited by Timo Myllyntaus, 1–14. Cambridge, UK: White Horse Press, 2011.

Nankervis, Max. "The Effect of Weather and Climate on Bicycle Commuting". *Transportation Research Part A: Policy and Practice* 33, no. 6 (1999): 417–31.

National Association of City Transportation Officials. *Cities for Cycling*. Washington, DC: National Association of City Transportation Officials, 2010.

*National Bicycling and Walking Study: Ten Year Status Report October 2004*. Washington, DC: Federal Highway Administration 2004.

*National Cycling Strategy 2006–2015*. Oslo: Norwegian Public Roads Administration, 2003.

*Nationaler Radverkehrsplan 2002–2012, FahrRad! Maßnahmen zur Förderung des Radverkehrs in Deutschland*. Berlin, Bonn: Bundesministerium für Verkehr, Bau- und Wohnungswesen, 2002.

Nelson, Arthur C., and David Allen. "If You Build Them, Commuters Will Use Them: The Association between Bicycle Facilities and Bicycle Commuting". *Transportation Research Record* 1578 (1997): 79–83.

Newson, Alex. *Fifty Bicycles That Changed the World*. London: Design Museum & Conran/Octopus, 2103.

Nihon Jitensha Shinkō-kai [Japan Bicycle Promotion Association], ed. *Keirin ju-nen shi* [A Ten-Year History of *Keirin*]. Tokyo: Nihon Jitensha Shinkō-kai, 1959.

Nihon Jitensha Shinkō-kai [Japan Bicycle Promotion Association], ed, *Keirin niju-nen shi* [A Twenty-Year History of *Keirin*]. Tokyo: Nihon Jitensha Shinkō-kai, 1969.

Nihon Jitensha Shinkō-kai [Japan Bicycle Promotion Association], ed. *Keirin sanju-nen shi* [A Thirty-Year History of *Keirin*]. Tokyo: Nihon Jitensha Shinkō-kai, 1979.

Nihon Jitensha Shinkō-kai [Japan Bicycle Promotion Association], ed. *Keirin yonju-nen shi* [A Forty-Year History of *Keirin*]. Tokyo: Nihon Jitensha Shinkō-kai, 1989.

Nihon Jitensha Shinkō-kai [Japan Bicycle Promotion Association], ed. *Keirin goju-nen shi* [A Fifty-Year History of *Keirin*]. Tokyo: Nihon Jitensha Shinkō-kai, 1999.

Nihon Jitensha Shinkō-kai [Japan Bicycle Promotion Association], ed. *Keirin rokuju-nen shi* [A Sixty-Year History of *Keirin*]. Tokyo: Nihon Jitensha Shinkō-kai, 2009.

*Sankō shiryō* [Reference materials on *Keirin* finances]. Tokyo: Zenkoku Keirin Shikōsha Kyōgikai, 1969.

Nilsson, Lars. "The Return to the City: Twentieth Century Urban Development in Sweden". In *Reclaiming the City: Innovation, Culture, Experience*, edited by Marjaana Niemi and Ville Vuolanto, 47–62. Helsingfors: Finnish Literature Society, 2003.

Noland, Robert B., and Howard Kunreuther. "Short-Run and Long-Run Policies for Increasing Bicycle Transportation for Daily Commuter Trips". *Transport Policy* 2, no. 1 (1995): 67–79.

Norcliffe, Glen. "Popeism and Fordism: Examining the Roots of Mass Production". *Regional Studies* 31, no. 3 (1997): 267–80.

Norcliffe, Glen. *The Ride to Modernity: The Bicycle in Canada, 1869–1900*. Toronto, Buffalo, and London: University of Toronto Press, 2001.

Norcliffe, Glen. "Associations, Modernity and the Insider-citizens of a Victorian Highwheel Bicycle Club". *Journal of Historical Sociology* 19, no. 2 (2006): 121–50.

Norcliffe, Glen. *Critical Geographies of Cycling: History, Political Economy and Culture*. Farnham: Ashgate, 2015.

Norton, Peter. *Fighting Traffic: The Dawn of the Motor Age in the American City*. Cambridge, MA: MIT Press, 2008.

Norton, Peter. "Four Paradigms: Traffic Safety in the Twentieth-Century United States". *Technology and Culture* 56, no. 2 (2015): 319–34.

Norwegian Public Roads Administration. *National Cycling Strategy: Making Cycling Safe and Attractive: National Transport Plan 2006–2015*. Oslo: Norwegian Public Roads Administration, 2003.

Oakley, William. *Winged Wheel*. Godalming: CTC, 1977.

O'Brien, Flann. *The Third Policeman*. London: MacGibbon & Kee, 1967.

O'Connell, Sean. *The Car and British Society: Class, Gender and Motoring, 1896–1939*. Manchester: Manchester University Press, 1998.

Oddy, Nicholas. "Cycling's Dark Age? The Period 1900–1920 in Cycling History". In *Cycle History 15: Proceedings of the 15th International Cycling History Conference*, edited by Rob van der Plas, 79–86. San Francisco: Van der Plas and Cycle, 2005.

Oddy, Nicholas. "The Flaneur on Wheels?" In *Cycling and Society*, edited by Dave Horton, Paul Rosen, and Peter Cox, 97–112. Aldershot, Burlington: Ashgate, 2007.

Oddy, Nicholas. "Cycling, A Game for all Players". In *Cycle History 17: Proceedings of the 17th International Cycling History Conference*, edited by Glen Norcliffe, 11–16. San Francisco: Van der Plas/Cycle Publishing, 2007.

Oddy, Nicholas. "Let Bygones be Bygones". In *Cycle History 21: Proceedings of the 21st International Cycling History Conference*, edited by Andrew Ritchie, 172–78. Birmingham: Cycle History Publishing, 2012.

Oddy, Nicholas. "The Roaster—Iconic Bygone of the Future". In *Cycle History 24: Proceedings of the 24th International Cycling History Conference*, edited by Andrew Ritchie and Gary Sanderson, 82–87. Birmingham: Cycle History Publishing, 2014.

Oddy, Nicholas. "This Hill is Dangerous". *Technology and Culture* 56, no. 2 (2015): 335–69.

Oddy, Nicholas. "Signing Off". In *Cycle History 25: Proceedings of the 25th International Cycling History Conference*, edited by Andrew Ritchie and Gary Sanderson, 165–73. Birmingham: Cycle History Publishing, 2015.

Swiss Federal Council. *Stratégie pour le développement durable: Lignes directrices et plan d'action (2008–2011)—Guide*. Bern: Federal Office for Spatial Development, 2008.

Ogilvie, David, Matt Egan, Val Hamilton, and Mark Petticrew. "Promoting Walking and Cycling as an Alternative to Using Cars: Systematic Review". *British Medical Journal* 329, no. 7469 (2004): 763–66.

Olde Kalter, Marie-José. *Vaker op de fiets? Effecten van overheidsmaatregelen*. The Hague: KiM Netherlands Institute for Transport Policy Analysis, 2007.

Oldenziel, Ruth, and Adri Albert de la Bruhèze. "Contested Spaces: Bicycle Lanes in Urban Europe, 1900–1995". *Transfers* 1, no. 2 (2011): 29–49.

Oldenziel, Ruth, and Mikael Hård. *Consumers, Tinkerers, Rebels: The People Who Shaped Europe*. Hampshire: Palgrave Macmillan, 2013.

Oldenziel, Ruth, Martin Emanuel, Adri Albert de la Bruhèze, and Frank Veraart, eds. *Cycling Cities: The European Experience*. Eindhoven: Foundation for the History of Technology, 2016.

Oosterhuis, Harry, and Stoffers, Manuel. "A Strong Presence, but a Weak History: The Bicycle in Dutch Historiography". In *Cycle History 21: Proceedings of the 21st International Cycling History Conference*, edited by Andrew Ritchie, 166–71. Birmingham: Cycling History Publishing, 2012.

Oosterhuis, Harry. "Bicycle Research between Bicycle Policies and Bicycle Culture". *Mobility in History* 5 (2014): 20–36.

Opie, John. "Environmental History: Pitfalls and Opportunities". *Environmental History Review* 7, no. 1 (1983): 8–16.

Papon, Francis. "The Evolution of Bicycle Mobility in France". In *Cycle History 22: Proceedings of the 22nd International Cycling History Conference*, edited by Andrew Ritchie, 189–99. San Francisco: Cycling History, 2012.

Papon, Francis. "Historiographical Needs in the Study of Bicycling Mobility in France". *Mobility In History* 5, no. 1 (2014): 43–48.

Parkin, John, Tim Ryley and Tim Jones. "Barriers to Cycling: An Exploration of Quantitative Analyses". In *Cycling and Society*, edited Dave Horton, Paul Rosen, and Peter Cox, 67–82. Aldershot: Ashgate, 2007.

Parkin, John, Mark Wardman, and Matthew Page. "Estimation of the Determinants of Bicycle Mode Share for the Journey to Work Using Census Data". *Transportation* 35, no. 1 (2008): 93–109.

Parkin, John, ed. *Cycling and Sustainability*. Bingley: Emerald, 2012.

Parkin, John, and Glen Koorey. "Network Planning and Infrastructure Design." In *Cycling and Sustainability*, edited by John Parkin, 131–60. Bingley: Emerald, 2012.

Patton, David L. "Aspects of a Historical Geography of Technology: A Study of Cycling, 1919–1939". In *Cycle History 5: Proceedings of the 5th International Cycling History Conference*, edited by Rob van der Plas, 21–28. San Francisco: Bicycle Books, 1995.

Pelzer, Peter. "Bicycling as a Way of Life: A Comparative Case Study of Bicycle Culture in Portland and Amsterdam". Master's thesis, University of Amsterdam, 2010.

Pelzer, Peter. "Fietsmulticulturalisme." *Agora. Magazine voor sociaalruimtelijke vraagstukken* 26, no. 4 (2010): 17–19.

Peters, Peter F. *De haast van Albertine. Reizen in de technologische cultuur. Naar een theorie van passages*. Amsterdam: Uitgeverij De Balie, 2003.

Peters, Peter F. *Time, innovation and mobilities: Travel in technological cultures*. London, New York: Routledge, 2006.

Petty, Ross D. "Peddling the Bicycle in the 1890s: Mass Marketing Shifts into High Gear". *Journal of Macromarketing* 15, no. 1 (1995): 32–46.

Pikora, Terri, Billy Giles-Corti, Fiona Bull, Konrad Jamrozik, and Rob Donovan. "Developing a Framework for Assessment of the Environmental Determinants of Walking and Cycling". *Social Science and Medicine* 56, no. 8 (2003): 1693–1703.

Pinch, Trevor J., and Wiebe E. Bijker. "The Social Construction of Facts and Artefacts: Or How the Sociology of Science and the Sociology of Technology Might Benefit Each Other". *Social Studies of Science* 14, no. 3 (1984): 399–441.

Pinkerton, John. "Who Put the Working Man on a Bicycle?" In *Cycle History 8: Proceedings of the 8th International Cycling History Conferences*, edited by Nicholas Oddy and Rob van der Plas, 101–06. San Francisco: Van der Plas Publications, 1998.

Pivato, Stefano. "The Bicycle as a Political Symbol: Italy, 1885–1955". *International Journal of History of Sport* 7, no. 2 (1990): 172–87.

Pivato, Stefano. "Italian Cycling and the Creation of a Catholic Hero: The Bartali Myth". *International Journal of the History of Sport* 13, no. 1 (1996): 128–38.

Ploeger, Jan. "Designing for Cycling: The New Dutch Design Manual". In *The Greening of Urban Transport: Planning for Walking and Cycling in Western Cities*, edited by Rodney Tolley, 397–402. Chichester: John Wiley & Sons, 1997.

Plowden, William. *The Motor Car and Politics, 1896–1970*. London: Bodley Head, 1971.

Pohl, Michael, ed. *HPV Chronik. 10 Jahre HPV—Eine Chronik und mehr!* Erlangen: HPV Deutschland e.V., 1995.

Pooley, Colin G., and Jean Turnbull. "Modal Choice and Modal Change: The Journey to Work in Britain since 1890". *Journal of Transport Geography* 8, no. 1 (2000), 11–24.

Pooley, Colin G., Jean Turnbull, and Mags Adams. *A Mobile Century? Changes in Everyday Mobility in Britain in the Twentieth Century*. Aldershot: Ashgate, 2005.

Pooley, Colin. "Landscapes without the Car: A Counterfactual Historical Geography of Twentieth-Century Britain". *Journal of Historical Geography* 36, no. 3 (2010): 266–75.

Pooley, Colin G., Dave Horton, Griet Scheldeman, Caroline Mullen, Tim Jones, Miles Tight, Ann Jopson, and Alison Chisholm. "Policies for Promoting Walking and Cycling in England: A View from the Street". *Transport Policy* 27 (2013): 66–72.

Porter, Christopher, John Suhrbier, and William L. Schwartz. "Forecasting Bicycle and Pedestrian Travel: State of the Practice and Research Needs". *Transportation Research Record* 1674 (1999): 94–101.

Poyer, Alex. *Les premiers temps des vélo-clubs. Apparition et diffusion du cyclisme associative francais entre 1867 et 1914*. Paris: L'Harmattan, 2003.

*Promenadstaden: Översiktsplan för Stockholm*. Stockholm: Stadsbyggnadskontoret, 2010.

Pucher, John. "Urban Travel Behavior as the Outcome of Public Policy: The Example of Modal-Split in Western Europe and North America". *Journal of the American Planning Association* 54, no. 4 (1988): 509–20.

Pucher, John. "Bicycling Boom in Germany: A Revival Engineered by Public Policy". *Transportation Quarterly* 51, no. 4 (1997): 31–46.

Pucher, John, Charles Komanoff, and Paul Schimek. "Bicycling Renaissance in North America? Recent Trends and Alternative Policies to Promote Bicycling". *Transportation Research Part A* 33, nos. 7–8 (1999): 625–54.

Pucher, John, and Lewis Dijkstra. "Making Walking and Cycling Safer: Lessons from Europe". *Transportation Quarterly* 54, no. 3 (2000): 25–50.

Pucher, John. "Cycling Safety on Bikeways vs. Roads". *Transportation* Quarterly 55, no. 4 (2001): 9–11.

Pucher, John, and Lewis Dijkstra. "Promoting Safe Walking and Cycling to Improve Public Health: Lessons from the Netherlands and Germany". *American Journal of Public Health* 93, no. 9 (2003): 1509–16.

Pucher, John, and Ralph Buehler. "Cycling Trends and Policies in Canadian Cities". *World Transport Policy and Practice* 11, no. 1 (2005): 43–61.

Pucher, John, and Ralph Buehler. "Why Canadians Cycle More than Americans: A Comparative Analysis of Bicycling Trends and Policies". *Transport Policy* 13, no. 3 (2006): 265–79.

Pucher, John, and Ralph Buehler. "Making Cycling Irresistible: Lessons from the Netherlands, Denmark and Germany". *Transport Reviews* 28, no. 4 (2008): 495–528.

Pucher, John, and Ralph Buehler. "At the Frontiers of Cycling: Policy Innovations in the Netherlands, Denmark, and Germany". *World Transport Policy* 1313 (2008): 8–57.

Pucher, John, and Ralph Buehler. "Cycling for a Few or For Everyone: The Importance of Social Justice in Cycling Policy". *World Transport Policy & Practice* 15, no. 1 (2009): 57–64.

Pucher, John, Jennifer Dill, and Susan Hardy. "Infrastructure, Programs, and Politics to Increase Bicycling: An International Review". *Preventive Medicine* 50 (2010): 106–25.

Pucher, John, and Ralph Buehler. "Walking and Cycling for Healthy Cities". *Built Environment* 36, no. 4 (2010): 391–414.

Pucher, John, Ralph Buehler, and Mark Seinen. "Bicycle Renaissance in North America? An Update and Re-appraisal of Cycling Trends and Policies". *Transportation Research Part A: Policy and Practice* 45, no. 6 (2011): 451–75.

Pucher, John, Jan Garrard, and Stephen Greaves. "Cycling Down Under: A Comparative Analysis of Bicycling Trends and Policies in Sydney and Melbourne". *Journal of Transport Geography* 19, no. 2 (2011): 332–45.

Pucher, John, Ralph Buehler, D. Merom, and A. Baumann. "Walking and Cycling in the United States: Evidence From the National Household Travel Surveys". *American Journal of Public Health* 101, no. 1 (2011): 310–17.

Pucher, John, and Ralph Buehler. *City Cycling*. Cambridge, MA: MIT Press, 2012.

Pye, Denis. *Fellowship Is Life: The Story of the Clarion Cycling Club*. Bolton: Clarion, 2004.

Rabenstein, Rüdiger. *Radsport und Gesellschaft. Ihre sozialgeschichtlichen Zusammenhänge in der Zeit von 1867 bis 1914*. Hildesheim, München, and Zürich: Weidmann, 1991.

Rabenstein, Rüdiger. "The Vienna–Berlin Race of 1893 and Its Influence on the Cycling Movement in Germany and Austria". In *Cycle History 4: Proceedings of the 4th International Cycle History Conference*, edited by Rob van der Plas, 113–20. San Francisco: Bicycle Books, 1995.

Radkau, Joachim. "Das Fahrrad in den Technikvisionen der Jahrhundertwende, oder: das Erlebnis in der Technikgeschichte". In *Wege zur Fahrradgeschichte*, edited by Volker Briese, Wilhelm Matthies, and Gerd Renda, 9–32. Bielefeld: BVA Bielefelder Verlag, 1995.

Rajan, Sudhir C. "Automobility and the Liberal Disposition". *Sociological Review* 54, no. 1 (2006): 113–29.

Rautio, Anna-Maria, and Lars Östlund. "'Starvation Strings' and the Public Good: Development of a Swedish Bike Trail Network in the Early Twentieth Century". *The Journal of Transport History* 33, no. 1 (2012): 42–63.

"Regional cykelplan för Stockholms län 2014–2030". Stockholm: Trafikverket Region Stockholm, 2014.

*Regional utvecklingsplan för Stockholmsregionen: RUFS 2010*. Stockholm: Regionplanenämnden, Stockholms läns landsting, 2010.

Reid, Carlton. *Roads Were Not Built for Cars: How Cyclists Were the First to Push for Good Roads & Became the Pioneers of Motoring*. Washington, DC: Island Press, 2015.

Replogle, Michael A. *Non-Motorized Vehicles in Asia: Lessons for Sustainable Transport Planning and Policy*. Washington, DC: World Bank, 1991.

Reynolds, Conor C.O., M. Anne Harris, Kay Teschke, Peter A. Cripton, and Meghan Winters. "The Impact of Transportation Infrastructure on Bicycling Injuries and Crashes: A Review of the Literature". *Environmental Health* 8, no. 47 (2009). https://ehjournal.biomedcentral.com/articles/10.1186/1476-069X-8-47.

Rhoads, Edward J.M. "Cycles of Cathay: A History of the Bicycle in China". *Transfers* 2, no. 2 (2012): 95–120.

Richardson, A.J. *Seasonal and Weather Impacts on Urban Cycling Trips*. TUTI Report 1–2000. Victoria, Australia: The Urban Transport Institute, 2000.

Richardson, Tim, Karolina Isaksson, and Anders Gullberg. "Changing Frames of Mobility through Radical Policy Interventions? The Stockholm Congestion Tax". *International Planning Studies* 15, no. 1 (2010): 53–67.

Rieß, Falk. "Leichtfahrzeuge für den Verkehr der Zukunft: Vom Fahrrad zum Alltags-Vector". *Provelo* 10 (1987): 23–32.

Rietveld, Piet, and Vanessa Daniel. "Determinants of Bicycle Use: Do Municipal Policies Matter?" *Transportation Research Part A: Policy and Practice* 38, no. 7 (2004): 531–50.

Rissel, Chris, Catriona Bonfiglioli, Adriana Emilsen, and Ben J. Smith. "Representations of Cycling in Metropolitan Newspapers: Changes over Time and Differences between Sydney and Melbourne, Australia". *BMC Public Health* 10, no. 371 (2010): 1–8.

Ritchie, Andrew. *King of the Road: An Illustrated History of Cycling*. London: Wildwood House, 1974.

Ritchie, Andrew. *Major Taylor: The Extraordinary Career of a Champion Bicycle Racer*. San Francisco: Bicycle Books, 1988.

Ritchie, Andrew. *Quest for Speed: A History of Early Bicycle Racing, 1868–1903*. San Francisco: Cycle Publishing/Van der Plas Publications, 2011.

Rivers, Patrick. *Restless Generation: A Crisis in Mobility*. London: Davis-Poynter, 1972.

Roberts, Derek. *Cycling History, Myths and Queries*. Birmingham: Pinkerton, 1991.

Rodgers, Gregory B. "The Characteristics and Use Patterns of Bicyclists in the United States". *Journal of Safety Research* 25, no. 2 (1994): 83–96.

Rogers, Everett. *Diffusion of Innovations*. 5th ed. New York: Simon and Schuster, 2003.

Rosa, Hartmut, and William E. Scheuerman. *High-Speed Society: Social Acceleration, Power, and Modernity*. University Park: Penn State University Press, 2008.

Rosen, Paul. "The Social Construction of Mountain Bikes". *Social Studies of Science* 23, no. 3 (1993): 479–513.

Rosen, Paul. *Framing Production: Technology Culture and Change in the British Bicycle Industry*. Cambridge, MA: MIT Press, 2002.

Rosen, Paul. *How Can Research into Cycling Help Implement the National Cycling Strategy? Review of Cycling Research Findings and Needs.* Report of Whitehall Summer Placement in the Department for Transport, CLT3 and CLT4. York: University of York, Science and Technology Studies Unit, 2003.

Rosen, Paul. "Up the Vélorution: Appropriating the Bicycle and the Politics of Technology". In *Appropriating Technology: Vernacular Science and Social Power*, edited by Ron Eglash, Jennifer L. Croissant, Giovanna di Chiro, and Rayvon Fouché, 365–89. Minneapolis: University of Minnesota Press, 2004.

Rubenson, Paul. "Missing Link: The Case for Bicycle Transportation in the United States in the Early 20th Century". In *Cycle History 16: Proceedings of the 16th International Cycling History Conference*, edited by Andrew Ritchie, 3–84. San Francisco: Cycling History, 2006.

Rubinstein, David. "Cycling in the 1890s". *Victorian Studies* 21, no. 1 (1977): 47–71.

Sabelis, Ida H.J. "Diversity in Cycle Policies". In *Cycling Cultures*, edited by Peter Cox, 43–62. Chester: University of Chester Press, 2015.

Sabir, Muhammad, Mark J. Koetse, and Piet Rietveld. *The Impact of Weather Conditions on Mode Choice Decisions: Empirical Evidence for the Netherlands.* Tinbergen Institute Discussion Paper. Amsterdam: Vrije Universiteit, 2008.

Saelens, Brian E., James F. Sallis, and Lawrence D. Frank, "Environmental Correlates of Walking and Cycling: Findings From the Transportation, Urban Design, and Planning Literatures". *Annals of Behavioral Medicine* 25, no. 2 (2003): 80–91.

Saelensminde, Kjartan. "Cost-Benefit Analyses of Walking and Cycling Track Networks Taking into Account Insecurity, Health Effects and External Costs of Motorized Traffic". *Transportation Research Part A: Policy and Practice* 38, no. 8 (2004): 593–606.

Sagaris, Lake. "Lessons from 40 Years of Planning for Cycle-Inclusion: Reflections from Santiago, Chile". *Natural Resources Forum* 39, no. 1 (2015): 64–81.

Sahlqvist, Shannon, and Kristiann Heesch. "Characteristics of Utility Cyclists in Queensland, Australia: An Examination of the Associations between Individual, Social and Environmental Factors and Utility Cycling". *Journal of Physical Activity and Health* 9, no. 6 (2011): 818–28.

Salomon, Warren. "How Australian Cyclists Got Their New Helmet Laws". In *The Bicycle: Global Perspectives: Papers Presented at the Conference vélo mondiale Pro Bike–Velo-city, Montreal, Canada, 13–17 September 1992*, edited by Robert Boivin and Jean-François Pronovost, 366–69. Québec: Vélo Québec, 1992.

Sandgruber, Roman. "Cyclisation und Zivilisation. Fahrradkultur um 1900". In *Glücklich ist, wer vergisst …? Das andere Wien um 1900*, edited by Hubert Ch. Ehalt, Gernot Heiss, and Hannes Stekl, 285–303. Vienna, Cologne, and Graz: Hermann Böhlaus, 1986.

Scheepers, Caroline E. "Opportunities to Stimulate Active Transport". Dissertation, Free University Amsterdam, 2016.

Scheiner, Joachim. "Mobility Biographies: Elements of a Biographical Theory of Travel Demand". *Erdkunde* 61, no. 2 (2007): 161–73.

Scheiner, Joachim and Christian Holz-Rau. "Travel Mode Choice: Affected by Objective or Subjective Determinants?" *Transportation* 34, no. 4 (2007): 487–511.

Scheiner, Joachim. "Interrelations between Travel Mode Choice and Trip Distance: Trends in Germany 1976 to 2002". *Journal of Transport Geography* 18, no. 1 (2010): 75–84.

Schimmelpennink, Luud. "Solo: de 365-dagen fiets". *HPV Nieuws* no. 2 (1990): 15–17.

Schmitz, Arnfried. "Why Your Bicycle Hasn't Changed for 106 Years". *Human Power* 13, no. 3 (1994): 4–9.

Schmitz, Arnfried, and Tony Hadland. *Human Power: The Forgotten Energy, 1913–1992*. Coventry: Le Thor, 2000.

Schneider, Robert J. "Theory of Routine Mode Choice Decisions: An Operational Framework to Increase Sustainable Transportation". *Transport Policy* 25 (2013): 128–37.

Schöndorf, Paul. "20 Jahre Fahrradforschung an der FH Köln". In *HPV Chronik. 10 Jahre HPV—Eine Chronik und mehr!* edited by Michael Pohl, 31–33. Erlangen: HPV Deutschland e.V., 1995.

Schot, Johan W., and Adri A. Albert de la Bruhèze. "The Mediated Design of Products, Consumption and Consumers in the Twentieth Century". In *How Users Matter: The Co-Construction of Users and Technology*, edited by Nellie Oudshoorn and Trevor Pinch, 228–45. Cambridge, MA: MIT Press, 2003.

Schulte Behrbühl, Margrit. "Migration, Transfer and Appropriation: German Pork Butchers in Britain". *Transfers* 2, no. 3 (2012): 97–119.

Schwanen, Tim. "Urban Form and Commuting Behavior: A Cross European Comparison". *Tijdschrift voor Economische en Sociale Geografie* 93, no. 3 (2002): 336–43.

Schwedes, Oliver. "The Field of Transport Policy: An Initial Approach". *German Policy Studies* 7, no. 2 (2011): 7–41.

Sener, Ipek N., Naveen Eluru, and Chandra R. Bhat. "An Analysis of Bicyclists and Bicycling Characteristics: Who, Why, and How Much Are They Bicycling". *Transportation Research Record* 2134 (2009): 63–72.

Seray, Jacques. *Deux Roues: La véritable histoire du vélo*. Rodez: Éditions de Rouergue, 1988.

Shafizadeh, Kevan, and Debbie Niemeier. "Bicycle Journey-to-Work: Travel Behavior Characteristics and Spatial Analysis". *Transportation Research Record* 1578 (1997): 84–90.

Sheller, Mimi, and John Urry. "The New Mobilities Paradigm". *Environment and Planning A: Economy and Space* (2006): 207–26.

Sheller, Mimi. "The New Mobilities Paradigm for a Live Sociology". *Current Sociology* 62, no. 6 (2014): 789–811.

Sick Nielsen, Thomas A., Hans Skov-Petersen, and Trine A. Carstensen. "Urban Planning Practices for Bikeable Cities—The Case of Copenhagen". *Urban Research & Practice* 6, no. 1 (2013): 110–15.

Simon, David. *Transport and Development in the Third World.* New York: Taylor & Francis, 1996.

Simonsen, Dorthe Gert. "Accelerating Modernity: Time–Space Compression in the Wake of the Aeroplane". *Journal of Transport History* 26, no. 2 (2005): 98–117.

Skinner, David, and Paul Rosen. "Hell Is Other Cyclists: Rethinking Transport and Identity". In *Cycling and Society*, edited by Dave Horton, Paul Rosen, and Peter Cox, 83–96. Aldershot: Ashgate, 2007.

Sluyterman, Keetie E. *Dutch Enterprise in the 20th Century: Business Strategies in a Small Open Country*. London: Routledge, 2005.

Smart, Michael. "US Immigrants and Bicycling: Two-Wheeled in Autopia". *Transport Policy* 17, no. 3 (2010): 153–59.

Smethurst, Paul. *The Bicycle: Towards a Global History*. New York: Palgrave Macmillan, 2015.

Smith, Robert A. *A Social History of the Bicycle: Its Early Life and Times in America*. New York: American Heritage Press, 1972.

Spencer, Stephanie. *Gender, Work and Education in Britain in the 1950s.* Basingstoke: Palgrave Macmillan, 2005.

Spinney, Justin. "A Place of Sense: A Kinaesthetic Ethnography of Cyclists on Mont Ventoux". *Environment and Planning D: Society and Space* 24, no. 5 (2006): 709–73.

Spinney, Justin. "Cycling the City: Movement, Meaning and Method". *Geography Compass* 3, no. 2 (2009): 817–35.

Spolander, Krister. "Cykel i huvudstad". In *1000 meter cykelfält som skakade Stockholm*, edited by Erik Beckman and Svante Linusson, 59–69. Bromma: Forum08, 2009.

Staal, Peter Eloy. *Automobilisme in Nederland: Een geschiedenis van gebruik, misbruik en nut*. Zutphen: Walburg Pers, 2003.

Stahre, Ulf. *Den alternativa staden: Stockholms stadsomvandling och byalagsrörelsen*. Stockholm: Stockholmia, 1999.

Stahre, Ulf. *Den gröna staden: Stadsomvandling och stadsmiljörörelse i det nutida Stockholm*. Stockholm: Atlas, 2004.

Statistisches Reichsamt. *Der auswärtige Handel Deutschlands*. Berlin, 1897–1913.

Statistisches Reichsamt. *Der auswärtige Handel Deutschlands*. Berlin, 1922–1927.

Steele, M. William. "The Speedy Feet of the Nation: Bicycles and Everyday Mobility in Modern Japan". *Journal of Transport History* 32, no. 1 (2010): 182–209.

Steele, M. William. "The Making of a Bicycle Nation: Japan". *Transfers* 2, no. 2 (2012): 70–94.

Steele, M. William. "Betting on the Wheel: The Bicycle and Japan's Postwar Recovery, 1945–1958". Paper presented at the ICOHTEC, Manchester, 22–28 July 2013.

Stehlin, John. "Regulating Inclusion: Spatial Form, Social Process, and the Normalization of Cycling Practices in the USA". *Mobilities* 9, no. 1 (2014): 21–41.

Steinbach, Rebecca, Judith Green, Jessica Datta, and Phil Edwards. "Cycling and the City: A Case Study of How Gendered, Ethnic and Class Identities Can Shape Healthy Transport Choices". *Social Science & Medicine* 72, no. 7 (2011): 1123–30.

Stellner, Norbert. *Radfahrervereine in der bayerischen Provinz: Raum Mühldorf/Altötting 1882–1994*. Regensburg: EditionVulpis, 2000.

Sterling, Bruce. *A Good Old-fashioned Future*. New York: Bantam Books, 1999.

Stinson, Monique A., and Chandra R. Bhat. "An Analysis of the Frequency of Bicycle Commuting Using an Internet-Based Survey". *Transportation Research Record* 1878 (2004): 122–30.

Stinson, Monique A., and Chandra R. Bhat. "Commuter Bicyclist Route Choice: Analysis Using a Stated Preference Survey". *Transportation Research Record* 1828 (2003): 107–15.

Stoffers, Manuel, and Harry Oosterhuis. "Ons populairste vervoermiddel. De Nederlandse fietshistoriografie in internationaal perspectief". *Bijdragen en Mededelingen betreffende de Geschiedenis der Nederlanden/The Low Countries Historical Review* 124, no. 3 (2009): 390–418.

Stoffers, Manuel, Harry Oosterhuis, and Peter Cox. "Bicycle History as Transport History: The Cultural Turn". In *Mobility in History: Themes in Transport: T2M Yearbook 2011*, edited by Gijs Mom, Peter Norton, Georgina Clarsen, and Gordon Pirie, 265–74. Neuchatel: Alphil, 2010.

Stoffers, Manuel. "Cycling as Heritage: Representing the History of Cycling in the Netherlands". *Journal of Transport History* 33, no. 1 (2012): 92–114.

Stoffers, Manuel. "The Human Powered Vehicle Movement and the Changing Image of the Bicycle at the End of the Twentieth Century". In *Cycle History 22: Proceedings of the 22nd International Cycling History Conference*, edited by Andrew Ritchie, 211–19. San Francisco: Cycling History, 2012.

Stoffers, Manuel, and Anne-Katrin Ebert. "New Directions in Cycling Research: A Report on the Cycling History Roundtable at T2M Madrid". *Mobility in History* 5, no. 1 (2014): 9–19.

Stoffers, Manuel. "Exoot wordt icon: de inburgering van de fiets. Van gewild buitenlands product naar nationaal vervoermiddel". In *Ons stalen ros. Nederland wordt een land van fietsers 1820 tot 1920*, edited by Kaspar Hanenbergh and Michiel Röben, 290–99. Utrecht: Uitgeverij Ons Stalen Ros, 2015.

Stoffers, Manuel. "The Politics of Bicycle Innovation: Comparing the American and Dutch Human-Powered Vehicle Movements, 1970s–present". In *Cycling and Recycling. Histories of Sustainable Practices*, edited by Ruth Oldenziel and Helmuth Trischler, 33–57. New York, Oxford: Berghahn, 2016.

*Storstadstrafik 2: Bakgrundsmaterial*, edited by Storstadstrafikkommittén. Stockholm: Allmänna förl., 1989.

Sudjic, Deyan. "Megalopolis Now: Hong Kong, Shanghai, Jakarta". *City* 1, no 1–2 (1996): 30–37.

*Supreme Command of the Allied Powers Declassified Document 775019*. National Diet Library of Japan, Tokyo.

Susilo, Yusak, and Kees Maat. "The Influence of Built Environment to the Trends in Commuting Journeys in the Netherlands". *Transportation* 34, no. 5 (2007): 589–609.

Takeuchi, Johzen. "Historical Features of Japanese Small and Medium Enterprises: A Comparative Economic Approach". Ch. 8 in *Small Firms, Large Concerns: The Development of Small Business in Comparative Perspective*, edited by Odaka, Konosuke and Minoru Sawai, 197–216. Oxford: Oxford University Press, 1999.

Tengström, Emin. *Bilismen—i kris? En bok om bilen, människan, samhället och miljön.* Stockholm: Rabén & Sjögren, 1991.

Tengström, Emin. "Transport Sustainability in Denmark, Sweden and the Netherlands". In *Making Urban Transport Sustainable*, edited by Nicholas Low and Brendan Gleeson, 132–48. Basingstoke: Palgrave Macmillan, 2003.

Tetz, John. "AHPV: Assisted Human Powered Vehicle". In *Fourth International Human Powered Vehicle Scientific Symposium August 6, 1992*, edited by Chester Kyle, Jean A. Seay, and Joyce S. Kyle, 21–26. Indianapolis: IHPVA, 1994.

Teuteberg, Hans-Jürgen. "Entwicklung, Methoden und Aufgaben der Verkehrsgeschichte". *Jahrbuch für Wirtschaftsgeschichte* (1994): 173–94.

*The Swedish National Strategy for More and Safer Cycle Traffic*. Stockholm: N/A, 2000.

Thompson, Christopher. "Un troisième sexe? Les bougeoisies et la bicyclette dans la France fin de siècle". *Le Mouvement Social* 192 (2000): 9–39.

Thompson, Christopher S. "Bicycling, Class, and the Politics of Leisure in Belle Epoque France". In *Histories of Leisure*, edited by Rudy Koshar, 131–46. Oxford and New York: Berg, 2002.

Thompson, Christopher S. *Tour de France: A Cultural History*. Berkeley: University of California Press, 2006.

Thompson, Flora. *Over to Candleford*. Oxford: Oxford University Press, 1941.

Tilahun, Nebiyou Y., David M. Levinson, and Kevin J. Krizek. "Trails, Lanes, or Traffic: Valuing Bicycle Facilities with an Adaptive Stated Preference Survey". *Transportation Research Part A: Policy and Practice* 41, no. 4 (2007): 287–301.

Timmerman, Nico. *Rijwielfabriek Union: Een eeuw in beweging*. Eindhoven: N.B.E. Timmerman, 2004.

Titze, Sylvia, Willibald J. Stronegger, Susanne Janschitz, and Pekka Oja. "Association of Built-Environment, Social-Environment and Personal Factors with Bicycling as a Mode of Transportation among Austrian City Dwellers". *Preventive Medicine* 47, no. 3 (2008): 252–59.

Tjong Tjin Tai, Sue-Yen, Frank Veraart, and Mila Davids. "How the Netherlands Became a Bicycle Nation: Users, Firms and Intermediaries, 1860–1940". *Business History* 57, no. 2 (2015): 257–89.

Tjong Tjin Tai, Sue-Yen. "Hoe werd de fiets een Nederlands product? Hollandse degelijkheid vormt de fiets voor dagelijks gebruik". In *Ons stalen ros. Nederland wordt een land van fietsers 1820 tot 1920*, edited by Kaspar Hanenbergh and Michiel Röben, 192–97. Utrecht: Uitgeverij Ons Stalen Ros, 2015.

Tobin, Gary A. "The Bicycle Boom of the 1890s: The Development of Private Transportation and the Birth of the Modern Tourist". *Journal of Popular Culture* 7, no. 4 (1974): 838–49.

*Tōkyō zaisei-shi, gekan* [Tokyo Fiscal History]. Vol. 3. Tokyo: Tokyo Metropolis, 1969.

"Trafikplan 87: Diskussionsunderlag." Stockholm: Gatukontoret, Stockholms kommun, 1987.

Transport for London. *Cycling in London: Final Report*. London: Transport for London, 2008.

Trapp, Georgina S.A., Billie Giles-Corti, Hayley E. Christian, Max Bulsara, Anna F. Timperio, Gavin R. McCormack, and Karin P. Villaneuva. "On Your Bike! A Cross-Sectional Study of the Individual, Social and Environmental Correlates of Cycling to School". *International Journal of Behavioral Nutrition and Physical Activity* 8, no. 1 (2011): 123–33.

Tschofen, Bernard, and Yonca Krahn, eds. *Velo: Erkundungen zu Zürcher Fahrradkultur*. Zürich: University of Zürich, 2016.

Turpin, Robert J. "'Our Best Bet is the Boy': Bicycle Marketing Schemes and American Culture after World War I". In *Cycle History 22: Proceedings of the 22nd International Cycling History Conference*, edited by Andrew Ritchie, 159–70. San Francisco: Cycling History, 2012.

Ueda, Tatsuzō. *The Development of the Japanese Bicycle Industry in Japan after World War II*. Tokyo: United Nations University Press, 1981. http://darch.ide.go.jp/je_archive/english/society/wp_je_unu38.html.

Uyl, Bob den. *Wat fietst daar?* Den Haag and Rotterdam: Nijgh & Van Ditmar, 1970.

Van Acker, Veronique, Bert van Wee, and Frank Witlox. "When Transport Geography Meets Social Psychology: Toward a Conceptual Model of Travel Behaviour". *Transport Reviews* 30, no. 2 (2010): 219–40.

Vandenbulcke, Grégory, Claire Dujardin, Isabelle Thomas, Bas de Geus, Bart Degrauewe, Romain Meeusen, and Luc Int Panis. "Cycle Commuting in Belgium: Spatial Determinants and 'Re-Cycling' Strategies". *Transportation Research Part A: Policy and Practice* 45, no. 2 (2011): 118–37.

Vannini, Phillip. "Mobile Cultures: From the Sociology of Transportation to the Study of Mobilities". *Sociology Compass* 4, no. 2 (2010): 111–21.

Vanysacker, Dries. "Le cyclisme en Wallonnie jusqu'à la Seconde Guerre mondiale: Une histoire sociale comparable à celle des Flandriens (1860–1945)". In *Images et paysages mentaux des 19e et 20e siècles, de la Wallonie à l'outre-mer*, edited by Luc Courtois, Jean-Pierre Delville, Francoise Rosart, and Guy Zelis, 147–72. Louvain-la-Neuve: Presses universitaires de Louvain, 2007.

Velox (pseud.). *Velocipedes: Bicycles and Tricycles: How to Make and How to Use Them: With a Sketch of Their History, Invention and Progress*. London: G. Routledge, 1869.

Veraart, Frank C.A. "Geschiedenis van de fiets in Nederland 1870–1940. Van sportmiddel naar massavervoermiddel". Master's thesis, Technische Universiteit Eindhoven, 1995.

Verhoeven, Rik, and Pieter M. Schrijnen. "Allochtonen onderweg. Fietsgebruik onder immigranten". Paper presented at Colloquium Vervoersplanologisch Speurwerk, Roermond, The Netherlands, 25–26 November 2010.

Verplanken, Bas, Henk Aarts, Ad van Knippenberg, and Carina van Knippenberg. "Attitude versus General Habit: Antecedents of Travel Mode Choice". *Journal of Applied Social Psychology* 2, no. 44 (1994): 285–300.

Verplanken, Bas, Henk Aarts, and Ad van Knippenberg. "Habit, Information Acquisition, and the Process of Making Travel Mode Choices". *European Journal of Social Psychology* 27, no. 5 (1997): 539–60.

Ververs, Ruud, and Arnold Ziegelaar. *Verklaringsmodel voor fietsgebruik gemeenten. Eindrapport*. Leiden: Research voor Beleid, 2006.

Vinne, Vincent van der. *Eysink: van fiets tot motorfiets. Ondernemen tijdens de opkomst van het gemotoriseerd verkeer*. Amsterdam: Bataafsche Leeuw, 2001.

Vinne, Vincent van der. *De trage verbreiding van de auto in Nederland 1896–1939*. Amsterdam: Bataafsche Leeuw, 2007.

Vigarello, Georges. "Le Tour de France". In *Les lieux de mémoire*, vol. 3, edited by Pierre Nora, 885–925. Paris: Gallimard, 1992.

Vivanco, Luis. *Reconsidering the Bicycle: An Anthropological Perspective on a New (Old) Thing*. New York and London: Routledge, 2013.

Wachs, Martin. "Creating Political Pressure for Cycling". *Transportation Quarterly* 52, no. 1 (1998): 6–8.

Waerden, Theodorus van der. *Geschooldheid en techniek. Onderzoek naar den invloed van arbeidssplitsing en machinerie op de mate van vereischte oefening en bekwaamheid der arbeiders*. Amsterdam: Van Rossen, 1911.

Walle, Frederick van de. "The Velomobile as a Vehicle for more Sustainable Transportation. Reshaping the Social Construction of Cycling Technology". Master's thesis, Kungliga Tekniska Högskolan, Stockholm, 2004.

Wardlaw, Malcom J. "Assessing the Actual Risks Faced by Cyclists". *Tec*, December (2002): 352–56.

Wardman, Mark, Richard Hatfield, and Matthew Page. "The UK National Cycling Strategy: Can Improved Facilities Meet the Targets?" *Transport Policy* 4, no. 2 (1997): 123–33.

Webb, Sydney, and Beatrice. *English Local Government: The Story of the King's Highway.* London: Longmans Green, 1913.

Welleman, Anthony G. "Why a Bicycle Policy in the Netherlands?" In *The Bicycle: Global Perspectives: Papers Presented at the Conference vélo mondiale Pro Bike–Velo-city, Montreal, Canada, 13–17 September 1992*, edited by Robert Boivin and Jean-François Pronovost, 275–77. Québec: Vélo Québec, 1992.

Westall, Sylvia. "Rickshaws Pick up Speed in Europe's Top Cities". *www.reuters.com*, last modified 3 October 2007. http://www.reuters.com/article/us-europe-rickshaws-idUSL0343085520071003.

Westwood, Kim. *The Courier's New Bicycle*. Sydney: Harper Voyager, 2011.

Whitt, Frank Rowland, and David Gordon Wilson. *Bicycling Science: Ergonomics and mechanics*. Cambridge, MA: MIT Press, 1974.

*Why Cyclists Object To Compulsory Rear Lights*. London: CTC, 1927.

Williamson, John A. *The Motor Car Acts 1896 & 1903*. London: Autocar, 1903.

Willis, Devon, Kevin Manaugh, and Ahmed El-Geneidy. "Cycling under Influence: Summarizing the Influence of Attitudes, Habits, Social Environments and Perceptions on Cycling for Transportation". *International Journal of Sustainable Transportation* 9, no. 8 (2015): 565–79.

Wilson, David G. "Man-Powered Land Transport". *Engineering* 207, no. 5372 (1969): 567–73.

Wilson, David G. "Human Muscle Power in History". In *Pedal Power: In Work, Leisure, and Transportation*, edited by James C. McCullagh, 1–36. Emmaus, PA: Rodale Press, 1977.

Wilson, David G. "The Future Potential for Muscle Power". In *Pedal Power: In Work, Leisure, and Transportation*, edited by James C. McCullagh, 106–23. Emmaus, PA: Rodale Press, 1977.

Wilson, David G. "Getting in Gear: Human-Powered Transportation". *Technology Review* 82, no. 1, October (1979): 42–54.

Wilson, David G., Richard Forrestall, and Dereck Henden. "Evolution of Recumbent Bicycles and the Design of the Avatar Bluebell". In *Second International Human Powered Vehicle Scientific Symposium: Proceedings*, edited by Allan V. Abbott, 92–103. Indianapolis: IHPVA, 1983.

Wilson, David G. "A Blueprint for an HPV Revolution". In *Third International Human Powered Vehicle Scientific Symposium, August 28 and 29, 1986: Proceedings*, edited by Allan V. Abbott, 79–83. Indianapolis: IHPVA, 1986.

Wilson, David G. "Editorial", *Human Power* 7, no. 2 (1988): 2.

Wilson, David G. "The Development of Modern Recumbent Bicycles". In *Human-Powered Vehicles*, edited by Allan V. Abbott and David G. Wilson, 113–27. Champaign: Human Kinetics, 1995.

Wilson, David G. "Human Power in the Future." *Human Power* 54 (2003): 13–14.

Wilson, David Gordon. "White Paper on Proposed Tax-and-Rebate Bill to Reduce Emissions and Nonrenewable Energy Use [etc.]". *lessgovletsgo.org*. 3 January 2011. http://lessgovletsgo.org/wp-content/uploads/2013/10/White_paper_on_energy_3Jan2011.pdf.

Winters, Meghan, Melissa C. Friesen, Mieke Koehoorn, and Kay Teschke. "Utilitarian Bicycling: A Multilevel Analysis of Climate and Personal Influences". *American Journal of Preventive Medicine* 32, no. 1 (2007): 52–58.

Winters, Meghan, Michael Brauer, Eleanor M. Setton, and Kay Teschke. "Built Environment Influences on Healthy Transportation Choices: Bicycling versus Driving". *Journal of Urban Health* 87, no. 6 (2010): 969–93.

Winters, Meghan, Gavin Davidson, Diana Kao, and Kay Teschke. "Motivators and Deterrents of Bicycling: Comparing Influences on Decisions to Ride". *Transportation* 38, no. 1 (2011): 153–68.

Withers, Jeremy. "Bicycles across the Galaxy: Attacking Automobility in the 1950s Science Fiction". *Science Fiction Studies* 44, no. 3 (2017): 417–36.

Wolmar, Christian. *Fire and Steam*. London: Atlantic Books, 2007.

Woodforde, John. *The Story of the Bicycle*. London: Routledge and Kegan Paul, 1971.

Woodward, Gilbert E. *Woodward's Road Traffic Acts and Orders 1930–1934*. London: Eyre & Spottiswoode, 1934.

Worster, Donald. *The Wealth of Nature: Environmental Histories and the Ecological Imagination*. New York and Oxford: Oxford University Press, 1993.

Wray J. Harry. *Pedal Power: The Quiet Rise of the Bicycle in American Public Life*. Boulder and London: Paradigm Publishers, 2008.

Yamazaki, Takuya, and Yuki Mabuchi. "Sports Betting and the Law in Japan". In *Sports Betting: Law and Policy*, edited by Paul M. Andersen, 512–15. Den Haag: TMC Asser Press, 2012.

Xing, Yan, Susan L. Handy, and Patricia L. Mokhtarian. "Factors Associated with Proportions and Miles of Bicycling for Transportation and Recreation in Six Small US Cities". *Transportation Research Part D: Transport and Environment* 15, no. 2 (2010): 73–81.

Yang, Lin, Shannon Sahlqvist, Alison McMinn, Simon J. Griffin, and David Ogilvie. "Interventions to Promote Cycling: Systematic Review". *British Medical Journal* 341, c5293 (2010.

Young, Michael, and Peter Wilmott. *Family and Kinship in East London.* Harmondsworth: Penguin, 1957.

Zahran, Sammy, Samuel D. Brody, Praveen Maghelal, Andrew Prelog, and Michael Lacy. "Cycling and Walking: Explaining the Spatial Distribution of Healthy Modes of Transportation in the United States". *Transportation Research Part D: Transport and Environment* 13, no. 7 (2008): 462–70.

*Zicht op Nederland Fietsland.* Amersfoort: Stichting Landelijk Fietsplatform, 2009.

# Index

www.ingramcontent.com/pod-product-compliance
Lightning Source LLC
Chambersburg PA
CBHW071353220926
51613CB00009B/182
*9789004289963*